Marine Microbiology

Marine Microbiology

Ecology and Applications

C.B. Munn

Dept. of Biological Sciences, University of Plymouth, Plymouth, UK

First published 2004

A CIP catalogue record for this book is available from the British Library.

ISBN 1 85996 288 2

Garland Science/BIOS Scientific Publishers
4 Park Square, Milton Park, Abingdon, Oxon, OX14 4RN, UK and 29
West 35th Street, New York, NY 10001-2299, USA
World Wide Web home page: www.bios.co.uk

Garland Science/BIOS Scientific Publishers is a member of the Taylor & Francis Group.

Distributed in the USA by
Fulfilment Center
Taylor & Francis
10650 Toebben Drive
Independence, KY 41051, USA
Toll Free Tel.: +1 800 634 7064; E-mail: taylorandfrancis@thomsonlearning.com

Distributed in Canada by
Taylor & Francis
74 Rolark Drive
Scarborough, Ontario M1R 4G2, Canada
Toll Free Tel.: +1 877 226 2237; E-mail: tal_fran@istar.ca

Distributed in the rest of the world by
Thomson Publishing Services
Cheriton House
North Way
Andover, Hampshire SP10 5BE, UK
Tel.: +44 (0)1264 332424; E-mail: salesorder.tandf@thomsonpublishingservices.co.uk

Library of Congress Cataloging-in-Publication Data

Munn, C. (Colin)
Marine microbiology : ecology and applications / C. Munn.
p. cm.
Includes bibliographical references and index.
ISBN 1-85996-288-2 (pbk. : alk. paper)
1. Marine microbiology. I. Title.

QR106.M86 2004
579'.177–dc22
2003018397

Production Editor: Andrew Watts
Typeset by Charan Tec Pvt. Ltd, Chennai, India
Printed and bound by Cromwell Press, Trowbridge, UK

Contents

List of research focus boxes

Preface

This book is intended for upper-level undergraduates and Master's degree students. University courses often include some element of marine microbiology as a specialist option for major marine biology or oceanography students who have little previous knowledge of microbiology. Marine microbiology is poorly covered in most marine biology courses and textbooks, despite its importance in ocean processes and interactions with other marine life – I hope that this book may play some small part in rectifying this deficiency. I also hope that the book will be useful to major microbiology students studying courses in environmental microbiology, who may have little knowledge about ocean processes, nor about the applications of the study of marine micro-organisms. I have attempted to make the book sufficiently self-contained to satisfy the various potential audiences, with the overall aim of bringing together an understanding of microbial biology and ecology with consideration of the applications for environmental management, human welfare, health and economic activity. I am particularly interested to receive comments from students and instructors – please e-mail me at cmunn@plymouth.ac.uk

Each chapter focuses on a particular aspect of marine microbiology. As will become evident, many common themes and recurring concepts link the activities, diversity, ecology and applications of marine microbes. I have attempted to summarize the current state of knowledge about each aspect with extensive cross-linking to other sections. To improve readability, I avoid the use of references in the main text, but each chapter contains one or more *Research Focus* boxes, which explore in more detail some topical areas of investigation. The choice of these topics is entirely my personal whim – they represent subjects that I think are particularly exciting, intriguing or controversial. They are intended, as far as possible, to be relatively self-contained 'mini-essays', which can be read in almost any order. I hope that students will be stimulated to read some of the original research papers suggested and use these as a starting point for investigations and seminar discussions. I have also included a glossary of some of the most important key terms, which I hope will be a useful learning aid.

The book begins with an introduction to the marine environment, the various habitats that exist and the role of microbes within them. In *Chapter 2*, I describe some of the methods used in marine microbiology. It is essential that the student appreciates the principles of methodology underpinning our current views about the role of microbes in the sea, which will enable better appreciation of research papers in this fast-moving field. I do not include technical details, but the recent book *Marine Microbiology* (*Methods in Microbiology Volume 30*) edited by John Paul (Academic Press, 2001) is essential reading for students undertaking practical work. *Chapters 3 and 4* discuss the structure, growth and physiology of marine *Bacteria* and *Archaea*. Students with a strong background in microbiology will find some of this material familiar, but I emphasize aspects of particular relevance to marine microbes that are often missing from general microbiology texts. *Chapters 5 and 6* consider the diversity of *Bacteria* and *Archaea*. Space does not permit detailed taxonomic or biochemical coverage and the interested student should consult one of the major microbiology texts for further information. These chapters emphasize the fact that much of our knowledge of prokaryotic diversity comes from the recent use of molecular methods to investigate organisms that cannot currently be cultured. I hope that you will get a flavor of the exciting developments in this fast-moving field. *Chapter 7* describes the main properties and activities of marine eukaryotic microbes, especially the recently discovered role of protists in ocean food webs. *Chapter 8* shows how the importance of viruses infecting marine bacteria and protists is emerging as a major area of research, in view of the impact of viruses on population structures and ocean processes. *Chapter 9* brings together the concepts from earlier chapters to provide a consideration of the overall importance of microbes in nutrient cycling and biogeochemical cycles. This area is one of the most active branches of marine science, asking questions about the 'big issues' such as productivity, gaseous exchange between the atmosphere and oceans and the role of microbes in climate control. Theoretical considerations and modeling play a big part in such research, but I do not attempt to cover these aspects here. Students who wish to explore this topic in more detail are advised to consult David Kirchman's excellent book *Microbial Ecology of the Oceans* (Wiley, 2000), which contains chapters by world experts on these issues. Symbiotic interactions between different organisms are of major significance in marine ecology and *Chapter 10* gives examples of recent discoveries that show just how widespread the symbiotic mode of life is. Unravelling the molecular basis of these interactions and thinking about how they evolved is a fascinating area of research. At this point in the book, you should be aware of the highly important role of microbes in ecosystems, which mostly occur unperceived by ordinary human experience. The remaining chapters emphasize aspects of marine microbiology that have a more obvious impact or direct application in human affairs. *Chapter 11* describes the importance of bacteria and viruses in human disease, including both indigenous marine organisms and those introduced by pollution. One of the most interesting aspects to emerge here is the evidence that some indigenous aquatic bacteria have evolved through genetic exchange to become human pathogens. *Chapter 12* describes the increasing importance of harmful algal blooms caused by toxic dinoflagellates and diatoms, which many view as a sign of the deteriorating condition of our oceans. The chapter includes some developments that have caused great controversy in the scientific world. *Chapters 13–15* deal with microbial diseases of marine mammals, fish and invertebrates respectively. A strong link between disease and environmental change emerges and the economic importance of disease and its control in aquaculture is emphasized. *Chapter 16* considers other detrimental activities, but emphasizes the beneficial effects of marine microbes of direct importance in human affairs. Marine microbial biotechnology holds great promise for mitigation of environmental damage and the development of new products and processes. The final chapter attempts to link the major recurring themes in our coverage of marine microbiology and finishes with a 'crystal ball' look at some of the likely developments in the next few years.

Acknowledgments

I gratefully acknowledge support from the Leverhulme Trust for a Study Abroad Fellowship, during which I undertook much of the writing. This would not have been possible without the help of many colleagues at the University of Plymouth, especially Graham Bradley, David Gaudie, Martyn Gilpin, Awadhesh Jha and Christine King, who covered responsibilities during my absence. Many experts kindly commented on individual chapters or gave me valuable ideas during visits to their laboratories, including Rita Colwell, Fenny Cox, Rocky de Nys, Terry Done, Jed Fuhrman, Paul Jepson, Ian Joint, Dave Karl, JoAnn Leong, Teresa Lewis, Lyndon Llewellyn, Margaret McFall-Ngai, Madeleine van Oppen, Ned Ruby, Bette Willis and Willie Wilson. I also thank Brian Austin for his encouragement and helpful comments during the initial stages and for review of the manuscript. To Geoff Wigham, Clare Morrall and friends at St Georges University and True Blue – thank you for support and friendship during my stay in the beautiful island of Grenada. I am especially grateful to Leigh Owens (James Cook University), David Bourne (Australian Institute of Marine Science), Jan Smith and Lone Høj – thank you for your splendid hospitality, scientific support and the wonderful experience of Tropical North Queensland. Finally, to Sheila – thank you for your constant support and love.

Colin Munn
Plymouth, UK

Abbreviations

AAnP	aerobic anoxygenic photosynthesis
AFLP	amplified fragment length polymorphism
AHL	acyl homoserine lactone
AMP	adenosine monophosphate
AO	acridine orange
APB	acid-producing bacteria
ASP	amnesic shellfish poisoning
atm	unit of atmospheric pressure (=0.103 Mpa)
ATP	adenosine triphosphate
BAC	bacterial artificial chromosome
BBD	black band disease
BBL	benthic boundary layer
BKD	bacterial kidney disease
BMNV	baculoviral midgut gland necrosis virus
BP	baculovirus penaei
cDNA	complementary DNA
CDV	canine distemper virus
CFB	*Cytophaga – Flavobacterium – Bacteroides* group
CFP	ciguatera fish poisoning
CFU	colony-forming units
CLSM	confocal laser scanning microscopy
CM	cytoplasmic membrane
CMV	cetacean morbillivirus
CTC	5-cyano-2,3-dilotyl tetrazolium chloride
CZCS	Coastal Zone Color Scanner
DAPI	4′6′- diamido-2-phenylindole
ddNTP	dideoxyribonucleotide triphosphate
DDT	dichloro-diphenyl-trichloroethane
DGGE	denaturing gradient gel electrophoresis
DHA	docosahexanoic acid
DIN	dissolved inorganic nitrogen
DMS	dimethyl sulfide
DMSP	dimethylsulfide propionate
DNA	deoxyribonucleic acid
dNTP	deoxyribonucleotide triphosphate
DOC	dissolved organic carbon
DOM	dissolved organic material
DSP	diarrhetic shellfish poisoning
ELISA	enzyme-linked immunosorbent assay
ELM	epifluorescence light microscopy
EPS	exopolymeric substances
FACS	fluorescent-activated cell sorting (sorter)
FAME	fatty acid methyl esters

FAT	fluorescent antibody technique
FC	fecal coliforms
FCM	flow cytometry
FISH	fluorescence *in situ* hybridization
FITC	fluoroscein isothiocyanate
FS	fecal streptococci
GC or G+C	guanine+cytosine base pair
GCAT	glycerophospholipid:cholesterol acyltransferase
GFP	green fluorescent protein
HNHC	high nutrient, high chlorophyll
HNLC	high nutrient, low chlorophyll
HPLC	high performance liquid chromatography
HPV	hepatopancreatic parvovirus
IFAT	indirect fluorescent antibody technique
IHHNV	infectious hypodermal and hematopoietic necrosis virus
IHN (IHNV)	infectious hematopoietic necrosis (virus)
IPN (IPNV)	infectious pancreatic necrosis (virus)
IROMPs	iron-restricted outer membrane proteins
ISA (ISAV)	infectious salmon anemia (virus)
kDa	kilodalton
LAB	lactic acid bacteria
LAL	*Limulus* amoebocyte assay
LNHC	low nutrient, high chlorophyll
LNLC	low nutrient, low chlorophyll
LPS	lipopolysaccharide
Mb	megabase (10^6 nucleotides)
MBV	monodon baculovirus
MIC	microbiologically influenced corrosion
MPN	most probable number
mRNA	messenger RNA
MUG	methylumbiliferyl-β-glucuronide
NADPH	nicotinamide adenine dinucleotide phosphate (reduced)
NAG	N-acetyl glucosamine
NSP	neurotoxic shellfish poisoning
OM	outer membrane (of Gram-negative bacteria)
ONPG	o-nitrophenol-β-galactopyranoside
ORF	open reading frame
PAGE	polyacrylamide gel electrophoresis
PAH	polyaromatic hydrocarbon
PCB	polychlorinated biphenyl
PCR	polymerase chain reaction
PDV	phocine distemper virus
PFGE	pulsed field gel electrophoresis
POM	particulate organic material
POP	persistent organic pollutant
PSP	paralytic shellfish poisoning
PST	paralytic shellfish toxin
PUFA	polyunsaturated fatty acid
PyMS	pyrolysis mass spectroscopy
QS	quorum sensing
RAPD	random amplified polymorphic DNA
rDNA	DNA encoding ribosomal RNA

RFLP	restriction fragment length polymorphism
RNA	ribonucleic acid
rRNA	ribosomal RNA
RT	reverse transcriptase
RTN	rapid tissue necrosis
RubisCO	ribulose bisphosphate carboxylase
RUBP	ribulose bisphosphate
S	Svedberg unit
SA	surface area
SDS-PAGE	sodium dodecyl sulfate polyacrylamide gel electrophoresis
SeaWiFS	Sea-viewing Wide Field-of-view Sensor
SEM	scanning electron microscopy
SMV	spawner mortality virus
SRB	sulfate-reducing bacteria
SRSV	small round structured viruses
SSU	small subunit
SWI	sediment–water interface
TBT	tributyl tin
TCA	tricarboxylic acid
TCBS	thiosulfate-citrate-bile-sucrose
Tcp	toxin coregulated pili
TDH	thermostable direct hemolysin
TEM	transmission electron microscopy
TGGE	denaturing gradient gel electrophoresis
Tm	dissociation temperature ('melting point') of DNA
TRFLP	terminal restriction fragment length polymorphism
TRH	thermostable related hemolysin
tRNA	transfer RNA
USEPA	United States Environment Protection Agency
UV	ultraviolet
V	volume
VAI	vibrio autoinducer
VBNC	viable but nonculturable
VHML	*Vibrio harveyi* myovirus like
VHS (VHSV)	viral hemorhagic septicemia (virus)
VLP	virus-like particles
VPI	*Vibrio* pathogenicity island
w/v	weight per volume
WBD	white band disease
YHV	yellow head virus

Glossary of key terms

adaptive bleaching hypothesis hypothesis that corals that have expelled zooxanthellae due to stress (bleaching) are recolonized by genetically different types with greater tolerance of the stress condition

aerobic anoxygenic photosynthesis (AAnP) process of photosynthesis occurring under aerobic conditions in which electron donors such as sulfide or organic matter are used, without evolution of oxygen

algae common name for polyphyletic group of unicellular or multicellular protists usually obtaining nutrition by photosynthesis (may be mixotrophic)

Archaea domain of prokaryotes characterized by isoprenoid glycerol diether or diglycerol tetraether membrane lipids, archaeal rRNA, complex RNA polymerase and other distinctive properties

autotroph organism using CO_2 as principle source of carbon

bacteria general term for prokaryotes of the domains *Bacteria* and *Archaea*

Bacteria domain of prokaryotes characterized by diacyl glycerol diester membrane lipids, bacteria rRNA, simple RNA polymerase and other distinctive properties

bacteriophage (phage) virus that infects bacteria

bacterioplankton free-floating aquatic bacteria

barophile (piezophile) organism that requires high pressures for growth (usually >400 atm)

barotolerant organism that can grow at high pressures, but is not dependent on them

benthic organisms living on or in the sediment of the ocean floor

bioaugmentation modification of microbial community composition by addition of specific microbes to improve the rate of bioremediation

biodegradation breakdown of complex organic compounds

biodeterioration damage to natural or fabricated materials through microbial activities

biofilm organized structure of microbial cells, extracellular products and associated substances formed on surfaces

biofouling colonization of marine surfaces by microbes with successive colonization by algae and animals

biogeochemical cycles movements through the Earth system of key elements essential to life, such as carbon, nitrogen, oxygen, sulfur and phosphorus

bioinformatics the computational storage, retrieval and analysis of information about biological structure, sequence and function

biological pump process by which CO_2 at the ocean–atmosphere interface is fixed by photoautotrophs into organic matter, redistributing carbon throughout the oceans and sediments

bioluminescence production of light by living organisms

biomimetics design process mimicking processes or principles of assembly found in living organisms

bioremediation biological process to enhance the rate or extent of naturally occurring biodegradation of pollutants; removal or degradation of pollutants using biological processes

biotechnology application of scientific and engineering principles to provide goods and services through mediation of biological agents

carbon cycle the flux of carbon through interconnected reservoirs (atmosphere, terrestrial biosphere, oceans, sediments and fossil fuels)

carbon fixation incorporation of CO_2 into cellular organic material

carbon sequestration uptake and storage of carbon via biological and geological processes

chemolithoautotroph microbe using CO_2 as carbon source, deriving energy and electrons from oxidation of reduced inorganic compounds

chemoorganotrophic heterotroph organism using organic compounds as a source of carbon, energy and electrons

chemotaxis microbial behavior in which microbes move towards attractant chemicals or away from repellents

clade a monophyletic group or lineage of organisms which share common inherited characteristics

climate change significant changes in global climate patterns, often synonymous with 'global warming'

cloning (molecular) isolation of a DNA sequence and propagation of multiple copies in a host organism, usually a bacterium, for production of large quantities of DNA for molecular analysis

coccolithophores unicellular marine bloom-forming algae with cell surface covered by calcified plates

community fingerprinting analysis of the genetic sequences in an assemblage of different types of microbes, usually achieved via DGGE of PCR-amplified genes

conditioning film layer of proteins and polysaccharides that coats surfaces within a short period of immersion in seawater; an essential first step in biofilm formation

confocal laser scanning microscopy (CLSM) microscopic methods in which laser light scans the specimen at one level, yielding an image with high contrast and resolution; especially valuable for examining biofilms and biological tissues

Cyanobacteria large group of oxygenic photosynthetic *Bacteria*

denaturing gradient gel electrophoresis (DGGE) technique for the separation of PCR products with different sequences

denitrification reduction of nitrate of N_2 during anaerobic respiration

diatoms unicellular or chain-forming marine and freshwater algal protists; major contributors to primary productivity

dimethylsulfide (DMS) volatile compound produced from DMS propionate, a major component of marine algae; DMS has important effects on climatic processes

dinoflagellate unicellular algal protists with two flagella and spinning motion; photosynthetic, phagotrophic or mixotrophic; some are parasites and pathogens

DNA vaccination direct administration of DNA encoding antigenic proteins into tissue, such that the recipient produces an effective immune response

DNA–DNA hybridization method for determining the relatedness of genetic sequences by determining hybridization of single-stranded DNA extracted from two organisms

El Niño extended warming of the central and eastern Pacific that leads to a major shift in ocean currents and weather patterns across the Pacific; occurs at irregular intervals of 2–7 years

endosymbiont microbe that lives symbiotically within the body of another organism (usually used for intracellular associations)

environmental genomics direct extraction and sequencing of nucleic acids from environmental samples, without the need for isolation or culture of the constituent organisms

epibiotic microbe that lives on the surface of another organism

epifluorescence light microscopy (ELM) method for visualizing bacteria, viruses and other particles on the surface of a membrane after staining with a fluorochrome, which emits light at a particular wavelength after illumination

eukaryotic cells with a membrane-bound nucleus and organelles (the spelling eucaryotic is also used)

eutrophic environment enriched by high levels of nutrients

export production amount of fixed organic matter produced in the photic zone of the oceans which is exported to deeper waters (carbon flux)

fermentation energy-yielding process in which the substrate is oxidized without an exogenous electron acceptor; organic molecules usually serve as both electron donors and acceptors

flow cytometry (FCM) method for quantifying and determining the properties of particles passed in a 'single-file' flow through a laser beam and detected by their fluorescent properties; cells or viruses are usually tagged with a fluorochrome, which is often attached via specific oligonucleotides or antibodies

Fungi monophyletic group of heterotrophic eukaryotes with absorptive nutrition; unicellular (yeasts) or mycelial

Gaia hypothesis concept that temperature, gaseous composition and oxidation state of the atmosphere, oceans and Earth's surface are actively controlled by its living organisms behaving as a system to maintain a stability conducive to the maintenance of life

gene probe oligonucleotide sequence used in hybridization methods to detect organisms belonging to a specific group

genome complete complement of genetic information in a cell or virus

genomic fingerprinting method for distinguishing closely related individuals based on small differences in DNA sequences

genomics study of the molecular organization of genomes and gene products using sequence information in coding and noncoding regions

glycocalyx layer of interconnected polysaccharides surrounding bacterial cells; important in cell interactions and biofilm formation

greenhouse effect natural warming of the atmosphere by absorbance and re-emission of infrared radiation while allowing shortwave radiation to pass; usually refers to the enhanced effect due to elevated levels of gases such as water vapor, CO_2, methane, nitrous oxide and chlorofluorocarbons

harmful algal bloom (HAB) unusual excessive growth of cyanobacterial, microalgal or macroalgal species resulting in toxin production, mortalities of marine life, foaming or other nuisance effects

halophilic microbe that requires high levels of sodium chloride for growth

heat-shock response expression of a range of proteins in response to sudden exposure to elevated temperatures or other stressful conditions; the response helps to protect cells from damage

heterotroph see chemoorganotrophic heterotroph

HNLC high nutrient, low chlorophyll; ocean regions characterized by low phytoplankton growth despite ample concentrations of nitrate and in which iron is probably the limiting nutrient

horizontal (lateral) gene transfer transfer to genes from one independent, mature organism to another; occurs via physical contact (conjugation), transfer of naked DNA (transformation) or via bacteriophages (transduction)

hyperthermophile extremely thermophilic bacterium (optimum growth above 80°C)

iron acquisition active mechanism by which ocean microbes and pathogens must obtain iron for cellular processes via secretion of siderophores and/or by surface components

iron hypothesis hypothesis that concentration of free iron in oceans plays a major regulatory role in phytoplankton productivity; most offshore waters contain very low levels since they are remote from land-masses and so receive low inputs of iron from terrestrial sources

limiting nutrient nutrient in shortest supply, the concentration of which limits growth and reproduction of particular types of organisms (includes N, P, Si and Fe)

live attenuated vaccine a virus or bacterium in which virulence has been eliminated; used to stimulate immunity against infection

lysogeny incorporation of a phage genome into a bacterial genome so that it replicates without initiating a lytic cycle unless stimulated to do so by inducing conditions

marine snow particles composed of aggregated cellular detritus, polymers and living microbes

mesocosm experimental system holding large volumes of seawater to stimulate open water conditions

methanogens group of strictly anaerobic *Archaea* that obtain energy by producing methane from CO_2, H_2, acetate and some other compounds

methanotrophs subset of methylotrophic organisms capable of oxidizing methane to CO_2

methylotrophs group of aerobic *Bacteria* that oxidize organic compounds without carbon–carbon bonds, including methanol, methylamine and sometimes methane, as a sole source of carbon and energy (cf. methanotrophs)

microalgae microscopic (mostly unicellular) protists, traditionally classified as algae

microarray technology method for determining gene expression by the binding of mRNA or cDNA from cells to an array of oligonucleotides immobilized on a surface

microbial loop process by which organic matter synthesized by photosynthetic organisms is remineralized by the activity of bacteria and protists enabling reuse of minerals and CO_2 by primary producers

microbial mat complex layered community of microbes on aquatic surfaces, characterized by chemical gradients and associated physiological activities

mixed layer uppermost layer of the ocean, mixed by wind; depth varies in different regions, depending on temperatures, upwelling and seasonal effects containing the surface waters

mixotrophy combination of autotrophic and heterotrophic processes

monophyletic a lineage of organisms belonging to the same phylogenetic cluster or clade

most probable number (MPN) method for determining microbial population in a liquid based on statistical probability of growth after inoculation of media with various dilutions of the sample

mutualism symbiotic association in which both partners benefit

nanotechnology construction of materials and functional objects assembled from basic molecular building blocks

neuston the surface film between water and the atmosphere

nitrification oxidation of ammonia to nitrate by certain chemolithotrophic bacteria

nitrogen fixation conversion of atmospheric N_2 to ammonia, carried out by *Cyanobacteria* and some other prokaryotes

oligotrophic environment with very low nutrient levels; also used to describe an organism adapted to such low-nutrient conditions

open reading frame (ORF) arrangement of nucleotides in triplet codons in DNA which does not contain a stop codon; sequences larger than 100 are considered to be potential protein-coding regions

osmoprotectant compatible solute accumulated in the cytoplasm to protect cells from loss of water to the external environment

osmotrophy feeding by absorption of soluble nutrients

pelagic water column in the open ocean; also used to describe organisms in this habitat

phagotrophy feeding by ingestion of particles into vacuoles

photic zone upper layer of ocean water penetrated by light (of appropriate wavelengths to permit phototrophy by different organisms)

photoautotroph organism that grows using light energy, inorganic compound(s) as a source of electrons and CO_2 as a carbon source

photoheterotroph (photoorganotroph) organism that grows using light energy, with organic compounds as source of electrons and carbon.

photoreactivation DNA repair process for excision of pyrimidine dimers by an enzyme activated by blue light

photosynthesis utilization of light as a source of energy for the formation of organic compounds from CO_2

phototrophy utilization of light as a source of energy for metabolism

phylogenetic classification a system of classification based on the genetic relatedness and evolutionary history of organisms, rather than similarity of current characteristics

phytoplankton free-floating photosynthetic algae and *Cyanobacteria*

picoplankton plankton microbes in the 0.2–2.0 μm size range (*Bacteria, Archaea* and some flagellates)

piezophile see barophile

plankton general term for free-floating microscopic organisms in water

plasmid a double-stranded DNA molecule in bacteria, carrying genes for specialized functions that replicates independently or can be integrated into the chromosome

polymerase chain reaction (PCR) *in vitro* amplification of DNA fragments employing sequence-specific oligonucleotide primers and thermostable polymerases

primary productivity rate of carbon fixation by autotrophic processes in the oceans; either gross (total biomass) or net (gross productivity less the respiration rate of producing organisms)

prokaryotic cells without a true membrane-bound nucleus (the spelling procaryotic is also used)

proteomics analysis of the complete protein complement of a cell using two-dimensional electrophoresis, mass spectrometry and other techniques

protist simple eukaryotes, usually unicellular or may be colonial without true tissues

protozoa common name for a polyphyletic group of unicellular protists that lack cell walls, usually feeding by phagotrophy (may be mixotrophic)

psychrophilic microbe adapted to growth at low temperatures (grows well around 0°C; temperature maximum usually 15–20°C)

pycnocline a zone having a marked change in density of water as a function of depth

quorum sensing mechanism of regulating gene expression in which bacteria measure their population density by secretion and sensing of signaling molecules; when these reach a certain critical level, the bacterium will express specific genes

radioisotope unstable isotope of an element that decays or disintegrates spontaneously, emitting radiation; compounds labeled with ^{14}C, ^{3}H, ^{32}P or ^{35}S are commonly used to study metabolic pathways and the rate of reactions

recombinant DNA technology insertion of a gene into a cloning vector such as a plasmid and transfer into another organism to produce a recombinant molecule (genetic engineering)

Redfield ratio the relatively constant proportions maintained between the elements C, N, P and O taken up during synthesis of cellular material by marine organisms and released by subsequent remineralization

remote sensing collection and interpretation of information about an object without being in physical contact; usually refers to measurement of physical and chemical properties of the oceans via satellite instruments

respiration energy-yielding metabolic process in the substrate is oxidized by transfer in an electron-transport chain to an exogenous terminal electron acceptor such as O_2, nitrate or certain organic compounds

rRNA analysis analysis of the nucleotide sequence of ribosomal RNA molecules or of the genes that encode them together with noncoding spacer regions (this may be designated rDNA); the main method used in phylogenetic classification and environmental genomics

siderophore organic molecule excreted by bacteria, which complexes with iron in the environment and transports it into the cell

signature sequence oligonucleotide sequence (e.g. in rRNA) that characterizes a particular group of organisms and used to design gene probes

solubility pump ocean process producing a vertical gradient of dissolved inorganic carbon due to increased solubility of CO_2 in cold water; deep water with high carbon content is formed in high latitudes and is transported by deep ocean currents, upwelling at low latitudes

Southern blotting techniques for transferring denatured DNA fragments from an agarose gel to a nitrocellulose sheet for identification using a hybridization probe

species (of bacteria) a collection of strains that share many relatively stable common properties and differ significantly from other collections of strains; attempts to define bacterial species by genetic methods are problematic but DNA–DNA hybridization is commonly employed

stable isotope analysis analysis of 'heavy' and 'light' forms of an element used to distinguish biological from purely geochemical processes; e.g. by measuring the ratio of ^{13}C:^{12}C or ^{34}S:^{32}S in organic material

strain (of bacteria) a population descended from an isolate in pure culture

sulfate-reducing bacteria (SRB) bacteria able to use sulfate or elemental sulfur as a terminal electron acceptor during anaerobic respiration; usually applied specifically to a group of δ-*Proteobacteria*, although some *Archaea* can also reduce sulfate

symbiosis living together; close association of two different organisms (usually interpreted as mutualism)

syntrophic association between microbes in which the growth of one or both partners depends on the provision of nutrients or growth factors from the activities of the other

TCA cycle series of metabolic reactions in which a molecule of acetyl coA is completely oxidized to CO_2, generating precursors for biosynthesis and NADH and $FADH_2$, which are oxidized in the electron transport chain; also known as Krebs cycle and citric acid cycle

thermocline boundary layer in ocean water separating water of different temperatures

thermophilic organism that grows at temperatures above 55°C

ultramicrobacteria term used to describe very small bacteria (as low as 0.2 μm diameter) occurring naturally in seawater or formed by miniaturization under low-nutrient conditions

upwelling upward motion of deep water, driven by effects of winds, temperature and density, bringing nutrients to the upper layers of the ocean

viable but nonculturable (VBNC) different interpretations have been made of this term; in this book, used to describe a physiological state in which it is not possible to culture an organism on media that normally supports its growth, although cells retain indicators of metabolic activity

virulence the degree of pathogenicity of a microorganism as indicated by the production of disease signs

Western blotting technique for transferring protein bands from a polyacrylamide gel to a nitrocellulose sheet for identification using an enzyme-linked antibody

zoonosis disease generally associated with animals that can be transmitted to humans

zooplankton free-floating small animals and nonphotosynthetic protists

Microbes in the marine environment 1

1.1 What is marine microbiology?

Ever since a detailed study of the microbial world began at the end of the nineteenth century, microbiologists have asked questions about the diversity of microbial life in the sea, its role in ocean processes, its interactions with other marine life and its importance to humans. However, despite excellent work by pioneering scientists, progress in understanding these issues was often slow and most microbiologists remained unaware of this field of study until recently. Towards the end of the twentieth century, a number of factors conspired to propel marine microbiology to the forefront of 'mainstream' science and it is now one of the most exciting and fast-moving areas of investigation. Powerful new tools (especially in molecular biology, remote sensing and deep-sea exploration) have led to astonishing discoveries of the abundance and diversity of marine microbial life and its role in global ecology. We now realize the vital role that marine microbes play in the maintenance of our planet, a fact that will have great bearing on our ability to respond to problems such as human population increase, overexploitation of fisheries, climate change and pollution. Study of the interactions of marine microbes with other organisms is providing intriguing insights into the phenomena of food webs, symbiosis and pathogenicity. Since some marine microbes produce disease or damage, we need to study these processes and develop ways to overcome them. Finally, marine microbes have beneficial properties such as the manufacture of new products and development of new processes in the growing field of marine biotechnology.

Defining the terms *microbiology* and *microorganism* is surprisingly difficult! Microbiologists study very small organisms that are too small to be seen clearly with the naked eye (i.e. less than about 1 mm diameter), but many aspects of microbiology are concerned with the activities or molecular properties of microbial communities rather than viewing individual cells with a microscope. As usually defined, microbiology encompasses the study of bacteria, viruses and fungi, although the study of each of these groups is a specialized field. The term microorganism simply refers to forms of life that fall within the microscopic size range, but there is a huge spectrum of diversity concealed by this all-encompassing term. Indeed, some 'microorganisms' are large enough to see without using a microscope, so this is not entirely satisfactory either. Some scientists would argue that the distinguishing features of microorganisms are small size, cellularity and osmotrophy (feeding by absorption of nutrients). The osmotrophic characteristic is important, because diffusion processes are a major limitation to cell size, as discussed in the next section. However, this characteristic would exclude many microscopic protists (algae and protozoa), many of which feed by phagotrophy (engulfment of particles). These 'plant-like' or 'animal-like' groups are most commonly studied by specialists who traditionally have a background in botany or zoology. However, many marine protists are mixotrophic and can switch from photosynthesis to phagotrophic feeding, so the plant or animal similarity is meaningless. Additionally, viruses are microscopic and are obviously included in the remit of microbiologists, but they are not cellular *organisms* and, arguably, they are not *living* either. There is a huge diversity of interconnected microbial life forms in the marine environment and problems of definition and artificial divisions are often unhelpful. In this book, I use the term *microbe* as a generic term, which includes all microscopic organisms, i.e. the *Bacteria, Archaea, Fungi,* protists (protozoa and the unicellular algae) and the viruses.

The book's subtitle is an important component of my approach. *Ecology* in the subtitle refers to the interaction of marine microbial life with the environment. We shall encounter a huge diversity of marine microbes and see that their activities have enormous importance in marine ecosystems. As more scientists become involved in this field and our fundamental knowledge increases, we see an increasing number of important *applications* for environmental management, human health and welfare and economic activity. This provides the second theme indicated by the subtitle.

This chapter introduces the microbes and considers modern approaches to classifying them within the living world. This is followed by a description of the major habitats for marine microbes, which serves as the foundation for detailed consideration in later chapters. What do we mean by the marine environment? It obviously includes the open ocean, but this book also considers coastal and estuarine waters as well as surfaces and sediments. However, specialized coastal habitats such as mangroves and salt marshes are only covered briefly.

1.2 Biological organization and the evolution of life

1.2.1 Cells

All living organisms are composed of one or more basic units of structure and function – the cell. The key feature of all cells is a cell membrane (composed of lipids and proteins), which encloses the cytoplasm containing the structures, macromolecules and chemicals that enable it to carry out the functions of metabolism, growth and reproduction, together with the informational molecules (DNA and RNA) that direct cellular processes, replication and evolution. Cells are highly dynamic structures, with a constant exchange of materials between the cytoplasm and the external environment; the cell membrane is critical in this process. All cells contain ribosomes, structures composed of protein and RNA that carry out the synthesis of proteins. The metabolic activities and the information processing functions of the cell are coordinated to ensure that the cell shows balanced growth and reproduces with high fidelity. This is achieved by regulation of enzyme activity and genetic expression.

Although all cellular life on Earth shares the common features referred to above, there are two fundamentally different levels of cellular organization, which are obvious when sections of cells are examined in the electron microscope. Prokaryotic cells have a simple internal structure and usually carry out all their metabolic activities in an undifferentiated cytoplasm and the genetic material is not bound by a nuclear membrane. Eukaryotic cells are usually larger and more complex and contain a membrane-bound nucleus and organelles with specific functions. Mitochondria occur in all eukaryotic cells, with the exception of a few anaerobic protozoa, and carry out the processes of respiratory electron transport. In photosynthetic eukaryotes, chloroplasts carry out reactions for the transfer of light energy for cellular metabolism. The organelles of eukaryotes are of similar size and have many features of prokaryotic cells, factors that (together with other evidence) suggest that they have evolved from prokaryotic ancestors. Most prokaryotes are unicellular whereas many eukaryotes are multicellular. Marine microbiology provides numerous examples that emphasize the fact that the distinction between prokaryotic and eukaryotic cells is not clear-cut. Some marine prokaryotes are much larger than eukaryotic cells and some show obvious multicellularity or differentiation during their lifecycle, or have complex intracellular structures.

1.2.2 The nature of viruses

Viruses are *not* cellular. Virus particles consist of a protein capsid containing the viral genome composed of either RNA *or* DNA. In order to replicate, viruses must infect living cells and take over the host's cellular machinery. It is often thought that viruses could have evolved (perhaps from bacteria) as obligate parasites that have progressively lost genetic information until they consist of only a few genes. A more likely explanation for their origin is that they represent fragments of host-cell RNA or DNA that has gained independence from cellular control. The main evidence in support of this is that the genome of viruses often contains sequences that are equivalent to specific sequences in the host cell. Viruses exist for every major group of cellular organisms (*Bacteria, Archaea,* protists, *Fungi,* plants and animals), but we currently have knowledge of only a tiny proportion of the viruses infecting marine life. Viruses have undoubtedly played a major role in evolution through the phenomenon of gene transfer.

1.2.3 Phylogenetic approaches to classifying the living world

Biologists usually rely on the study of morphology and physiological properties to classify living organisms, but these characteristics have always proved frustratingly unhelpful when dealing with microbes. We need a way of measuring the relatedness of organisms in evolutionary terms, i.e. a phylogenetic system of classification. The background to this method has its origins in the ideas of Zuckandl and Pauling in the 1960s, who suggested that we could study evolutionary relationships between organisms by looking at the information content of their macromolecules (especially nucleic acids and proteins). If two organisms are very closely related, we expect the sequence of the individual units in a macromolecule to be more similar than they would be in two unrelated organisms. Initial studies were based on the protein-sequencing methods that were then available and progress was slow. However, in the 1970s Carl Woese and colleagues pioneered the use of ribosomal RNA (rRNA) sequencing in order to develop a better view of prokaryotic diversity. Our view of the living world has since been revolutionized by advances in this approach, made possible because of the parallel advances in molecular biological techniques and computer processing of the large amounts of information generated.

Ribosomes are composed of two subunits that can be separated by high-speed centrifugation. Based on the sedimentation rates, a prokaryotic ribosome is termed 70S and is composed of 50S and 30S subunits. The eukaryotic ribosome is 80S overall, with 60S and 40S subunits (see *Figure 2.4*). For various reasons discussed in *Section 2.6.1,* the small subunit (SSU) rRNAs (16S in prokaryotes and 18S in eukaryotes) have become the molecules of choice for phylogenetic comparisons. Because the secondary structure of rRNA is so important in the ribosome and the vital cell function of protein synthesis, base sequence changes in the rRNA molecule occur quite slowly in evolution. In fact, some parts of SSU rRNA are highly conserved and sequence comparisons can be used to ascertain the similarity of organisms on a broad basis. The methods and applications of this major technique are described in *Section 2.6.*

1.2.4 The three-domain tree of life

Using SSU rRNA sequencing, Woese identified three distinct lineages of cellular life, referred to as *domains.* The domains *Bacteria* and *Archaea* have a prokaryotic cell structure, whilst the domain *Eukarya* has the more complex eukaryotic cell structure.

By constructing a phylogenetic 'tree of life', we can assume that the three domains diverged from an original 'universal ancestor' (*Figure 1.1*). From this hypothetical ancestor, it appears that life developed in two main directions. One division evolved into the *Bacteria,* whilst the other division split again, with one subdivision forming the *Archaea* and the other forming the *Eukarya.* The three-domain system of classification is now almost universally adopted by microbiologists. Apart from our preference for a phylogenetic system, it allows microbiologists to say that we study two entire domains of life, and a significant proportion of the third! The most important consequence of the three-domain tree of life is that we now realize that the *Archaea* are not a peculiar, specialized group of bacteria as originally thought (for many years they were called the archaebacteria), but are in fact closer phylogenetically to the *Eukarya* than they are to the *Bacteria.* The *Bacteria* and *Archaea* are completely different groups, each with their own evolutionary history. They share the fact that they have a simple cellular organization, but have

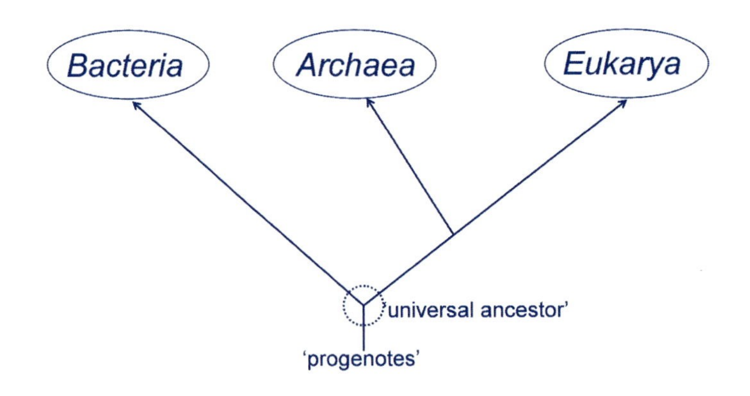

Figure 1.1

The three domains of life. The root of the tree is the hypothetical universal ancestor that evolved from precellular life (progenotes). This simplified representation does not take account of extensive lateral gene transfer between the domains.

Table 1.1 Principal cellular features distinguishing the three domains of life

Property	*Bacteria*	*Archaea*	*Eukarya*
Cell structure	Prokaryotic	Prokaryotic	Eukaryotic
Covalently closed circular DNA	Yes	Yes	No
Histone proteins in DNA	No	Yes	Yes
Plasmid DNA	Yes	Yes	Rarely
Membrane lipids	Ester linked	Ether linked	Ester linked
Membrane enclosed nucleus	No	No	Yes
Peptidoglycan in cell wall	Usually	No	No
Ribosome structure	70S	70S	80S (70S in organelles)
Initiator tRNA in protein synthesis	N-formyl methionine	Methionine	Methionine (N-formyl methionine in organelles)
Sensitivity of elongation factor to diphtheria toxin	No	Yes	Yes
Poly A tailing of mRNA	No	No	Yes
RNA polymerases	One type (four subunits)	Several types (8–12 subunits)	Three types (12–14 subunits)
Promoter structure	Pribnow box	TATA box	TATA box
Transcription factors required	No	Yes	Yes
Sensitivity of protein synthesis to chloramphenicol, streptomycin, kanamycin	Yes	No	No

many fundamental differences. However, for most marine scientists the term 'bacteria' (lower case) is deeply embedded and is still useful to describe the range of organisms that share the same basic unit of cellular organization and similar basic physiological properties. Thus, in this book 'bacteria' is used to mean the same as 'prokaryotes' and '*Bacteria*' (capitalized, italics) is only used when describing the domain of evolutionarily related prokaryotes that forms a distinct lineage from the *Archaea*.

Various lines of evidence (especially the molecular analysis of the nucleic acids and proteins in eukaryotic cellular organelles) support the theory of serial endosymbiosis to account for the evolution of eukaryotes from primitive prokaryotic cells, first proposed by Lynn Margulis in 1970. Following the discovery of the phylogenetic relationships of the three domains, Wolfram Zillig has suggested that the original eukaryotic cell evolved through a fusion event between a cell of a pre-*Bacteria* type and a pre-*Archaea* type. As shown in *Table 1.1,* cells of *Eukarya* share several features with cells of *Archaea* (notably the protein synthesis machinery) and some with *Bacteria* (notably the nature of the cell membrane lipids). According to the Margulis hypothesis, subsequent symbiotic events led to the evolution of the chloroplast (perhaps from a cyanobacterial ancestor) and mitochondria (perhaps from a protobacterial ancestor). In modern marine systems, there is much evidence to support this hypothesis. The processes of endosymbiosis, retention of organelles by phagotrophs and mixotrophy are very common in marine protists (*Section 10.6*).

An important factor, which confounds our understanding of evolution, is the increasing recognition of horizontal (lateral) gene transfer. As our knowledge of genomes increases, we find increasing evidence of extensive gene transfer within, and between, domains. *Bacteria* and *Archaea* contain genes with very similar sequences, and *Eukarya* contain genes from both prokaryotic domains. Some bacteria have even been shown to contain eukaryotic genes. We will probably be forced to abandon the simple tree of life with its single trunk and three main branches depicted in *Figure 1.1,* in favor of a network of branches linking various trunks.

1.3 The importance of microbes in the living world

Probably the most important overriding features of microbes are their exceptional diversity and ability to occupy every conceivable habitat for life. Indeed, what we consider 'conceivable' is challenged constantly by the discovery of new microbial communities in habitats previously thought of as inhospitable, or carrying out processes which we had no idea were microbial in nature. Prokaryotes were the first life forms to evolve over 3.5 billion years ago and they have shaped the subsequent development of life on Earth ever since. The metabolic processes that they carry out in the transformation of elements, degradation of organic matter and recycling of nutrients play a central role in innumerable activities that affect the support and maintenance of all other forms of life. Microbial life and the Earth have evolved together and the activities of microbes have affected the physical and geochemical properties of the planet. Indeed, as we shall see in later parts of the book, they are actually the driving forces responsible for major planetary processes like changes in the composition of the atmosphere, oceans, soil and rocks. This is especially relevant to our consideration of the marine environment, in view of the huge proportion of the biosphere that this constitutes. Despite the preponderance of microbes and the importance of their activities, they are unseen in everyday human experience (*Box 1.1*). Microbes live and grow almost everywhere, using a huge range of resources, whereas plants and animals occupy only a small subset of possible environments and have a comparatively narrow range of lifestyles.

1.4 The importance of size

Even by the usual standards of microbiology, most microbes found in seawater are *exceptionally* small. Their very small size is the main reason why appreciation of their abundance eluded us until quite recently. As described in *Chapter 2,* this realization depended on the development of fine-pore filters, direct counting methods using epifluorescence microscopy and flow cytometry. Small cell size has great significance in terms of the physical processes that affect life. At the microscale, the rate of molecular diffusion becomes the most important mechanism for transport of substances into and out of the cell. Small cells feeding by absorption (osmotrophy) can take up nutrients more efficiently than larger cells. The surface area:volume ratio is the critical factor

Box 1.1 RESEARCH FOCUS

Prokaryotes – the unseen majority

The title of this box comes from a keynote paper by William Whitman, David Coleman and William Wiebe (Whitman *et al.* 1998), who used data from various sources to estimate the number of prokaryotes in the biosphere. The dominance of *Bacteria* and *Archaea* is confirmed by their conclusion that there are $4–6 \times 10^{30}$ prokaryotes on Earth, with a total cellular carbon content of $3.5–5.5 \times 10^{11}$ tonnes. The carbon content in prokaryotes is between 60 and 100% of that found in plants, and the protoplasmic biomass of prokaryotes is more than half that of all other life forms.

Estimated numbers of prokaryotes in various habitats (adapted from Whitman *et al.* 1998)

Habitat	Estimated total number of prokaryotes ($\times 10^{28}$)
Continental shelf	1.00×10^{26}
Open ocean. Water, upper 200 m	3.60×10^{28}
Open ocean. Water, below 200 m	6.50×10^{28}
Open ocean. Sediment, top 10 cm	1.70×10^{28}
Total marine	$\mathbf{1.18 \times 10^{29}}$
Saline lakes	1.00×10^{26}
Freshwater lakes and rivers	1.00×10^{26}
Soil	2.56×10^{28}
Subsurface sediments	3.80×10^{30}
Animal and plant surface	Estimated $10^{23}–10^{24}$
Atmosphere (lower 3 km)	Estimated 10^{19}
Grand total	$\mathbf{4.17 \times 10^{30}}$

The table shows that the deep subsurface sediments (of which 67% occur beneath the deep seas) contain over 90% of all prokaryotes. Marine habitats, which are mostly very low in nutrients, contain about half as many prokaryotes as terrestrial soils. Since the introduction of direct counting by epifluorescence microscopy (see *Section 2.2.4*) the density of prokaryotes in seawater is reported to be in the range $10^5–10^7\,ml^{-1}$. Numbers vary according to location and depth, but an average of about $10^6\,ml^{-1}$ seems to be generally accepted. In their calculations, Whitman *et al.* use values of $5 \times 10^5\,ml^{-1}$ for the upper ocean and $0.5 \times 10^5\,ml^{-1}$ for water below 200 m; these may be underestimates, so the total number of marine prokaryotes may be even higher. Calculating the number of other marine microbes is much more difficult. Estimates for the number of marine protists vary from about 10^1 to 10^5 per ml in surface waters, but much less is known about their distribution in deep water and sediments.

Another recent paper of enormous importance is that by Markus Karner, Ed DeLong and David Karl (Karner *et al.* 2001). They used the FISH technique (see *Section 2.6.10*) to provide the first quantitative estimates of the number of *Archaea* in the deep ocean. During a year-long study in the Pacific Ocean, they found that *Bacteria* dominate the top 150 m of water, representing about 90% of all cells visualized by direct microscopic counts. Below 150 m, *Bacteria* decrease in abundance and below 1000 m, they represent only 35–40% of cells (see *Figure 6.2*). By contrast, *Archaea* are a small proportion of the total count in the upper layer, but increase sharply at 250 m, so that below 1000 m they are as common as *Bacteria*. Karner *et al.* extrapolate from their data, using the known volumes of water at different depths. They conclude that the world's oceans contain 1.3×10^{28} archaeal cells and 3.1×10^{28} bacterial cells. The total estimated number of marine prokaryotes is thus within the same order of magnitude as that obtained by Whitman *et al.* The number 4.4×10^{28} is unimaginably large. To make the point, it is quite instructive to write it in the form 44 thousand, million, million, million, million.

because, as cell size increases, volume (V) increases more quickly than surface area (SA). In the case of a spherical cell, volume is a function of the cube of the radius ($V = \frac{4}{3}\pi r^3$), whereas surface area is a function of the square of the radius ($SA = 4\pi r^2$). Prokaryotic cells with large SA/V ratios are more efficient at obtaining nutrients and will grow more rapidly and reach higher cell densities. Most ocean prokaryotes have very small cell volumes and large SA/V ratios. The vast majority are smaller than about 0.6 μm in their largest dimension, and many are less than 0.3 μm, with cell volumes as low as 0.03 μm^3. Since the first description of such small cells, termed ultramicrobacteria, their size has provoked considerable controversy. Such extremely small cells could result from a genetically fixed phenotype maintained throughout the cell cycle or because of physiological changes associated with starvation. The latter explanation is supported by the fact that some cultured bacteria become much smaller when starved. Because most naturally occurring bacteria have been impossible to grow in culture, it has been difficult to determine whether small size is a genotypically determined condition for marine bacteria. However, studies with some recently cultured marine bacteria from low-nutrient (oligotrophic) ocean environments show that addition of nutrients does not cause an increase in cell size. If nutrients are severely limiting, as they are in most of the oceans, selection will favor small cells. It is quite possible that both explanations could be true for different members of the population and these points are considered further in *Section 4.5.* Small cell size also has important implications for mechanisms of active motility and chemotaxis, due to the microscale effects of Brownian movement (bombardment by water molecules) and shear forces. As discussed in *Box 3.1,* small marine bacteria have mechanisms of motility and chemotaxis quite unlike those with which we are familiar in 'conventional' microbiology.

Cells use various strategies to increase the SA/V ratio and thus improve efficiency of diffusion and transport. In fact, spherical cells are the least efficient shape for nutrient uptake and many marine bacteria are long, thin filaments (e.g. spirilla and many cyanobacteria) or have appendages (e.g. the budding and prosthecate bacteria). Many organisms have extensive invaginations of the cytoplasmic membrane, which results in increased SA. *Table 1.2* shows the dimensions and volumes of some representative marine prokaryotes. Although the majority of marine bacteria are very small, there are some notable exceptions. The bacteria *Epulopiscium fishelsoni*

Table 1.2 Size range of some representative marine prokaryotes

Organism	Characteristics	Size (μm)[1]	Volume (μm^3)
Thermodiscus sp.	Disk-shaped. Hyperthermophilic *Archaea*.	0.08 × 0.2	0.003
'Pelagibacter' (SAR11)	Crescent-shaped. *Bacteria* ubiquitous in ocean plankton.	0.1 × 0.9	0.01
Prochlorococcus sp.	Cocci. Dominant photosynthetic ocean *Bacteria*.	0.6	0.1
Vibrio sp.	Curved rods. *Bacteria* common in coastal environments and associated with animal tissues.	1 × 2	2
Staphylothermus marinus	Cocci. Hyperthermophilic *Archaea*.	15	1800
Thiploca auracae	Filamentous. Sulfur *Bacteria*.	30 × 43	40 000
Beggiatoa sp.	Filamentous. Sulfur *Bacteria*.	50 × 160	1 000 000
Epulopiscium fishelsoni	Rods. *Bacteria* symbiotic in fish gut.	80 × 600	3 000 000
Thiomargarita namibiensis	Cocci. Sulfur *Bacteria*.	750	200 000 000

[1]Where one value is given, this is the diameter of spherical cells.

Table 1.3 Classification of plankton by size

Size category	Size range (μm)	Microbial groups
Femtoplankton	0.01–0.2	Viruses
Picoplankton	0.2–2	*Bacteria*[1], *Archaea,* some flagellates
Nanoplankton	2–20	Flagellates, diatoms, dinoflagellates
Microplankton	20–200	Ciliates, diatoms, dinoflagellates, other algae

(SIZE increases downward; ABUNDANCE increases upward.)

[1]Some filamentous *Cyanobacteria* and sulfur-oxidizing bacteria occur in larger size classes, see *Table 1.2.*

(*Figure 5.13*) and *Thiomargarita namibiensis* (*Figure 5.6*) are the largest prokaryotic cells known; in fact, they are bigger than many eukaryotic cells. How they overcome the diffusional limitations is not fully understood, but *E. fishelsoni* is thought to have an unusual surface membrane, which is folded in such a way that the SA is increased considerably. *T. namibiensis* contains large sulfur granules and the cell contains a large vacuole, so that the volume of active cytoplasm is much lower than the cell dimensions would suggest.

As shown in *Table 1.3,* marine eukaryotic microbes show a considerable variation in size. Many of the heterotrophic protists, especially flagellates, are in the 1–2 μm range, cellular dimensions that are more typical of many familiar prokaryotes. Again, the realization that such small cells play a vital role in ocean processes escaped attention until quite recently. Many small protists appear capable of engulfing bacteria of almost the same size as themselves or can prey on much larger organisms (*Chapter 7*). Many groups of the flagellates, ciliates, diatoms and dinoflagellates are somewhat larger, reaching sizes up to 200 μm and amoeboid types (radiolarians and foraminifera) can be even larger and visible to the naked eye.

1.5 The world's oceans and seas

The marine ecosystem is the largest on the planet and accounts for over 90% of the biosphere. The oceans cover 3.6×10^8 km^2 (71% of the Earth's surface) and contain 1.4×10^{21} litres of water (97% of the total on Earth). The average depth of the oceans is 3.8 km, with a number of deep-sea trenches, the deepest of which is the Marianas Trench in the Pacific (11 km). The ocean floor contains large mountain ranges, and is the site of over 90% of volcanic activity on Earth.

There are five oceans, of which the Pacific Ocean is the largest. This single body of water has an area of 1.6×10^8 km^2 and covers about 28% of the Earth's surface, more than the total land area. The ocean floor in the eastern Pacific is dominated by the East Pacific Rise, while the western Pacific is dissected by deep trenches. The Atlantic Ocean is the second largest with an area of 7.7×10^7 km^2 lying between Africa, Europe, the Southern Ocean, and the Americas. The mid-Atlantic Ridge is an underwater mountain range stretching down the entire Atlantic basin. The deepest point is the Puerto Rico Trench (8.1 km). The Indian Ocean has an area of 6.9×10^7 km^2 and lies between Africa, the Southern Ocean, Asia, and Australia. A series of ocean ridges cross the basin and the deepest point is the Java Trench (7.3 km). The Southern Ocean is the body of water between 60°S and Antarctica. It covers 2.0×10^7 km^2 and has a fairly constant depth of 4–5 km, with a continual eastward water movement called the Atlantic Circumpolar Current. The Arctic Ocean, lying north of the Arctic Circle, is the smallest ocean, with an area of 1.4×10^7 km^2. About half of the ocean floor is continental shelf, with the remainder a central basin interrupted by three submarine ridges. As well as the major oceans, the topography of many bodies of water leads to their designation as marginal seas, including the Mediterranean, Caribbean, Baltic, Bering, South China Seas and many others.

At the margins of major land masses, the ocean is shallow and lies over an extension of the land called the continental shelf. This extends offshore for a distance ranging from a few km to several hundred km and slopes gently to a depth of about 100–200 m, before there is a steep drop-off to become the continental slope. The abyssal plain covers much of the ocean floor. This is a mostly flat surface with few features but is broken in various places by ocean ridges, deep-sea trenches, undersea mountains and volcanic sites.

The upper surface of the ocean is in constant motion due to winds, which generate waves and currents. Wind belts created by differential heating of air masses generate the major surface current systems. Rotation of the Earth deflects moving water in a phenomenon known as the Coriolis Effect, which results in boundary currents between large water masses. This leads to large circular gyres that move clockwise in the northern hemisphere and anticlockwise in the southern hemisphere. Such gyres and currents affect the distribution of nutrients and marine organisms. On the basis of surface ocean temperatures, the marine ecosystem can be divided into four major biogeographical zones, namely polar, cold temperate, warm temperate (subtropical) and tropical (equatorial). The boundaries between these zones are not absolute and vary with season. As noted below, water beneath a certain depth is not affected by these wind-generated currents and mixing, but there are major subsurface circulation systems, which move at different depths.

1.6 Chemical and physical factors in the marine environment

1.6.1 Properties of seawater

Seawater is a slightly alkaline (pH 7.5–8.4) aqueous solution of over 80 solid elements, gases and dissolved organic substances. The concentration of these varies considerably according to geographic and physical factors and it is customary to refer to the salinity of seawater in parts per thousand (‰) to indicate the concentration of dissolved substances. The open ocean has a constant salinity in the range 34–37‰, with differences due to dilution by rainfall and evaporation. Oceans in subtropical latitudes have the highest salinity due to higher temperatures, whilst temperate oceans have lower salinity due to less evaporation. Around the equator, salinity is at the lower end of this range due to increased dilution by rainfall. In coastal regions, seawater is diluted considerably by freshwater from rivers and terrestrial runoff and is in the range 10–32‰. In enclosed areas such as the Red Sea and Arabian Gulf, the salinity may be as high as 44‰. In polar areas, the removal of freshwater by the formation of ice also leads to increased salinity. The major ionic components of seawater are sodium (55% w/v), chloride (31%), sulfate (8%), magnesium (4%), calcium (1%) and potassium (1%). Together, these constitute over 99% of the weight of salts. There are four minor ions, namely bicarbonate, bromide, borate and strontium, which together make up just less than 1% of seawater. Many other elements are present in trace amounts (less than 0.01%), including key nutrients such as nitrate, phosphate, silicate and iron. The concentration of these is crucial in determining the growth of marine microbes and the net productivity of marine systems, as discussed in *Chapter 9*.

The concentration of salts has a marked effect on the physical properties of seawater. The freezing point of seawater at 35‰ is −1.9°C, and seawater increases in density up to this point. This results in the formation of masses of cold, dense water in polar regions. These sink to the bottom of the ocean basins and are dispersed by deep-water circulation currents. The sinking water displaces water that returns to the ocean surface in a process known as upwelling, bringing nutrients to the surface (*Figure 9.1*). Differences in density of seawater create a discontinuity called the pycnocline, which separates the top few hundred meters of the water column from deeper water. This has great significance, because the gases oxygen and carbon dioxide are more soluble in cold water. Oxygen is at its highest concentrations in the top 10–20 m of water, due to exchange with the atmosphere and production of oxygen by photosynthesis. Concentration decreases with distance from the surface until it reaches a minimum between

200 and 1000 m, and biological activity may create conditions that are almost anoxic. Below this, the oxygen content increases again due to increased solubility at lower temperature. This oxygen gradient differs greatly in different regions. The solubility of carbon dioxide is an important factor in controlling the exchange of carbon between the atmosphere and the oceans, as discussed in *Section 9.2*. Carbon dioxide reacts with water to form carbonic acid, which rapidly dissociates to bicarbonate and carbonate in the reactions

$$CO_2\text{ (gas)} + H_2O \rightleftharpoons H_2CO_3 \rightleftharpoons H^+ + HCO^{3-} \rightleftharpoons 2H^+ + CO_3^{2-}$$

These reactions tend to stay in equilibrium, buffering the pH of seawater within a narrow range and constraining the amount of CO_2 taken up from the atmosphere.

1.6.2 Solar radiation and temperature

Light is of fundamental importance in the ecology of photosynthetic microbes and possibly others that may use light energy for other functions (see *Box 2.2*), thus affecting primary productivity. The extent to which light of different wavelengths penetrates seawater depends on a number of factors, notably cloud cover, the polar ice caps, dust in the atmosphere and variation of the incident angle of solar radiation according to season and location on the Earth's surface. Light is absorbed or scattered by organisms or suspended particles. Even in the clearest ocean water, photosynthesis is restricted by the availability of light of sufficient intensity to the upper 150–200 m. This is termed the photic zone. Blue light has the deepest penetration, and photosynthetic microbes at the lower part of the photic zone have light-harvesting systems that are tuned to collect blue light most efficiently (see *Box 2.2* and *Section 5.4.4*). In turbid coastal waters, during seasonal plankton blooms, the photic zone may be only a few meters deep.

Solar radiation also leads to thermal stratification of seawater. In tropical seas, the continual input of energy from sunlight leads to warming of the surface waters to 25–30°C, causing a considerable difference in density from that of deeper waters. Thus, throughout the year, there is a marked thermocline at about 100–150 m, below which there is a sudden reduction in temperature to 10°C or less. Little mixing occurs between these layers. In polar seas, the water is permanently cold except for a brief period in the summer, when a slight thermocline results. Apart from this period, turbulence created by surface winds generates mixing of the water to considerable depths. Temperate seas show the greatest seasonal variation in the thermocline, with strong winds and low temperatures leading to strong mixing in the winter. The thermocline develops in the spring, leading to a marked shallow surface layer of warmer water in summer. As the sea cools and wind increases, the thermocline breaks down again in the autumn. Combined with seasonal variations in light intensity, these temperature stratification effects and vertical mixing have a great impact on rates of photosynthesis and other microbial activities.

1.7 Marine microbial habitats

1.7.1 The water column and marine snow

Habitats in the water column in the open sea are referred to as the *pelagic*. In coastal zones, the term neritic zone is used. The top 100–200 m corresponds to the photic zone and is termed epipelagic, whilst progressively deeper habitats are classified as shown in *Figure 1.2*. Microbes occur in all of these habitats. The surface interface (neuston) between water and atmosphere is rich in organic matter and often contains the highest levels of microbes. Plankton is a general term in marine biology, which refers to organisms suspended in the water column that do not have sufficient power of locomotion to resist large-scale water currents (in contrast to the nekton, which are strong-swimming animals). Traditionally, we refer to the phytoplankton (plants) and zooplankton (animals). Using this approach, we could add the terms bacterioplankton for bacteria and virioplankton for viruses. Traditional concepts of 'plant' and 'animal' are not very

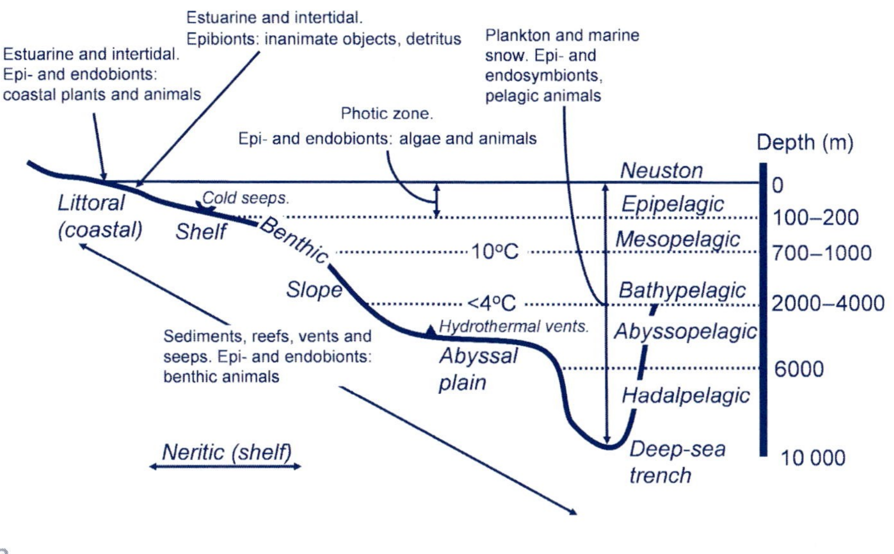

Figure 1.2

Ecological zones of the oceans and marine microbial habitats (not to scale).

satisfactory when we consider the role of photosynthetic bacteria or protistan microbes, but in common with general usage, the term phytoplankton in this book includes the *Cyanobacteria*. Another approach to classifying the plankton is in terms of size classes, for which a logarithmic scale has been devised. *Table 1.3* shows the size classes that encompass marine microbes. Thus, the viruses constitute the femtoplankton, bacteria constitute the picoplankton, whilst protists occur in the picoplankton, nanoplankton and microplankton.

It is tempting to think of seawater as a homogenous fluid, with planktonic microbes and nutrients evenly distributed within it. However, a growing body of evidence indicates that there is microscale heterogeneity in the distribution of nutrients around organisms and particles of organic matter. One of the pioneers of modern microbial oceanography, Farooq Azam (of the Scripps Institute, California), propounds the importance of studying microscale events. Large-scale processes like productivity, nutrient recycling and geochemical cycles are the result of microbial activity. In turn, large-scale physical processes like turbulence, photon flux and gas exchange are translated down to the microscale level, and affect microbial behavior and metabolism. Azam suggests that large quantities of transparent colloidal polymers structure seawater into a complex gel-like matrix. Physical factors such as diffusion, shear forces and viscosity must be considered in this context.

Marine snow is the name given to particles in the water column, so called by divers because of its resemblance to falling snowflakes when illuminated underwater. Marine snow consists of aggregates of inorganic particles, plankton cells, detritus from dead or dying plankton and zooplankton fecal material, glued together by a matrix of polymers released from phytoplankton and bacteria (*Figure 1.3*). Most particles are about 0.5 to a few μm in diameter, but they can grow to several cm in calm waters. Aggregates form as a result of collision and coagulation of primary particles and they increase in size as they acquire more material through these physical processes. The nucleus for snow formation is often the abandoned 'houses' of larvaceans, which are mucus-based feeding structures used by these zooplankton. Dying diatoms, at the end of a bloom, often precipitate large-scale snow formation due to the production of large amounts of mucopolysaccharides in their cell walls. The generation of water currents during feeding by flagellates and ciliates colonizing the aggregate also collects particles from the surrounding water and leads to growth of the snow particle. Marine snow is mainly produced in the upper

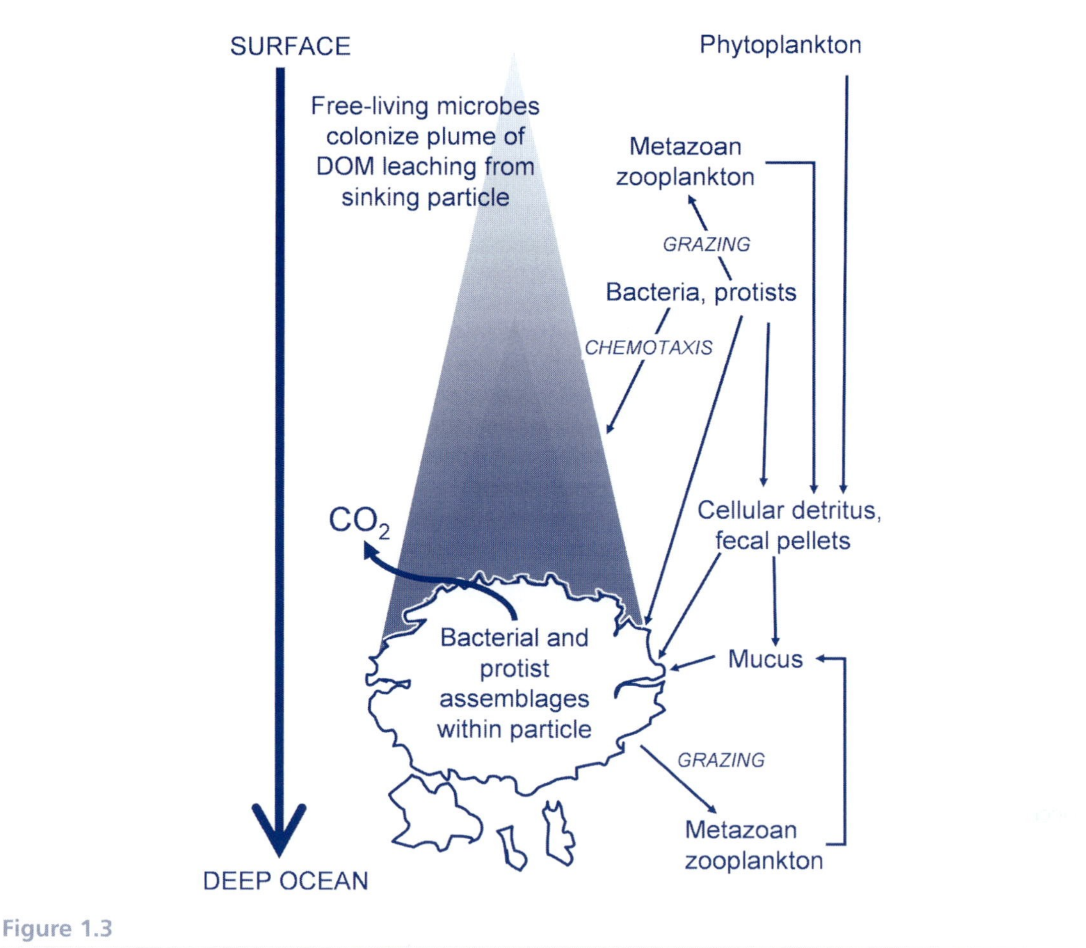

Figure 1.3

Formation and fate of marine snow particles (DOM = dissolved organic material).

100–200 m of the water column and large particles can sink up to 100 m per day, allowing them to travel from the surface to the ocean deep within a matter of days. This is the main mechanism by which a proportion of the photosynthetically fixed carbon is transported from the surface layers of the ocean to deeper waters and the seafloor. However, aggregates also contain active complex assemblages of bacteria and bacterivorous protists. Levels of microorganisms in marine snow are typically 10^8–10^9 ml^{-1}, which are about 100–10 000-fold higher than in the bulk water column. As particles sink, organic material is degraded by extracellular enzymes produced by the resident microbial population. Microbial respiration creates anoxic conditions, so that diverse aerobic and anaerobic microbes colonize different niches within the snow particle. The rate of solubilization exceeds the rate of remineralization, so dissolved material leaks from snow particles, leaving a plume of nutrients in its wake as it spreads by diffusion and advection. This may send chemical signals that attract small zooplankton, which may consume the particle. The trailing plume also provides a concentrated nutrient source for suspended planktonic bacteria, which may show chemotactic behavior in order to remain within favorable concentrations (*Box 3.1*). Thus, much of the carbon is recycled during its descent, but some material reaches the ocean bottom, where it is consumed by benthic organisms or leads to the formation of sediments. Photosynthesis by algae and bacteria leads to the formation of organic material through CO_2 fixation, but viruses, heterotrophic bacteria and protists all play a part in remineralizing some of this organic material. The discovery of this mechanism, termed the microbial loop, was one of the most important recent conceptual advances in biological oceanography and its significance is considered further in *Chapter 9*.

1.7.2 Sediments

Much of the continental shelf and slope is covered with terrigenous or lithogenous sediments, derived from erosion of the continents and transported into the ocean as particles of mud, sand or gravel. The mineral composition reflects the nature of the rocks and the type of weathering. Large rivers such as the Amazon, Orinoco or Ganges transfer millions of tonnes of fine sediments to the ocean each year. Most of this mud settles along the continental margins or is funneled by submarine currents as dilute suspensions.

In the deep ocean, about 75% of the deep ocean floor is covered by abyssal clays and oozes. Abyssal clays are formed by the deposition of wind-blown terrestrial dust from the continents, mixed with volcanic ash and cosmogenic dust (from meteor impact). These accumulate very slowly (less than 1 mm per 1000 years), whilst oozes accumulate at up to 4 cm per 1000 years. Biogenous oozes contain over 30% of material of biological origin, mainly shells of protistan plankton, mixed with clay. Oozes are usually insignificant in the shallow waters near continents. Calcareous oozes or muds cover nearly 50% of the ocean floor, especially in the Indian and Atlantic Oceans. They form by the deposition of the calcium carbonate shells (tests) of two main types of protist, the coccolithophorids (*Section 7.7*) and the foraminifera (*Section 7.9*). Siliceous oozes are formed from the shells (frustules) of diatoms (*Section 7.6*) and radiolarians (*Section 7.8*), which are composed of opaline silica ($SiO_2 \cdot nH_2O$). The rate of accumulation of biogenous oozes depends on the rate of production of organisms in the plankton, the rate of destruction during descent to the seafloor and the extent to which they are diluted by mixing with other sediments. In the case of coccolithophorids and foraminifera, depth has an important effect on dissolution of the tests. At relatively high temperatures near the surface, seawater is saturated with $CaCO_3$. As calcareous shells sink, $CaCO_3$ becomes more soluble due to the increased content of CO_2 in water at lower temperatures and higher pressures. The carbonate compensation depth is the depth at which carbonate input from the surface waters is balanced by dissolution in deep waters; this varies between 3 km (polar) and 5 km (tropical). For this reason, calcareous oozes tend not to form in waters more than 5000 m deep. Similarly, not all of the silica in the frustules of diatoms reaches the ocean floor. Recently, bacterial action has been shown to play a large part in the dissolution of diatom shells during their descent (*Box 7.1*). The rate of deposition of protist remains to the seabed is much more rapid than would be assumed from their small size. This is because they are aggregated into larger particles through egestion as fecal pellets after grazing by zooplankton and through the formation of marine snow as described above. In shallower waters near the continental shelf, the high input of terrigenous sediments mixes with, and dilutes, sediments of biogenous origin.

Remineralization of readily degradable organic matter in the water column though microbial action means that only a small fraction of fixed carbon (probably less than 10% of primary productivity) reaches the deep ocean floor; much of this is composed of more refractory particles. The sediments in much of the Atlantic and Pacific Ocean are typically 500–1000 m thick (in some places they are much thicker), below which they become heavily compacted. The microbiology of deep marine sediments and subsurface rocks is currently an area of active investigation using deep core drilling, but is beyond the scope of this book. Water saturation restricts O_2 penetration into sediments to a few cm, below which all microbial breakdown of organic matter is anaerobic. Indeed, the overlying water column may be completely anoxic in some regions where high nutrient concentrations promote high O_2 consumption by microbes. For example, the Black Sea contains no free O_2 from a depth of 150 m to the bottom at 2000 m, due to sulfate reduction. Large anoxic basins also exist off the coasts of Venezuela and Mexico. As organic material settles deeper into anoxic sediments, it accumulates more rapidly than it is degraded and thus joins the geological cycle, reemerging millions of years later when uplifted in continental rocks through tectonic processes. There is increasing recognition of the importance of microbial activities in the sediment–water interface (SWI) and deep-sea benthic boundary layer (BBL), which is a layer of homogeneous water 10 m or more thick, adjacent to the sediment surface. The SWI includes high concentrations of particulate organic debris and dissolved organic compounds that become adsorbed onto mineral particles. The structure and composition of the microbial habitat is modified by benthic 'storms' and the action of burrowing animals, which move and resuspend sediments. As well as the constant 'snowfall' of plankton-derived material, concentrated nutrient inputs reach the seabed in the form of large animal carcasses. For example, time-lapse photography has shown how quickly fallen whale carcasses attract colonies of animals and microbiological studies accompanying these investigations have yielded some novel bacteria with biotechnological applications. The sulfide-based microbial communities that develop are very similar to those found at hydrothermal vents and cold seeps (*Section 1.7.6*). Other types of sediment that provide special habitats for microbes include those in salt-marshes, mangroves and coral reefs.

Studies of the extent to which carbon fixed in the photic zone finds its way to the seabed, and its fate in sediments, are important in understanding the role of the oceans in the planetary carbon cycle. Microbial processes such as production and oxidation of methane and oxidation and reduction of sulfur compounds are of special interest. Studies of the diversity and activity of protistan and prokaryotic life in the various types of sediment are yielding many new insights, mainly because of the application of molecular techniques, and are described in subsequent chapters.

1.7.3 Microbial life at surfaces – biofilms and microbial mats

In the last two decades, the special phenomena that govern the colonization of surfaces by microbes have come under intense scrutiny, with the growing recognition that such biofilm formation involves complex physicochemical processes and community interactions. Biofilms consist of a collection of microbes bound to a solid surface by their extracellular products, which trap organic and inorganic components. In the marine environment, all kinds of surfaces including rocks, plants, animals and fabricated structures may be colonized by biofilms. As a result of metabolic processes, ecological succession results in the development of microenvironments and colonization by mixed communities of bacteria and protists to form layered structures known as microbial mats, which can be several mm thick. These are particularly important in shallow and intertidal waters. The composition of microbial mats is affected by physical factors such as light, temperature, water content and flow rate; and by chemical factors such as pH, redox potential, the concentration of molecular O_2 and other chemicals (especially sulfide, nitrate and iron) and dissolved organic compounds. Phototrophic bacteria and diatoms are major components of stratified microbial mats, and the species composition and zonation are determined by the intensity and wavelength of light penetration into the mat. Light normally only penetrates about 1 mm into the mat and, below this, anaerobic conditions develop. The formation and diurnal variations of oxygen and sulfide gradients have a major effect on the distribution of organisms in the mat. Detailed description of biofilm formation is in *Section 4.5.4* and the role of different bacteria in the composition of microbial mats is considered in *Chapter 5*. Biofilms are also of great economic importance in biofouling, as discussed in *Chapter 16*.

1.7.4 Sea ice

In the polar regions, the temperature is so low during the winter that large areas of seawater freeze to form sea ice, some of which forms adjacent to the coastal shoreline and some of which forms floating masses of pack ice. Sea ice forms when the temperature is less than −1.8°C, the freezing point of water at 35‰ salinity. The first stage in sea ice formation is the accumulation of minute crystals of frazil ice on the surface, which are driven by wind and wave action into aggregated clumps called grease ice. These turn into pancake-shaped ice floes that freeze together and form a solid ice cover. At the winter maxima, the combined coverage by sea ice at the north and south polar regions is almost 10% of the Earth's surface (1.8×10^7 km^2 in the Antarctic and 1.5×10^7 km^2 in the Arctic). During the formation of frazil ice, planktonic microbes become trapped between the ice crystals and wave motion transports further organisms into the grease ice during its formation. Near the ice–air interface, temperatures may be as low as -20°C during the depths of the polar winter, whilst the temperature at the ice–water interface remains fairly constant at about -2°C. When seawater freezes, it forms a crystalline lattice of pure water, excluding salts from the crystal structure. The salinity of the liquid phase increases and its freezing point drops still further. This very cold, high-density, high-salinity (up to 150‰) water forms brine pockets or channels within the ice, which can remain liquid to -35°C. The ice becomes less dense than seawater and rises above sea level, with the channels draining brine through the ice to the underlying seawater. Thus, sea ice is very different from freshwater glacial ice.

The structure of sea ice provides a labyrinth of different microhabitats for microorganisms, with variations in temperature, salinity, nutrient concentration and light penetration. This enables colonization and active metabolism by distinctive mixed communities of cold-adapted (psychrophilic) algae, protozoa and bacteria, as well as viruses. The properties of psychrophilic microbes are considered in *Section 4.6.1.* Microbial activities also alter the physicochemical conditions, mainly due to the production of large amounts of cryoprotectant compounds and extracellular polymers, leading to creation of further microenvironments. The dominant photosynthetic members of the sea ice microbiota are pennate diatoms, mainly in the lower part of the ice near the ice–sea interface, and small dinoflagellates in the upper part. The density of diatoms in sea ice may be up to 1000 times that in surface waters. In addition, a wide range of protozoa and heterotrophic bacteria has been found in sea ice, including new species with biotechnological potential because of their adaptation to high salinity and low temperature. Through photosynthesis, the microalgae make a small, but significant, contribution to primary productivity in the polar regions. For example, the contribution of sea ice to primary productivity in the Southern Ocean is only about 5% of the total, but it extends the short summer period of primary production and provides a concentrated food source that sustains the food web during the winter. During the Antarctic winter, microalgae on the undersurface of sea ice are the major source of food for grazing krill, shrimp-like crustaceans that are the main diet of fish, birds and mammals in the Southern Ocean.

1.7.5 Hydrothermal vents and cold seeps

Hydrothermal vents form a specialized and highly significant habitat for microbes. They occur mainly at the mid-ocean ridges at the boundary of the Earth's tectonic plates, where sea-floor spreading and formation of new ocean crust is occurring. Numerous such sites have been studied in the Pacific and Atlantic Oceans. Seawater permeates through cracks and fissures in the crust and interacts with the heated underlying rocks, thereby changing the chemical and physical characteristics of both the seawater and the rock. The permeability structure of the ocean crust and the location of the heat source determine the circulation patterns of hydrothermal fluids. As cold seawater penetrates into the ocean crust it is gradually heated along its flow path, leading to the removal of magnesium from the fluid into the rock, with production of acid during the process. This leads to the leaching of other major elements and transition metals from the rock into the hydrothermal fluid and sulfate in the seawater is removed by precipitation and reduction to hydrogen sulfide. As the percolating fluids reach the proximity of the magma heat source, extensive chemical reactions occur within the rock and the pressurized fluids are heated to over 350°C, becoming buoyant and rising towards the ocean floor. As they rise, the fluids cool slightly due to decompression, and precipitation of metal sulfides and other compounds occurs *en route*. The hydrothermal fluid is injected into the ocean as plumes of mineral-rich superheated water. The hottest plumes (up to 350°C) are generally black, due to the high

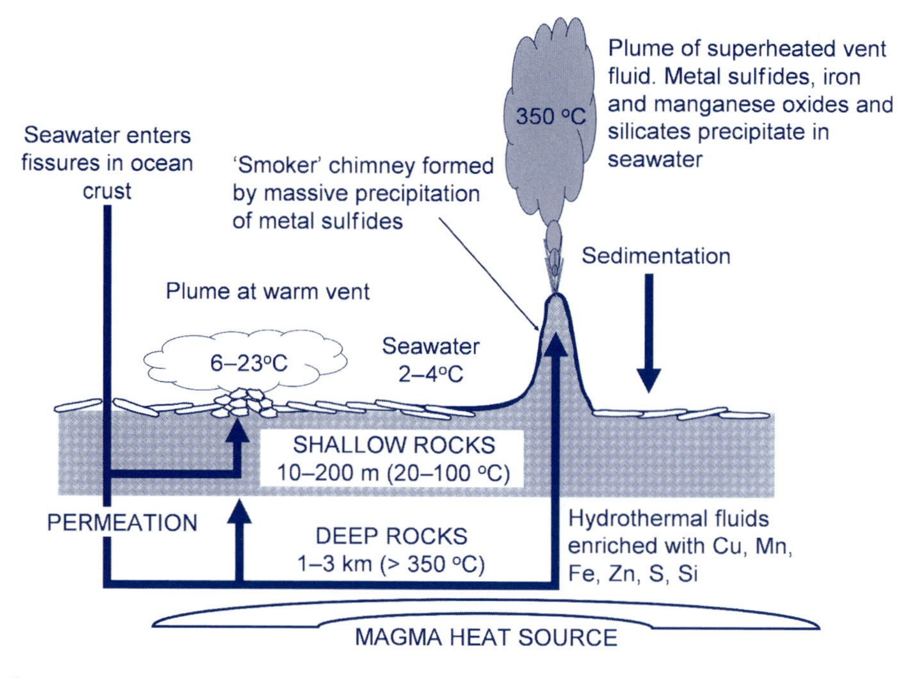

Figure 1.4

Processes occurring at a hydrothermal vent system. Gradients of minerals and temperature around the vents create a variety of habitats for diverse microbial and animal communities.

content of metal sulfide and sulfate particles and precipitation occurs as the hot plume mixes with the cold seawater. Some of these precipitates form chimney structures called 'black smokers', whilst others are dispersed through the water and form sediments in the vicinity. In other parts of the vent field, the circulation of hydrothermal fluid may be shallower, leading to diffuse plumes of water heated to 6–21°C (*Figure 1.4*). The gradients of temperature and nutrients that exist at hydrothermal systems provide a great diversity of habitats for microbes suspended in the surrounding heated waters, in sediments and attached to surfaces of the chimneys. Many of these are hyperthermophilic *Bacteria* and *Archaea,* which can grow at temperatures up to 113°C (*Section 4.6.2*), whilst others grow at lower temperatures further from the fluid emissions. Molecular studies are revealing an astonishing diversity of prokaryotes, many of which have biotechnological applications. The microbiology of the deep subsurface rocks beneath vents is also now under investigation, and many novel microbes and metabolic processes are being discovered. Microbial activity in the deep subsurface contributes to the chemical changes in composition during circulation of the hydrothermal fluids.

Hydrothermal vent systems were first described in 1977, when scientists aboard the submersible *Alvin,* from Woods Hole Oceanographic Institution, were exploring the seabed about 2500 m deep near the Galapagos Islands. The discovery of life around vents was totally unexpected. The *Alvin* scientists observed dense communities of previously unknown animals growing around the vents, including tubeworms, clams, anemones, crabs and many others. Subsequent research showed that the warm waters near hydrothermal vents contain large populations of chemolithotrophic bacteria, which fix CO_2 using energy from the oxidation of sulfides in the vent fluids. This metabolism supports a food chain with many trophic levels that is independent of photosynthesis. We now know that many of the animals at vent sites contain chemolithotrophic bacteria as symbionts within their tissues or on their surfaces and these relationships are discussed in *Section 10.3*. In addition, bacterial populations directly support the growth of animals by filter-feeding (e.g. molluscs) or grazing on microbial mats (e.g. crustacea), as well as indirectly by predators of these animals. Thus, hydrothermal vents are oases of life in the deep sea. Previously, life was thought to always rely ultimately on the fixation of CO_2 by photosynthesis, but the vent communities function without the input of material derived from the use of light energy. (However, note that sulfide oxidation depends on dissolved O_2 in the water, and the origin of this *is* photosynthetic.)

Cold seeps are abundant along the continental shelf and slope where the upwards percolation of fluids through the sediments is influenced by plate tectonics and other geological processes. At cold seeps, high concentrations of methane and sulfide support prolific chemosynthetic communities consisting of free-living *Bacteria* and *Archaea*, as well as those living symbiotically with invertebrates. The discovery of previously unknown syntrophic interactions between prokaryotes at seeps is highlighted in *Box 6.1.* Extensive populations of foraminifera also occur at cold seeps. Some sites are associated with seeps of hypersaline brines or hydrocarbons.

1.7.6 Living organisms as microbial habitats

Microbial biofilms also form on the surfaces of all kinds of animals, algae and coastal plants, which provide a highly nutritive environment through secretion or leaching of organic compounds. Many organisms appear to selectively enhance surface colonization by certain microbes and discourage colonization by others. This may occur by the production of specific compounds that inhibit microbial growth or interfere with microbial attachment processes. Once established, particular microorganisms may themselves influence colonization by other types. These processes offer obvious applications in the control of biofouling (*Section 16.2*). As well as surface (epibiotic) associations, microorganisms can form endobiotic associations within the body cavities, tissue or cells of living organisms.

Many microalgae such as diatoms and dinoflagellates harbor bacteria on their surfaces, or as endosymbionts within their cells. Seaweeds and seagrasses have dense populations (up to $10^6 cm^{-2}$) of bacteria on their surfaces, although this varies considerably with species, geographic location and climatic conditions. Microbial diseases of seaweeds and marine plants are being increasingly recognized, although this remains an area of limited investigation and is not covered in this book.

The surfaces and intestinal content of invertebrate animals provide a variety of habitats to a wide diversity of microbes. Such associations may be neutral in their effects, but commonly lead to some mutual benefit for host and microbe (*symbiosis*). Examples of symbiotic interactions between animals and microbes are considered in *Chapter 10,* whilst diseases of marine mammals, fish and invertebrates are discussed in *Chapters 13–15.*

References and further reading

Azam, F., and Long, R.A. (2001) Sea snow microcosms. *Nature* **414**: 495–498.

Deming, J.W. (2002) Psychrophiles and polar regions. *Curr Opin Microbiol* **5**: 301–309.

Doolittle, W.F. (2000) Uprooting the tree of life. *Sci Am* **282**: 90–95.

Fenchel, T. (2002) Microbial behaviour in a heterogeneous world. *Science* **296**: 1068–1070.

Gupta, R.S., and Golding, G.B. (1996) The origin of the eukaryotic cell. *Trends Biochem Sci* **21**: 166–171.

Hendrix, R.W., Lawrence, J.G., Hatfull, G.F., and Casjens, S. (2000) The origins and ongoing evolution of viruses. *Trends Microbiol* **8**: 504–508.

Howarth, R.W. (1993) Microbial processes in salt-marsh sediments. In: **Ford, T.E.** (ed.) *Aquatic Microbiology: an Ecological Approach.* Blackwell, Cambridge, MA, pp. 239–259.

Karner, M.B., DeLong, E.F., and Karl, D.M. (2001) Archaeal dominance in the mesopelagic zone of the Pacific Ocean. *Nature* **409**: 507–510.

Kjorbøe, T. (2001) Formation and fate of marine snow: small-scale processes with large-scale implications. *Sci Mar* **65 (S2)**: 57–71.
Li, L., Kato, C., and Horikoshi, K. (1999) Microbial diversity in sediments collected from the deepest cold-seep area, the Japan Trench. *Mar Biotechnol* **1**: 391–400.
Meyers, S.P. (2000) Developments in aquatic microbiology. *Intern Microbiol* **3**: 203–211.
National Academy of Science Space Studies Board (1999) *Size Limits of Very Small Microorganisms Proceedings of a Workshop*. Online @ http:// www7.nationalacademies.org/ssb/nanomenu.html
Pace, N.R. (1997) A molecular view of microbial diversity and the biosphere. *Science* **276**: 734–740.
Prieur, D. (1997) Microbiology of deep-sea hydrothermal vents. *Trends Biotechnol* **15**: 242–244.
Reysenbach, A.-L., and Cady, S.L. (2001) Microbiology of ancient and modern hydrothermal systems. *Trends Microbiol* **9**: 79–86.
Schulz, H.N., and Jørgensen, B.B. (2001) Big bacteria. *Ann Rev Microbiol* **55**: 105–137.
Sherr, E.B., and Sherr, B.F. (1991) Planktonic microbes: tiny cells at the base of the ocean's food webs. *Trends Ecol Evol* **6**: 50–54.
Sieburth, J.M. (1979) *Sea Microbes*. Oxford University Press, New York.
Staley, J.T., and Gosink, J.J. (1999) Poles apart: biodiversity and biogeography of sea ice bacteria. *Ann Rev Microbiol* **53**: 189–215.
Thomas, D.N., and Dieckemann, G.S. (2002) Antarctic sea ice – a habitat for extremophiles. *Science* **295**: 641–644.
Turley, C. (2000) Bacteria in the cold deep-sea benthic boundary layer and sediment–water interface of the NE Atlantic. *FEMS Microbiol Ecol* **33**: 89–99.
Valentine, D.L. (2002) Biogeochemistry and microbial ecology of methane oxidation in anoxic environments: a review. *Ant Van Leeuw Int J Gen Molec Microbiol* **81**: 271–282.
Van Dover, C. (2000) *Ecology of Hydrothermal Vents*. Princeton University Press, Princeton.
Whitman, W.B., Coleman, D.C., and Wiebe, W.J. (1998) Prokaryotes: the unseen majority. *Proc Natl Acad Sci USA* **95**: 6578–6583.
Williams, D.M., and Embley, T.M. (1996) Microbial diversity: domains and kingdoms. *Ann Rev Ecol Syst* **27**: 560–595.
Woese, C. (1998) The universal ancestor. *Proc Natl Acad Sci USA* **95**: 6854–6859.

Methods in marine microbiology

2.1 Sampling and experimental approaches

In studies of marine microbiology, it can be quite a challenge to collect samples in a form suitable for subsequent analysis together with environmental data about the sampling site. Because ocean water is highly heterogeneous, proper attention must be paid to the replication, frequency and location of sampling. Most microbes are too small for the traditional methods of examining plankton composition using nets. However, nets can be used to harvest phytoplankton and zooplankton which will have attached microbes and sometimes aggregates of free-living microorganisms may be large enough to collect in this way. Obtaining samples of seawater and sediments is relatively straightforward in shallow coastal waters. Samples may be collected from a boat or by SCUBA divers. For most microbiological work, samples are collected in sterile plastic bags or bottles and precautions are needed to ensure their aseptic collection. Careful attention is needed in the choice of construction materials for sampling containers, because many microbial processes can be affected by the presence of trace metals or rubber in hoses and stoppers. Triggering mechanisms can be used to open and reseal bottles, in order to collect water column samples from different depths. One of the most widely used systems is the Niskin sampler that can be operated individually or serially attached on a cable and activated to open a hose to draw water into a container. Some systems employ multiple sampling bottles that can be activated by pressure or remote control. For some investigations, it may be necessary to collect several hundred liters of water, which is concentrated by filtration or centrifugation. Various types of grabs and corers are available for sampling marine sediments. Sampling in the open ocean obviously requires the use of research vessels equipped with suitable sampling gear and on-board laboratory facilities. Research investigations are facilitated by the use of established long-term sampling programs combined with routine monitoring of oceanographic and atmospheric data. Well-known examples include the Pacific Ocean time-series site known as station ALOHA (A Long-term Oligotrophic Habitat Assessment) off Hawaii and the Bermuda Atlantic Time Series (BATS) site.

Exploration of the deep sea is extremely costly and is possible only with manned or remotely operated submersible vessels. The most famous of these is the deep submergence vehicle *Alvin,* owned by the US Navy and operated by the Woods Hole Oceanographic Institute. *Alvin* has been rebuilt numerous times and made nearly 4000 deep-sea dives since its first launch in 1964. A typical dive can take two scientists and a pilot down to 4500 m. At maximum depth, it takes about 2 hours for *Alvin* to reach the seafloor and another two to return to the surface, with 4 hours of carefully planned recording and sampling work on the sea floor. Viewing ports and video cameras allow direct observation so that specific samples can be taken. Other investigations are carried out using remotely operated vehicles. The Deepstar project in Japan uses a variety of submersible vessels and has made particular advances in the collection of deep-sea microbes under high pressure for biotechnological investigations. Other deep-sea submersibles include the *Nautile* operated by IFREMER in France, *MIR-1* and *MIR-2* operated by the Shirshov Institute of Oceanology in Russia and the *Johnson Sea Link* vessel operated by the Harbor Branch Oceanographic Institution, Florida. Special technology is needed to collect samples, especially from abyssal depths at great pressure and from high-temperature environments near hydrothermal vents.

Carrying out controlled experiments is essential in order to understand the effects of factors such as nutrient additions, light and temperature on community dynamics and microbial processes in the marine environment. Laboratory investigation of microbial processes in water uses 'bottle experiments' in which volumes of water up to a few liters can be handled. Use of these techniques led to great advances in our understanding of phototrophic and heterotrophic processes, for example by comparison of metabolic activities of samples incubated under light and dark conditions or the effects of temperature and nutrient additions. Such small-scale experiments are a vital first stage in the study of microbial processes in marine samples, but extending studies to the natural environment is difficult.

Mesocosms are containers holding large volumes of seawater under controlled conditions, which attempt to simulate open sea environments. Large tanks containing coastal seawater pumped in at controlled rates can be used in shore-based experimental stations. However, a more successful approach is the use of large enclosures constructed of polythene reinforced with vinyl and nylon (essentially, very big and very strong plastic bags immersed in the sea). The materials must be chosen carefully to simulate natural conditions (especially temperature and irradiances) as closely as possible. For example, the European Large Scale Facility for Marine Research (based at the University of Bergen, Norway) has operated a number of these mesocosm bags (about 4 m deep and 2 m in diameter) suspended at various depths in the relatively calm, deep waters of a fjord (*Figure 2.1*). The bags are filled by pumping in unfiltered fjord water so that natural communities of phytoplankton, zooplankton, bacteria and viruses are introduced. An example of the experimental use of such mesocosms is given in *Box 8.1* for the study of phytoplankton bloom dynamics and the effect of viruses.

Generally, it is impractical to conduct experiments in the open sea because of the continuous turbulence and dispersion of water and the organisms it contains. There are also considerable logistical problems in conducting research at sites sufficiently far from the continental

Figure 2.1

Mesocosm enclosures used to investigate responses of a bloom of *Emiliania huxleyi* to CO_2. The mesocosms (each with a volume of 4 m^3) are covered with plastic foil in order to control the partial pressure of CO_2 in the air overlying the water. Image courtesy of Jean-Pierre Gattuso, CNRS, Laboratoire d'Océanographie de Villefranche, France.

shelf to avoid land effects. However, a notable exception is the recent series of experiments to investigate the effect of iron additions on phytoplankton composition and productivity. The IRONEX experiments in the Pacific Ocean (1993, 1995) and the SOIREE (1999) and SOFeX (2002) experiments conducted in the Southern Ocean involved the release of iron over areas up to 100 km^2. These experiments were possible because of the deployment of an inert tracer compound (sulfur hexafluoride, SF_6). The dispersion of SF_6 can be measured analytically, thus establishing the degree of dilution of iron in the water mass under study. The significance of these experiments and plans for further large-scale investigations involving measurement of microbial parameters are discussed in *Box 9.2*

2.2 Microscopic methods

2.2.1 Light microscopy

The study of microorganisms began with the development of the light microscope and this remains the main method for the initial examination of plankton samples. Eukaryotic organisms such as diatoms, dinoflagellates and ciliates are large enough for microscopy to be useful in distinguishing morphological and structural features. In these groups, microscopic appearance is the main criterion for classification. Direct light microscopy is also used for enumeration of eukaryotic plankton in seawater samples. For prokaryotes, light microscopy reveals the general shape and morphology of the cells and internal structures are sometimes visible. However, the wavelength of visible light limits the effective magnification of the light microscope to about 1000–1500 times and it is not possible to resolve objects or structures smaller than about 200 nm. The use of special dyes and illumination techniques can improve the amount of information revealed by light microscopy. Viruses are below the limits of resolution of the light microscope, although they may be visualized by epifluorescence light microscopy (see *Figure 8.2*).

2.2.2 Electron microscopy

It was not until the development of the electron microscope that study of the ultrastructure of cells and viruses became possible. In the transmission electron microscope (TEM), a beam of electrons is focused onto an ultrathin section of the specimen in a vacuum, usually after staining with lead or uranium salts. Electrons are scattered as they pass through the specimen according to different densities of material in the cell. Because the wavelength of an electron beam is much smaller than that of light, the TEM has an effective magnification up to 1 million times and objects as small as 0.5 nm can be resolved. Various techniques such as shadowing, freeze-etching and negative staining are used to visualize the membranes, internal structures and surface appendages of cells. However, it must always be borne in mind that TEM images are only obtained after staining and fixing the cells and observing them in a vacuum. Therefore, the appearance of structures may not reflect their actual organization in the living cell. In the scanning electron microscope (SEM), a very fine electron beam scans the surface of the object, generating a three-dimensional image. SEM is, therefore, particularly useful for studying the structure of cell surfaces. Modern developments of electron microscopy include atomic force microscopy, in which a tiny probe is held in place very close to the surface of an object using weak atomic repulsion forces. Effective magnifications up to 100 million are possible and the atomic structure of molecules such as DNA or proteins can be visualized. A major advantage of this technique is that specimens do not need to be fixed or stained and can be examined in the living state.

2.2.3 Confocal laser scanning microscopy (CLSM)

CLSM uses a laser light source coupled to an optical microscope and computer-aided digital imaging system. The beam focuses on a narrow plane of the specimen. The advantage of CLSM lies in the ability to use it on living specimens and to generate a three-dimensional image. Because the magnification is much less than that of TEM or SEM, it is not as useful for revealing ultrastructural detail, but it is proving enormously valuable in ecological studies of microbial communities, especially when combined with fluorescent *in situ* hybridization (FISH) techniques (see *Section 2.6.10*). In particular, understanding of the structure of biofilms and microbial mats has advanced considerably using this technique.

2.2.4 Epifluorescence light microscopy (ELM)

Fluorescence occurs when material absorbs light at one wavelength (excitation or absorption spectrum) and then re-emits it at a different wavelength (emission spectrum). The specimen is illuminated with a tungsten-halogen or mercury vapor lamp. In direct fluorescence microscopy, a filter is placed between the light source and the specimen, allowing only light of the desired excitation wavelength to be transmitted, whilst a barrier filter placed between the specimen and the eyepiece transmits the emitted fluorescence and absorbs longer wavelengths. ELM depends on the use of dichroic mirrors as interference filters that transmit one set of wavelengths and reflect the others. Water samples are usually fixed with formalin and filtered through polycarbonate membrane filters of a pore size appropriate to the group under study (0.22 μm is most commonly used). ELM can therefore be used for observation and enumeration of all groups of marine microbes including microalgae, ciliates, flagellates, bacteria and viruses. The original ELM stain used in plankton studies was acridine orange (AO, 3,6-bis[dimethylamino]acridinium chloride), which binds to DNA and RNA. In comparison with more recent molecular methods, ELM seems a simple technique, so it is hard to believe that the first description of its use (by John Hobbie of the Woods Hole Laboratory in 1977) led to such a major upheaval in microbial oceanography. Estimates of the bacterial numbers in plankton were suddenly a thousand or more times higher than previously realized. Using AO, background fluorescence and distinguishing microbes from inanimate particles can be problematic and the most widely used stain today is DAPI (4′,6′-diamido-2-phenylindole) which binds to DNA and largely overcomes this problem. This fluorochrome is excited by UV-light and emits bright blue light. The use of incident light also permits observation of microbes attached to suspended particles. Different stains have advantages in particular applications; other stains include SYBR green and YoPro1. Some staining methods can distinguish between living and dead cells, based on differences in permeability; the most important of these is 5-cyano-2,3-dilotyl tetrazolium chloride (CTC) which is a fluorogenic redox dye that detects an active electron transport chain. Although there have been some criticisms of the technique, the CTC assay seems to be a reliable method for estimating the distribution of cell-specific metabolic activity in natural assemblages of marine bacteria.

The introduction of green fluorescent protein (GFP, a protein produced in the jellyfish *Aequorea victoria*) into other organisms by recombinant DNA technology has found wide application in cell biology. In marine microbiology, this marker is used to trace the fate of genetically modified organisms in the environment, for example in bioremediation applications. Another application is study of the fate of *gfp* marked bacteria following their ingestion by grazing ciliates.

2.3 Flow cytometry (FCM)

FCM was originally developed for biomedical use and was first applied to marine studies in the late 1970s. Since then, use of this technique has produced some of the most spectacular advances in enumeration and characterization of ocean microbes. A flow cytometer is an instrument that can separate and sort cells according to a specific fluorescent 'signature', leading to a wide range of physiological and ecological information. FCM simultaneously measures and analyzes multiple physical characteristics of single particles, ranging in size from 0.2 to 150 μm, as they flow in a fluid stream through a beam of light. The coupling of the optical

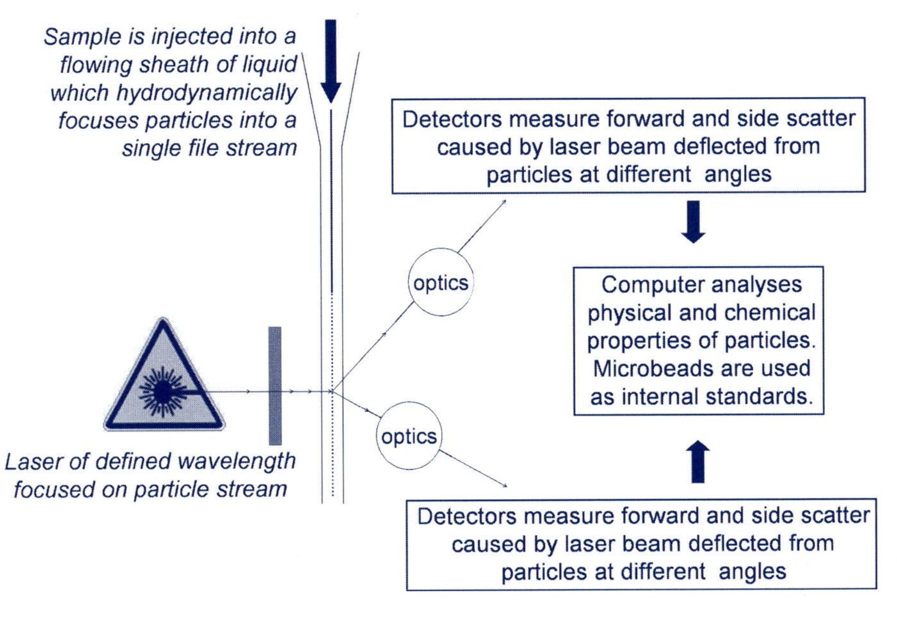

Figure 2.2

Schematic diagram of the components of a flow cytometer.

detection system to an electronic analyzer records how the particle scatters incident laser light and emits fluorescence. A flow cytometer is made up of three main components (*Figure 2.2*). The fluidics system injects particles into the instrument in a stream of liquid so that they pass in single file through a laser beam. The optics system consists of lasers (there may be four or more different wavelengths) to illuminate the particles in the sample stream and optical filters to direct the resulting light signals to the appropriate detectors. Each particle scatters light at different angles and may also emit fluorescence. The scattered and emitted light signals are converted to electronic pulses that can be processed by the electronics system. In a fluorescent-activated cell sorter (FACS), the electronics system can also transfer a charge and deflect particles with certain characteristics, so that different cell populations can be separated. FCM is used for a range of specific measurements in marine microbiology, including quantifying different microbial groups and their diversity, estimation of cell dimensions and volume, determination of DNA content and assessment of viability. FCM was first used for the analysis of phytoplankton cells, as they are naturally labeled with photosynthetic pigments such as chlorophyll and phycobilins, which are autofluorescent. However, all microbes can be detected in FCM by using the same principles described for epifluorescence microscopy, i.e. 'tagging' cells with fluorochromes that emit light of various wavelengths when illuminated by the appropriate laser. FCM has thus been particularly valuable in the analysis of bacteria and small eukaryotes in the picoplankton, especially following the development of portable instruments with stable optical and electronic systems that can be deployed on ships for direct examination of water samples. Recent improvements in the sensitivity of FCM instruments and the introduction of new dyes mean that FCM can now be used for detection and quantification of marine viruses. *Table 2.1* lists examples of fluorochromes used in FCM and ELM. Using an appropriate mixture of different fluorochromes and laser light of different wavelengths, different populations of microbes in aquatic ecosystems can be analyzed based on size, abundance and specific properties. Probably, the most dramatic impact of FCM was the discovery of *Prochlorococcus* in the late 1980s. Sallie Chisholm, of Woods Hole Laboratory noticed unusual fluorescence signals during one of the first research cruises to use a portable flow cytometer and subsequently identified the

Table 2.1 Some representative fluorochromes used in epifluorescence microscopy and flow cytometry

Fluorochrome	Excitation/emission (nm)	Target
Acridine orange	500/526; 460/650	DNA, RNA
DAPI	358/461	DNA
Ethidium bromide	518/605	Double-stranded DNA, RNA
FITC	495/520	General fluorescent label, proteins
Hoechst 33342	350/461	AT rich DNA
Mithramycin	425/550	GC rich DNA
Propidium iodide	535/617	Double-stranded DNA, RNA
Rhodamine 123	480/540	Membrane potential
SYBR Green	494/521	DNA, can be used for viruses

very small cyanobacterium *Prochlorococcus*. This is now considered to be one of the commonest photosynthetic organisms on Earth (*Section 5.4.4*) and it is remarkable that it escaped detection until so recently. Exciting new advances in FCM have been due to a fusion with FISH technology (*Section 2.6.10*), in which fluorescently labeled oligonucleotide probes can be used to discriminate specific taxa in heterogeneous natural marine microbial communities.

2.4 Antibody-labeling techniques

The surface proteins and polysaccharides of microbe are antigenic, that is they will induce the synthesis of a highly specific antibody if injected into an animal. If a particular organism has been isolated in sufficient quantities, it is possible to raise antibodies and then use these antibodies to probe a sample to determine if that organism is present. Antibody labeling is used particularly for identifying the presence of pathogens in water and in infected aquatic animals. Polyclonal antibodies are widely used; these are obtained by raising antisera in animals such as a rabbit or goat. Monoclonal antibodies, produced by culture of antibody-producing cells from mice, are preferred as they target specific antigens and are more easily standardized. In the direct fluorescent antibody technique (FAT), the antibody is conjugated to fluoroscein isothiocyanate (FITC), which forms strong bonds with amino groups in the antibody protein. The sample is then treated with the fluorescent antibody, washed carefully to remove unattached antibodies and examined in the fluorescence microscope. Preparing individual conjugated antibodies against each antigen of interest is expensive and time-consuming, and therefore an indirect method (IFAT) is often used. In this case, the interaction between the primary antibody and antigen is detected by adding a secondary anti-antibody that is conjugated to FITC. More importantly, IFAT can also be used to determine whether an animal has been infected with a particular pathogen by testing for the presence of antibodies in tissue or serum samples. A variation of this technique can be used in the electron microscope by conjugating antibodies to ferritin, which appears as dark spots in the TEM. As well as these microscopic techniques, a quantitative measurement of antigens or antibodies may be made using enzyme-linked immunosorbent assay (ELISA). This uses the same principles or either direct or indirect FAT, although in this case the primary or secondary antibodies are attached to an enzyme such as peroxidase or phosphatase, which gives rise to a colored product. *Figure 2.3* shows a summary of FAT and ELISA. Western blotting uses a similar principle to ELISA and can be used to identify specific antigens as components of microbial cells or viruses. In this case, proteins are separated by polyacrylamide gel electrophoresis (PAGE) and transferred to a sheet of nitrocellulose, which is then treated with enzyme-linked antibodies followed by addition of a chromogenic substrate. Specific protein bands can be excised from the original gel and subjected to further

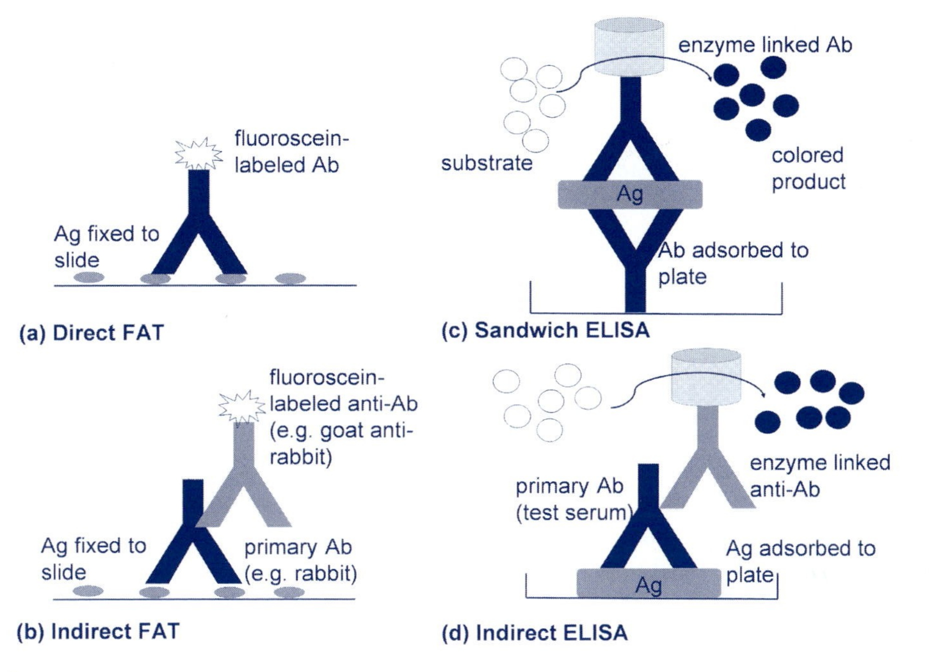

Figure 2.3

Antibody techniques. (a), (b) Fluorescent antibody techniques (FAT). A labeled antibody (Ab) reacts with an antigen (Ag) enabling specific microbes to be seen in the microscope, e.g. from tissue samples or on the surface of particles or biofilms. In the indirect technique (b), a stock anti-antibody is used to avoid the need for preparing fluorescently labeled primary antibodies. (c), (d) Enzyme-linked immunosorbent assay (ELISA). The antibody is conjugated to an enzyme (e.g. peroxidase) which catalyzes the conversion of a chromogen to a colored substrate. ELISA (c) can be used for detecting the presence of microbial antigens in animal serum or tissue to detect infection. The indirect ELISA (d) can be used to detect the presence of antibodies to a specific microbe as a way of monitoring immune responses to infection or vaccines. ELISA techniques are usually carried out in 96-well microtiter plates and the intensity of the enzymic reaction is measured in a special spectrophotometer.

analysis such as amino acid sequencing. Alternatively, Western blotting can be used in the reverse mode to screen serum or tissue extracts with a particular antigen, to see if an animal is producing antibodies in response to an infection. Such antibody techniques are widely used in the diagnosis and epidemiological studies of diseases of marine mammals and fish.

2.5 Laboratory culture

2.5.1 The importance of cultural conditions

A wide range of methods is used to isolate and culture microbes from marine samples. As will be referred to repeatedly in later sections, there is a large discrepancy between the number of microbes that can be observed in direct counts of marine samples and those that can be cultured in the laboratory. Even those organisms that can be successfully cultured differ enormously in their physiological properties (see *Chapter 4*), so the composition of the culture medium and environmental conditions (temperature, pH, atmospheric conditions and pressure) must be chosen with great care.

2.5.2 Enrichment culture

Since all marine habitats contain communities of microbes, investigation of their ecology often begins with enrichment culture, in which the provision of particular nutrients and incubation conditions will select for a certain group. For example, the addition of cellulose or chitin plus NH_4^+ salts to a marine sample under aerobic conditions will enrich for *Cytophaga*-like bacteria, whilst sulfates with acetic or propionic acid under strictly anaerobic conditions will favor growth of *Desulfovibrio* and related types. The Winogradsky column is especially useful in investigating communities in marine sediments. A tall glass tube is filled with mud and seawater and various substrates are added prior to incubation under appropriate conditions. A gradient of anaerobic to aerobic conditions will develop across the sediment and a succession of microbial types will be enriched. In the light, microalgae and cyanobacteria will grow in the upper parts of the column and will generate aerobic conditions by the production of O_2. The activity of various groups (depending on the substrate added) will lead to the production of organic acids, alcohols and H_2, which favor the growth of SO_4^- reducing bacteria. The sulfide produced leads to the development of anaerobic phototrophs that utilize H_2S as electron acceptor. A wide range of enrichment conditions may be adapted for particular groups, and the resulting communities may then be subject to further enrichment or selective plating to obtain pure cultures. The results of enrichment cultures must be interpreted carefully, because they often lead to over-representation of groups that may not be the dominant member of the natural community.

2.5.3 Isolation

Many autotrophic prokaryotes can be grown in a simple defined synthetic medium containing bicarbonate, nitrate or ammonium salts, sulfates, phosphates and trace metals. Many heterotrophs can be grown on a similar defined medium with the addition of appropriate organic substrates such as sugars, organic acids or amino acids. Usually however, isolation and routine culture are carried out using complex media made from semidefined ingredients such as peptone (a proteolytic digest of meat or vegetable protein, which provides amino acids) and yeast extract (which provides amino acids, purines, pyrimidines and vitamins). Many heterotrophs are more fastidious and require the addition of specific growth substances. An example of a widely used medium for routine culture of many marine bacteria is Zobell's 2216E medium, which contains low concentrations of peptone and yeast extract plus a mixture of various salts. It must be stressed that there is no single medium suitable for all types and microbiologists employ a very wide range of recipes for different purposes. Bacteria may be grown in liquid media (broth) or on media solidified with agar. Cultivation on agar plates (Petri dishes) permits samples to be streaked out to obtain single colonies, which can then be restreaked to form pure cultures. In the shake tube and roll tube techniques, samples are mixed with molten agar which allows culture of some anaerobic bacteria. Agar is a polysaccharide obtained from particular types of seaweed and has certain advantages for microbiological work. It is colorless, transparent and remains solid at normal incubation temperatures, but can be held in a molten state above 42°C. Its main advantage in microbiology is that it does not affect the nutritional status of the medium, because most bacteria and fungi do not possess extracellular enzymes for its degradation. However, not surprisingly, many marine microorganisms can degrade this algal compound. When marine samples are plated onto agar, craters or liquefaction of the agar are frequently observed surrounding the colonies. Furthermore, many marine bacteria are extremely oligotrophic and the addition of agar (even if purified) to seawater provides too high a concentration of nutrients for growth (see *Box 2.1*). Media for the culture of marine fungi usually contain complex substrates such as wood pulp or cellulose to encourage the growth of these organisms, which rely on the production of extracellular polymer-degrading enzymes. Many marine microalgae can be cultured in simple mixtures of salts supplemented with vitamins.

Selective media contain dyes, antibiotics or other chemicals that inhibit the growth of certain microorganisms, thus allowing only those of interest to multiply. They may also include certain

Box 2.1 RESEARCH FOCUS

Culturing the uncultured

New methods allow culture of marine microbes known only by their genes

As discussed in Chapters 5 and 6, our knowledge of microbial diversity in the oceans has been revolutionized by the application of techniques that allow recognition and relatedness of groups of organisms based solely on their genetic sequence. Most of the marine prokaryotes identified by these techniques (especially sequencing of 16s rRNA genes) are unculturable using the traditional microbiological methods. Our modern view of microbial diversity leads to the conclusion that we have little or no understanding of the properties of the great majority of microbes that inhabit our planet. Probably less than 1% of prokaryotic species have been cultured. Many major divisions of the *Bacteria* contain no known cultured species, so we have no idea what physiological or biochemical properties these possess and the roles they play in the ecosystem. Many free-living marine bacteria are extremely small ('ultramicrobacteria', which can pass through 0.45 μm membrane filters) and resemble the small cells induced when cultured bacteria are starved (see *Section 4.5.1*). For many years, it was assumed that most marine bacteria are perpetually starved and inactive. Consequently, as the number of previously unknown organisms identified by molecular analysis of the environment increased in the 1990s, many microbial ecologists adopted the pessimistic stance that these organisms were inherently unculturable and would never be grown. Perhaps this dogmatic view was also prompted by impatience generated by the quick generation of results produced by exciting new molecular biology techniques compared with the 'old-fashioned' traditional methods applied to the culture of microbes in the laboratory. Recent work shows that, provided sufficient effort and ingenuity are applied, there is every reason to be optimistic that we will be able to culture many more marine prokaryotes. So, we should regard these organisms as 'not yet cultured' rather than 'unculturable'. We now know that, far from being dormant, marine bacteria are very active in ocean processes.

Why are most marine prokaryotes so difficult to grow whereas others are easy? The most likely reason is that many pelagic microbes are obligate oligotrophs, adapted to growing at low cell densities with very low nutrient concentrations (see *Section 4.5.3*). It is very difficult to reproduce these conditions in the laboratory, especially if we try to culture organisms quickly at thousands of times the density at which they normally exist, for example on agar plates.

One way of overcoming the problems of unnaturally high cell densities and nutrient concentrations, pioneered by Button *et al.* (1993) and Schut *et al.* (1993) is to dilute samples of seawater containing microorganisms into sterile seawater, so that only one or two bacteria per tube are obtained. This leads to a successful enrichment of the dominant cell types and up to 50% of the indigenous bacterial population in seawater can be grown in dilution tubes containing only filtered, autoclaved natural seawater. Ultramicrobacteria with small genomes predominate in such cultures. Schut *et al.* (1997) used this technique to isolate a *Sphingomonas* sp. that possesses traits typical of a 'model oligotroph', having very low cell volumes, low DNA content, high protein content and only one copy of the rRNA operon. The cells are well adapted to the simultaneous utilization of mixed substrates.

A similar dilution approach was recently adapted by use of a high-throughput method in microtiter trays containing sterile seawater supplemented with phosphate, ammonia or low levels of organic compounds by Rappé *et al.* (2002), who succeeded in culturing representatives of the SAR11 clade. This group of α-*Proteobacteria* is the most common type identified by 16S rRNA sequencing and is ubiquitous in pelagic environments (*Section 5.2.2*). Novel microarray and FISH techniques were used to screen cultures. Bringing SAR11 into culture allows the physiological properties of this ubiquitous organism to be investigated for the first time and it has been given the proposed species name *Pelagibacter ubique*. Genome sequencing of SAR11 bacteria is now underway. Connon and Giovannoni (2002) used this method to culture additional bacteria from groups previously regarded as unculturable.

Zengler *et al.* (2002) used a novel method which involves encapsulation of single bacteria in microdroplets of gel to allow large-scale culture of individual cells. Concentrated seawater was emulsified in molten agarose to form very small droplets and FCM was used to sort droplets containing single bacteria. The droplets were transferred to growth columns (which filter out contaminating free bacteria) and various media and growth conditions were evaluated. Microbial diversity in the microcolonies within the droplets was examined using 16S rRNA typing. When organic media were used, the bacteria grown by this method resembled previously cultivated bacteria, but when unsupplemented filtered seawater was used as a growth substrate a broad range of bacteria, including previously uncultured types, were isolated and grown. Surprisingly, members of the SAR11 cluster (the most common type in environmental samples) did not grow. This technique permits physiological studies to be carried out on cultures as few as 100 cells, and further development will have major implications for microbial ecology and biotechnology.

A further recent breakthrough in culture methods is described by Kaeberlein *et al.* (2002), who attempted to mimic the natural environment of sandy sediments in an aquarium setting. Organisms were placed in diffusion chambers resting on the sand and covered with seawater. The walls of the chambers had permeable membranes which allow the flow of nutrients but contain the bacteria. Surprisingly, many microcolonies which developed were capable of growth on agar plates and some appear to be capable of growth only when cocultivated with other types.

ingredients that give a differential reaction when metabolized by bacteria belonging to particular groups. These media are used as an enrichment method in liquid culture or to select colonies of the desired type on agar plates. In marine microbiology, selective media are used mainly for the isolation and growth of animal and human pathogens and nonindigenous indicators of pollution. These media frequently contain bile or synthetic detergents, which select for enteric bacteria and related groups. For example, the medium thiosulfate-citrate-bile-sucrose (TCBS) agar is widely used for isolation of vibrios, whilst lauryl sulfate broth is used in the enumeration of coliform bacteria in assessing fecal pollution of seawater (*Section 11.3.3*). Selective anti-biotics may be added to suppress the growth of bacteria on fungal isolation plates.

As well as plating, successive dilution in liquid media can be used to separate individual microorganisms to obtain pure cultures. This method has been used successfully to grow previously nonculturable bacteria in seawater without the addition of nutrients (see *Box 2.1*). The dilution-to-extinction method is also used in the most probable number (MPN) technique for quantification of microbial populations. One widely used application of this is in the determination of coliforms as indicators of fecal pollution in seawater or shellfish (see *Figure 11.5*).

A recent advance in pure culture methods allows the isolation of individual cells viewed under a microscope, which can be trapped using a laser beam and separated off for subsequent pure culture. Such 'optical tweezers' provide a very powerful technique when used in conjunction with gene probes or antibody staining.

2.5.4 Biochemical methods for identification and taxonomy of bacteria

The morphology of bacteria and their appearance in the microscope helps in their initial characterization, but does not provide sufficient information for identification and taxonomy. With eukaryotes, the opposite is true and their classification is based largely on microscopic appearance. Bacteria are usually first identified and characterized using a combination of growth characteristics and tests for the production of particular enzymic activities, particularly those involved in carbohydrate and amino acid metabolism. Diagnostic tables and keys aid in the identification of unknown bacteria, although the inherent strain-to-strain variation in individual characteristics often causes problems for accurate identification and taxonomy. Commercial kits employing a battery of tests (e.g. API® or Micro-Bact® strips) offer advantages for standardization of methods and processing of results with the aid of databases developed using numerical taxonomic principles. However, most biochemical methods were developed for the identification of bacteria of medical importance and require adaptation for examination of marine environmental samples. These methods are most useful in areas such as identification of pathogens or fish spoilage organisms. For example, there are extensive databases and established diagnostic keys based on biochemical tests for the identification of *Vibrio* and related pathogens in aquaculture (*Section 14.3.2*). Some tests are very characteristic of particular groups. For example, the detection of the enzymes β-galactosidase and β-glucuronidase using fluorogenic substrates is important in the identification of coliforms and *E. coli* in polluted waters (*Section 11.3.3*). Of course, these biochemical test methods are suitable only for the small proportion of heterotrophic marine bacteria that can be easily cultured. The BIOLOG® system differentiates microbial taxa based on their nutritional requirements and incorporates up to 95 different substrates as a sole source of carbon or nitrogen, together with a tetrazolium salt. Different microbial groups utilize a range of substrates at characteristic levels and rates. Positive reactions are tested via a color reaction based on reduction of the tetrazolium salt. BIOLOG systems have been developed specifically for Gram-positive bacteria, Gram-negative bacteria and yeasts. Again, this method is not suitable for detection of nonculturable microorganisms but it provides a less stressful environment than the surface of solid growth media and some investigators have used the BIOLOG system for community characterization of metabolic activities in sediments or surface layers by direct inoculation without an intervening culture step.

Analysis of the composition of bacterial membranes via gas chromatography of fatty acid methyl esters (FAME) is a powerful technique that can detect very small differences between species and strains. Individual taxa have a distinct fatty acid fingerprint. However, careful standardization of growth conditions is needed because membrane composition is strongly affected by environmental conditions and the nature of the culture medium. Members of the *Archaea* cannot be identified with this method due to the nature of their membranes (see *Section 3.4*). Certain taxonomic groups of bacteria and fungi can be identified by the presence of specific biomarker fatty acids (e.g. isoprenoid quinones) and other molecules and this is used to monitor changes in microbial community composition as a result of environmental changes such as the introduction of pollutants into marine systems.

Analysis of the protein profile of microbes can also be used for comparison of microbial species. Extracted proteins are dissociated with the detergent sodium dodecyl sulfate and separated by polyacrylamide gel electrophoresis (SDS-PAGE). However, gene expression and the resulting protein pattern produced are extremely dependent on environmental conditions and the technique is of little use in identification and taxonomy. Indeed, its main use is in assessing the effect of various factors on the synthesis of particular proteins, such as bacterial colonization or virulence factors during infection of a host.

Pyrolysis mass spectrometry (PyMS) is a recently introduced technique that generates a chemical fingerprint of the whole microbe. This technique involves thermal degradation of complex organic material in a vacuum to generate a mixture of low-molecular-weight organic compounds that are separated by mass spectrometry. PyMS is easily automated and a high throughput of samples can be processed, but the equipment is expensive and is found only in a few specialized laboratories.

Table 2.2 shows a summary of the methods used for bacterial identification and classification. Some of these methods are applicable to fungi and microalgae. With the exception of the few applications mentioned for community analysis, all of these methods are suitable only for the easily cultured heterotrophs. They are of no use for studying the diversity and ecology of the great majority of marine microbes. For these, molecular methods are needed, as described in the next section.

Table 2.2 Summary of techniques used for the identification and classification of bacteria

Technique	Usefulness at different taxonomic levels
Nucleic acid techniques	
DNA sequencing	All levels, for phylogenetic analysis
DNA–DNA hybridization	All levels above species, 70% hybridization level used for species definition
GC ratios	Comparison of species known to be closely related
PCR based fingerprinting (e.g. RAPD, AFLP, rep-PCR)	Good discrimination at strain level
RFLP, ribotyping	Strain differentiation, but prone to variability
16S rRNA sequencing	Higher taxonomic levels, for phylogenetic analysis
Phenotypic characters	
Bacteriophage typing	Strain differentiation, highly specific
Biochemical tests (e.g. API, BIOLOG)	Species and strain differentiation, routine identification, community analysis
Morphology	Limited information except in some groups
Plasmid and protein profiles	Strain differentiation but highly variable
Serology	Strain (serotype) differentiation
Chemotaxonomic markers	
FAME	Rapid typing to species level
PyMS	Rapid typing to species level
Quinones and other biomarkers	Good differentiation above species level

2.6 Molecular methods

2.6.1 The impact of molecular tools in marine microbial diversity

Marine microbiologists use a wide range of techniques based on the isolation and analysis of nucleic acids. Many of these can be applied to the characterization of microbes in culture, for example for accurate identification and taxonomy or for diagnosis of disease. However, the most dramatic advances in marine microbial ecology have occurred due to the development of methods for the detection and identification of microbes based on direct extraction of genetic material from the environment, without the need for culture. This approach has been termed environmental genomics. Molecular techniques are very sensitive and can allow the detection of very small numbers of specific organisms or viruses among many thousands of others. Direct analysis of the genes present in an environmental sample allows us to make inferences about the diversity, abundance and activity of microbial communities.

The first attempts to characterize microbial diversity in marine samples by nucleic-acid-based methods were in the mid 1980s, using isolation of ribosomal RNA (rRNA). Recall from *Chapter 1* the pioneering work of Woese and colleagues, which led to revision of ideas about evolution of the major domains of life. The ribosome is composed of a large number of proteins and rRNA molecules, as shown in *Figure 2.4*. The small ribosomal subunit (SSU) rRNA molecules (16S in prokaryotes and 18S in eukaryotes) quickly became the first choice for diversity studies for a number of reasons. Firstly, rRNAs are universally present in all organisms, due to the role of the ribosome in the essential function of protein synthesis. The rRNA molecule has a complex secondary structure (see *Figure 2.9*) and mutations in parts of the gene that affect critical aspects of structure and function in the ribosome are often lethal; thus, changes in 16S rRNA occur slowly over evolutionary time. Some parts of the rRNA molecules (and the genes that encode them) are highly conserved whilst others have a high degree of variability. Growing cells contain multiple copies of rRNA, thus improving the efficiency of extraction. Following Woese's initial work, the parallel advances in nucleic acid sequencing technologies

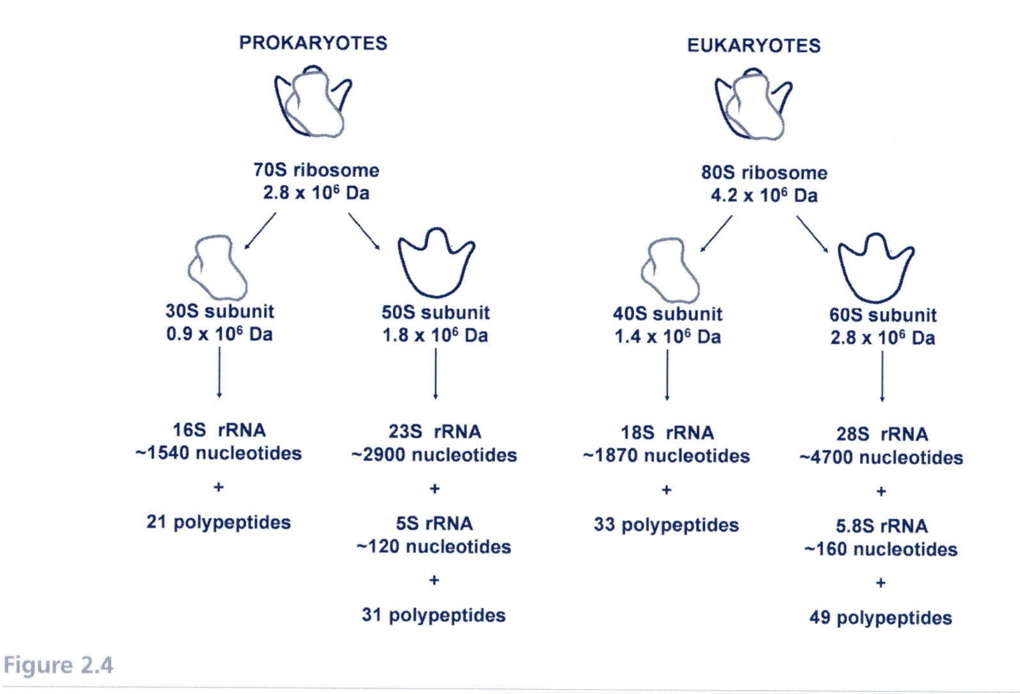

Figure 2.4

Structure of prokaryotic and eukaryotic ribosomes.

and computing led to the rapid generation of large databases of information linking sequences from known organisms to taxonomic position. Because many marine microbes cannot be cultured, microbial ecologists therefore asked whether it would be possible to obtain sequence information directly from environmental samples.

An outline of the methods used in this type of investigation is shown in *Figure 2.5*. Initial investigations relied on the isolation of rRNA and creation of a complementary DNA (cDNA)

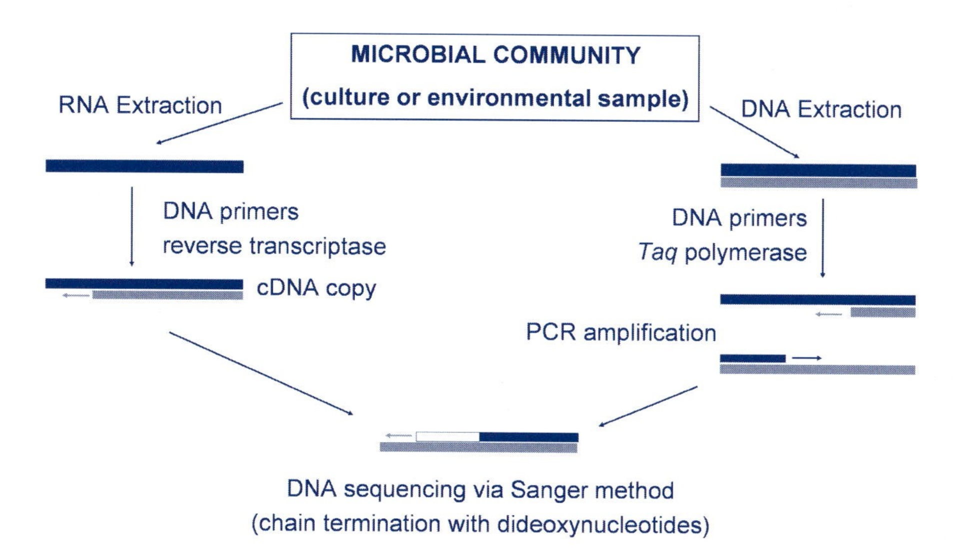

Figure 2.5

Outline of methods for sequencing ribosomal RNA or DNA from cultures or environmental samples. PCR amplification of DNA encoding rRNA genes is the most commonly employed method.

sequence using the enzyme reverse transcriptase (RT). However, when the polymerase chain reaction (PCR) was developed in 1988, direct amplification of 16S rDNA (i.e. the genes encoding rRNA) quickly became the preferred method. Furthermore, microorganisms often contain multiple copies of the rRNA genes, making it easy to amplify them from very small amounts of DNA. Early studies were laborious and labor-intensive, because they relied on 'shotgun' cloning, but by using PCR with primers directed against specific sequences in the 16S rRNA genes, gene fragments can be selectively amplified from mixed DNA in a sample from the marine environment. Most environmental analyses use 'universal' or broad-spectrum primers directed against short, highly conserved regions in the 16S rRNA genes and amplify the more variable sequences between the primer annealing sites. The PCR amplification products can be cloned to create a gene library. Alternatively, the PCR products can be separated by denaturing gradient gel electrophoresis (DGGE). After determination of DNA sequences, data are compared in order to generate phylogenetic trees. A brief description of each of the main stages in this process is given in the following sections.

Although tremendous advances have been made through the use of SSU rRNA sequencing and this method dominates much of the discussion about molecular tools in this book, it should be remembered that it is not the only approach and it has some inherent limitations. The highly conserved nature of the genes and its small size (1500–1800 bases) means that its useful information content is relatively low. SSU rRNA gene sequencing is best suited to comparison of higher-level microbial taxa and there can be drawbacks when it is used to delineate genera and species (see *Box 5.1*). In theory, genes encoding proteins contain more information, because the genetic code (based on four nucleotides) generates an amino acid sequence (based on 20 amino acids). Increasingly, the sequences of genes encoding key structural proteins (e.g. elongation factors in protein synthesis) and enzymes (e.g. ATPase) are being determined and this permits detailed phylogenetic comparisons. However, there is often a high degree of conservation in these genes, meaning that their real 'information content' is limited. In addition, the use of probes directed against specific metabolic genes can provide information about the importance of particular microbial activities in the environment (see *Box 9.1*). For very fine discrimination between closely related lower taxonomic levels (i.e. between species or between strains within a species), it is often preferable to use noncoding spacer regions between genes. These regions of DNA are not under the same constraints as functional genes and are often hypervariable.

The first application of direct sequencing of nucleic acids from the marine environment was in 1986 by Norman Pace and colleagues. Since then, the technique has been applied in many different studies of diversity with startling results that have transformed the science of marine microbiology. *Figure 2.6* shows an outline of the strategies used in the study of bacterioplankton diversity. A number of research teams, most notably including those led by Stephen Giovannoni (Oregon State University), Ed DeLong (Monterey Bay Aquarium Research Institute) and Jed Fuhrman (University of Southern California), pioneered research in this field. As an example, *Box 2.2* describes how molecular methods have led to revised views about diversity among phototrophic marine bacteria. The wider implications of these studies in our current view of prokaryotic diversity are described in *Chapters 5* and *6*. Whilst the greatest attention has been paid to prokaryotic diversity, there is a growing database of molecular information about eukaryotic marine microbes, described in *Chapter 7*.

2.6.2 Isolation of nucleic acids

The first step in all nucleic acid investigations involves the isolation of genomic DNA or RNA from the culture or community. Protocols for DNA or RNA extraction from cells in culture are usually straightforward and reproducible. By contrast, isolations from environmental samples are often problematic and require considerable optimization. Many methods are based on phenol-chloroform extraction or separation on solid-phase media. Commercially available kits provide advantages in speed and reproducibility, but are often not well suited to environmental

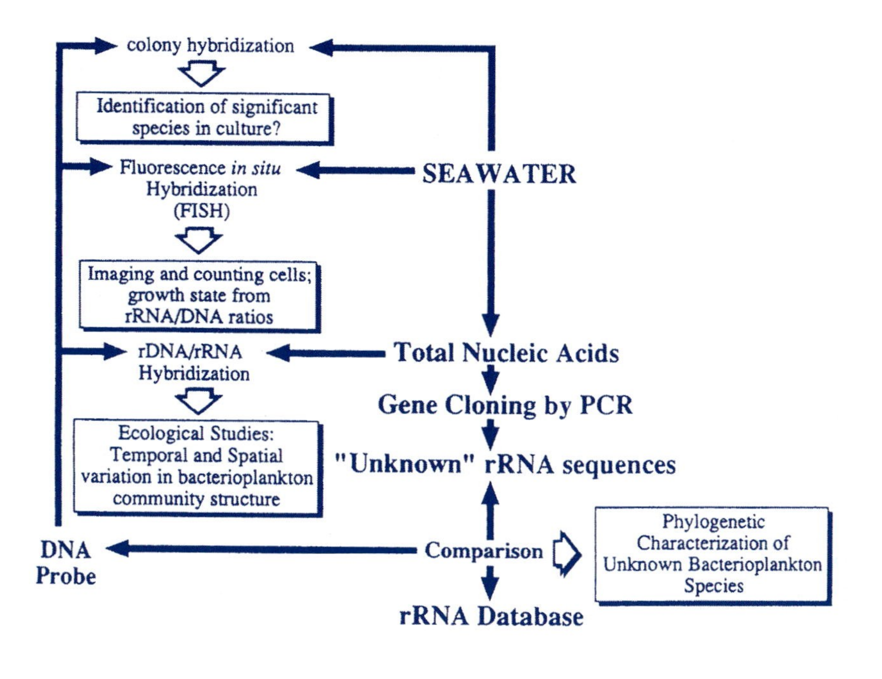

Figure 2.6

Strategy for culture-independent genetic investigations of bacterioplankton diversity. From Giovannoni S and Cary SC (1993), reproduced with permission, The Oceanography Society.

applications. With a few PCR techniques, preparation may involve simple procedures such as boiling a sample of culture, but good-quality purified RNA or DNA is usually needed. Samples must usually be processed rapidly or preserved in such a way that the nucleic acids are protected from degradation (e.g. immediate freezing in liquid N_2). Working with RNA is particularly prone to contamination by nucleases in the laboratory. For extraction from planktonic microbes, it may be necessary to concentrate biomass by centrifugation or tangential flow filtration of large volumes of water. All nucleic-acid-based methods for analysis of diversity in the environment depend on the assumption that DNA is extracted in a more or less pure state equally from the different members of the community, but there are several reasons why this may not always be the case. Microbes differ greatly in their sensitivity to the treatments used to break open the cells. In particular, the nature of the cell envelope varies enormously and cells may be protected by association with organic material or because they are inside the cells of other organisms (e.g. as symbionts or pathogens), meaning that DNA may be extracted preferentially from some types according to the treatment used. Techniques such as freeze–thawing, bead beating (rapid shaking of the mixture with tiny glass beads), ultrasonic disintegration or use of powerful lytic chemicals enhance the recovery of DNA, but can result in extensive degradation. Interference with the PCR often occurs because of the presence of inhibitors such as humic substances and metal ions and these can cause special problems in amplification of DNA from samples such as sediments or animal tissue.

2.6.3 The polymerase chain reaction (PCR)

PCR is a method of amplifying specific regions of DNA, depending on the hybridization of specific DNA primers to complementary sequences, as shown in *Figure 2.7*. The target DNA is mixed with a DNA polymerase, a pair of oligonucleotide primers and a mixture of the four deoxyribonucleoside triphosphates (dNTPs). The PCR depends on the use of a thermostable

Box 2.2 RESEARCH FOCUS

Shedding light on ocean processes

Genomic techniques reveal previously unknown types of phototrophic metabolism

Application of the molecular techniques known as environmental genomics, described in *Section 2.6,* have revealed unexpected microbial diversity and the existence of previously unknown ocean processes. Perhaps the most dramatic demonstration to date of the power of marine environmental genomics is the discovery of proteorhodopsin in marine *Bacteria* made in the laboratory of Ed DeLong and co-workers at the Monterey Bay Aquarium Research Institute, California. Rhodopsin is a light-absorbing pigment found in the eye of animals. It has been known for some time that membranes of the archaean genus *Halobacterium* contain analogs of this pigment, termed bacteriorhodopsins, which use light energy to synthesize ATP for generation of a proton pump for the maintenance of sodium/potassium balance in its cells (*Section 6.2.4*). This was thought to be a process unique to these extremely halophilic *Archaea.* How-ever, whilst carrying out sequence analysis of DNA from an uncultivated group of the γ-*Proteobacteria* known as SAR86 (see *Section 5.2.2*), Beja *et al.* (2000) found evidence of a genetic sequence with strong sequence homology to the archaeal rhodopsins. Using powerful bioinformatics tools, Beja and colleagues made a structural model of the protein believed to be encoded by the gene, and identified the transmembrane domains that are a critical feature associated with the rhodopsin function. Phylogenetic analysis showed that the gene for the proteobacterial protein (which they called proteorhodopsin) is phylogenetically distinct from the archaeal protein bacteriorhodopsin. By transferring the gene to bacterial artificial chromosomes (BACs, *Section 2.6.9*) the gene for the proteorhodopsin protein was cloned in *Escherichi coli.* Fortunately, the protein folded correctly in its recombinant host and was expressed as an active protein. Physiological experiments proved that it acts as a proton pump.

To determine whether this laboratory phenomenon was relevant in the natural environment, Beja *et al.* (2001) then used a technique known as laser-flash photolysis to measure proteorhodopsin-like properties of membranes isolated from marine bacterioplankton concentrated from seawater by filtration. Laser-flash photolysis measures conformational changes in the protein as it absorbs light. Results showed that marine bacteria do indeed possess a functional reaction cycle similar to that seen in *E. coli* expressing the recombinant SAR86 gene. Because bacteria of the SAR86 type are so abundant and widely distributed in marine surface waters, Beja *et al.* suggest that as much as 12% of surface bacterioplankton may contain proteorhodopsin and these bacteria might have a previously unrecognized form of photosynthesis. Beja *et al.* calculate that there are up to 20 000 copies of the protein in each cell, enough to cover a significant part of the cell surface. Further studies show that there is variation in sequence and biophysical properties of proteorhodopsin isolated from different regions and depths. For example, proteorhodopsin isolated from bacteria in surface waters absorbs light maximally at 530 nm, whereas at 75 m the absorption maximum is about 480 nm. Thus, there is an ecophysiological adaptation in SAR86 to ensure maximum absorption of blue light, which penetrates better in deeper waters. This is further evidence that light capture via proteorhodopsin is probably a significant source of energy uptake for cell maintenance and reproduction in the oceans, reducing requirements for energy from respiratory metabolism. However, we do not yet know if proteorhodopsin-containing bacteria can fix CO_2.

Other recent research demonstrates how rapidly our view of phototrophic mechanisms is changing. Most photosynthetic processes in the oceans were thought to be due to cyanobacteria, which carry out aerobic, oxygenic photosynthesis (see *Section 5.4*). However, Kolber *et al.* (2000) provided the first biophysical evidence demonstrating that aerobic anoxygenic photosynthesis (AAnP) by bacteria is widespread in surface waters of the Pacific and Atlantic oceans. Kolber *et al.* (2001) measured the vertical distribution of bacteriochlorophyll *a* and fluorescence signals and found that photosynthetically competent AAnP bacteria comprise at least 11% of the total microbial community. These organisms are photoheterotrophs, which utilize O_2 for respiratory metabolism of organic carbon and for the synthesis of bacteriochlorophyll *a*. However, AAnP bacteria metabolize more efficiently when light is available, suggesting that in oligotrophic waters they can use energy from both light and scarce nutrients simultaneously. Until recently, it was assumed that this type of metabolism was restricted to a very small number of types within the α-*Proteobacteria*. However, Beja *et al.* (2002) reported an unexpected diversity among AAnP bacteria. They prepared a BAC library from surface bacterioplankton and screened this for *puf* genes, which code for the subunits of the light harvesting and reaction center complexes (*Section 4.3.2*). To identify groups containing the photosynthetic genes in natural populations, Beja *et al.* used RT-PCR of mRNA (*Section 2.6.3*). Several different clones were identified and gene sequences were compared with those from known, cultured photosynthetic bacteria. New β-proteobacterial species were discovered that were genetically, and perhaps physiologically, different from previously known types. The discovery that currently uncultured bacteria may be contributing significantly to aerobic photosynthesis in the oceans necessitates a radical rethink of ocean energy budgets and cycling of organic and inorganic carbon.

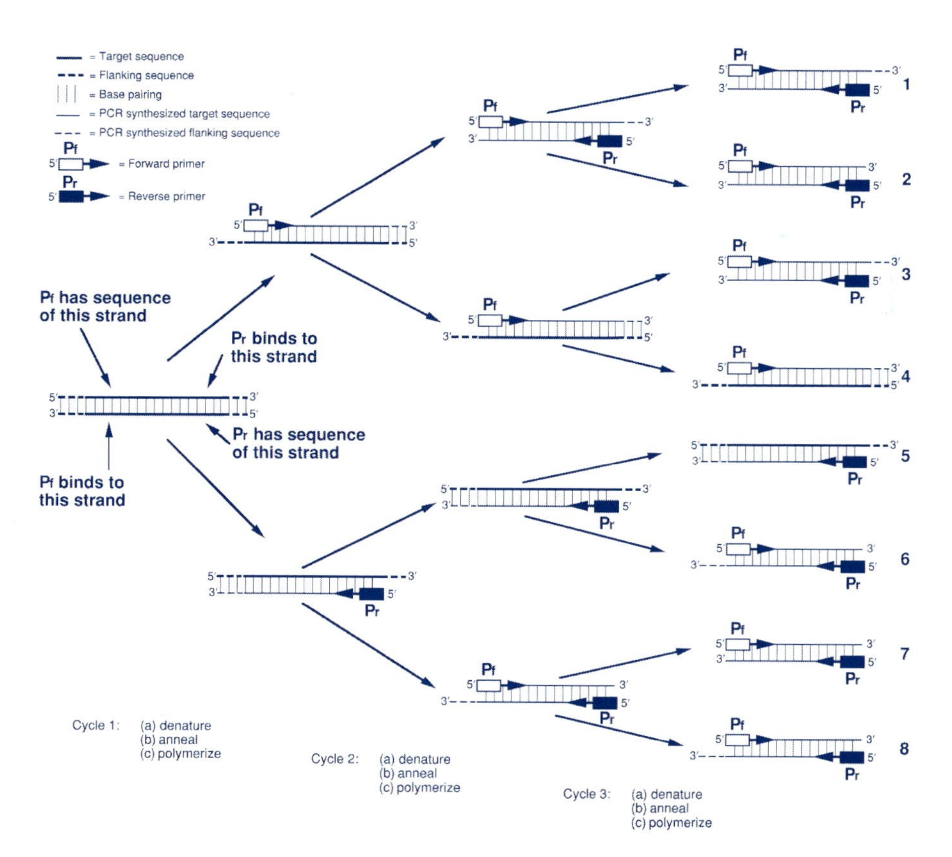

Figure 2.7

The polymerase chain reaction (first three cycles). Only after cycle 3 are there any duplex molecules of the exact length of the region to be amplified (molecules 2 and 7). After a few more cycles, these become the major product. From Turner PC *et al.* (2000), reproduced with permission, BIOS Scientific Publishers.

DNA polymerase, which is able to function during the repeated cycles of heating and cooling in the reaction. *Taq* polymerase is a thermostable enzyme isolated from the hot-spring thermophilic bacterium *Thermus aquaticus*. Interestingly, alternative thermostable polymerases from deep-sea thermophiles are now also used (see *Section 16.6*). The primers are designed to anneal to opposite strands of the target DNA so that the polymerase extends the sequences towards each other by addition of nucleotides from the 3′ ends, by base pairing using the target DNA sequence as a template for synthesis. The reaction is started by heating (usually to about 94°C for 1 min), so that the double-stranded DNA dissociates into single strands. The temperature is then lowered (usually to about 50–60°C for 0.5–1 min), so that the primers can attach to the complementary sequences on the DNA. The temperature is then raised to 72°C (the optimum for *Taq* polymerase), and primer extension begins. In the first stage of the reaction, the target DNA is copied for various distances until polymerization is terminated by raising the temperature again to 94°C. For the second cycle, the temperature is lowered and the other primer can now attach to the newly synthesized DNA strand. On raising the temperature again to 72°C, this strand is copied until the end of the first primer is reached. After three such cycles, the number of short duplexes containing the amplified sequence will increase exponentially; one target molecule will become 2^n molecules after n cycles. The heating and cooling cycles are carried out in specially designed electronically controlled thermal cyclers, in which the reaction tubes are placed in blocks allowing very rapid and precise heating and cooling. Usually, 30–40 cycles take place, resulting in amplification of a single target sequence to over 250 million PCR products. The outcome of the reaction is usually checked by agarose gel electrophoresis. The DNA molecules are separated according to size and visualized under UV light after staining the gel with ethidium bromide or SYBR green. Because of the high degree of conservation in the SSU gene, the PCR products will all have approximately the same size and should migrate as a single band. Suitable positive and negative controls are always included in the reaction and each application of the PCR must be carefully optimized. This is especially important when trying to amplify particular sequences from a large mixed population of DNA, as in direct examination of environmental samples. Conditions such as the annealing temperature, amount of template DNA and concentration of Mg^{2+} are critical. For example, *Taq* polymerase is inactive in the absence of Mg^{2+}, but at excess concentration the fidelity of the polymerase is impaired and nonspecific amplification can result.

There are many variations of the basic PCR technique. In nested PCR, the level of specificity and efficiency of amplification is improved by carrying out a PCR for 15–30 cycles with one primer set, and then a further 15–30 cycles with a second set of primers that anneal within the region of DNA amplified by the first primer set. Multiplex PCR involves the use of multiple sets of primers and results in the production of multiple products. This is often used as a quick screening method for the presence of certain organisms within water or infected tissue. The reverse transcriptase PCR (RT-PCR) is used to quantify messenger RNA (mRNA) and thus detects gene expression. This is a technique requiring great care in sample preparation and handling, since mRNA is very short-lived and the technique is prone to contamination from genomic DNA and nucleases. In quantitative real-time PCR, the formation of a fluorescent reporter is measured during the reaction process. By recording the amount of fluorescence emission at each cycle, it is possible to monitor the PCR reaction during the exponential phase where the first significant increase in the amount of PCR product correlates to the initial amount of target template. Real-time PCR offers significant advantages for the diagnostic detection of microbes in the environment and in infected tissue samples and hand-held instruments for use in the field are now becoming available.

2.6.4 DNA sequencing

The most commonly used method of obtaining the sequence of DNA is based on the Sanger 'dideoxy' technique of chain termination during synthesis of a copy of the DNA, as shown in

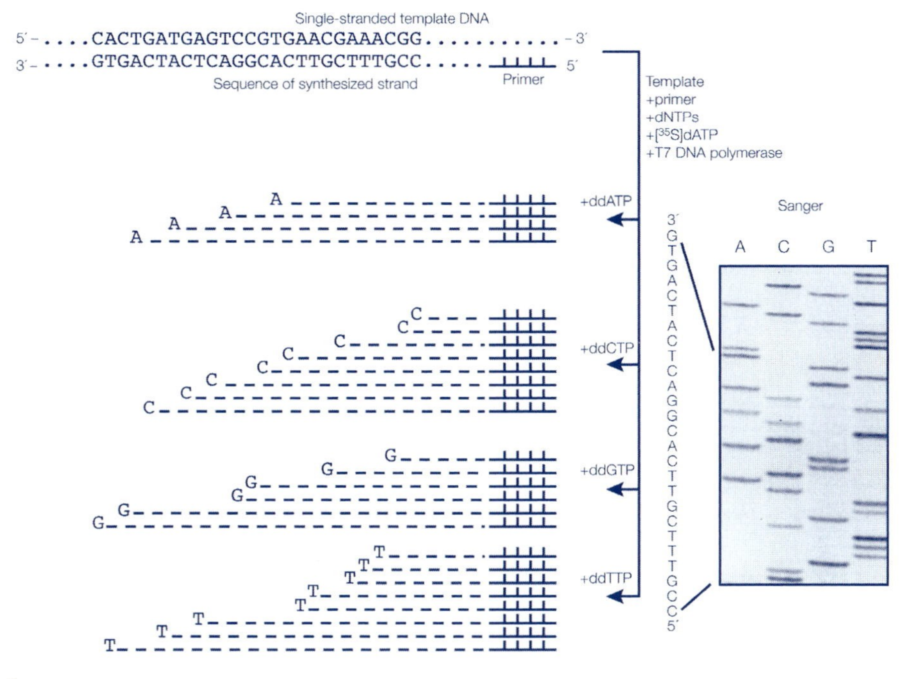

Figure 2.8

DNA sequencing by the Sanger dideoxy chain termination method. From Turner PC *et al.* (2000), reproduced with permission, BIOS Scientific Publishers.

Figure 2.8. A sequencing primer is added to a single-stranded DNA molecule in four separate reaction mixtures and chain extension begins using the normal dNTPs. The different reactions contain a small amount of dideoxy nucleotides (ddNTPs), which are synthetic analogs of the nucleotides lacking a –OH group on the sugar. Because of this, the DNA chain is extended only up to the point at which one of the ddNTPs is incorporated. The products from the four reactions are compared by running a polyacrylamide electrophoresis gel and the sequence of the DNA is determined by reading the sequence of nucleotides at which the chain extension is terminated. In the original method, a radioactive label is incorporated in the reaction mixture and the bands are visualized by autoradiography (exposure of an X-ray film). The basic technique has been modified to allow direct sequencing of PCR products as double-stranded DNA and most laboratories now rely on the use of dNTPs labeled with different fluorescent dyes, which can be read by lasers. Automated DNA sequencers allow robotic handling and high throughput of multiple samples, generating color-coded printouts of the sequence and data suitable for direct analysis by computer. For community analysis, it is often necessary to obtain only a partial sequence.

2.6.5 Phylogenetic analysis

Sequence data are used to determine the degree of phylogenetic relatedness between organisms. Data come both from microbes cultured in the laboratory and sequences obtained by direct community analysis of the environment. Data processing depends on the availability of publicly accessible databases (e.g. GENBANK, Ribosome Database Project and EMBL) and specialized computer software. A newly generated sequence is first compared and aligned to previously described sequences using programs such as SEQUENCE_MATCH, BLAST and CLUSTAL, available via the Internet. Various mathematical methods and computer algorithms are used for the construction of phylogenetic trees, but details of the theory of each technique and its merits are beyond the scope of this book. Briefly, in one method the different sequences are compared in order to determine the evolutionary distance between all the permutated pairs of sequences in the dataset by calculating the percentage of nonidentical sequences. Statistical corrections to allow for the possibility of multiple mutations at a given site are included in the calculation. A distance matrix is then constructed and a phylogenetic tree is drawn by grouping the most similar sequences and then adding less similar groups of sequences. The lengths of the lines in the tree are in proportion to their evolutionary distances apart (e.g. see *Figure 5.1*). In another approach called parsimony, the tree is constructed after calculating the minimum number of mutational events needed for each pair to transform one sequence into the other. The tree is then drawn using the minimum number of such parsimonious steps. This method relies on the inherent assumption that evolution proceeds in one direction by the fastest route; this is undoubtedly an oversimplification. The validity of trees can be tested by a computational technique known as 'bootstrapping'.

Sequence analysis shows that certain nucleotide sequences are highly conserved in particular regions of the rRNA molecule, the bases of which are numbered in a standard notation. These signature sequences are characteristic of groups of organisms at different taxonomic levels. For example, the sequence CACYYG occurs at approximate position 315 in the 16S rRNA molecule in almost all *Bacteria,* but does not occur in *Archaea* or *Eukarya.* Sequence CACACACCG at position 1400 is distinctive of *Archaea* and does not occur in *Bacteria* or *Eukarya.* Signature sequences can therefore be used to design domain-specific oligonucleotide probes for identification of any member of the respective group in a marine sample. Other sequences may be diagnostic for phyla, families or even genera (see *Figure 2.9*). This is a very powerful

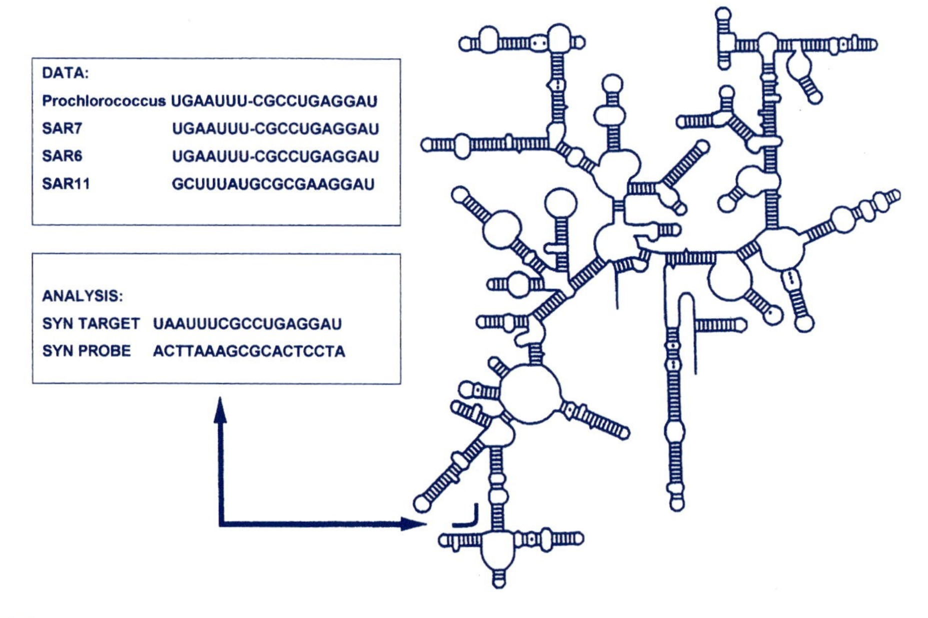

Figure 2.9

Simplified example of the procedure used to design phylogenetic DNA probes from databases of 16S rRNA sequences. In this example, a small sequence is identified which is shared by autotrophic picoplankton (marine *Synechococcus* and *Prochlorococcus*) but differs from homologous regions of the abundant bacterial clone SAR11, known only genetically. SAR7 and SAR6 are marine *Synechococcus* and *Prochlorococcus* genes respectively, which were cloned directly from seawater. The target site for the probe is indicated on the 16S rRNA structural model. From Giovannoni S & Cary SC (1993), reproduced with permission, The Oceanography Society.

technique for culture-independent community analysis and is usually achieved using FCM (*Section 2.3*) or FISH (*Section 2.6.10*).

2.6.6 Community fingerprinting

In denaturing gradient gel electrophoresis (DGGE), PCR products are run in a polyacrylamide gel under a linear gradient of denaturing conditions (7 M urea and 40% formamide at a temperature of 50–65°C). Small variations in sequence result in different migration of DNA. Double-stranded DNA migrates through the gel until it reaches denaturing conditions sufficient to cause separation of the double helix ('melting'). DGGE primers must contain a guanine- and cytosine-rich sequence (GC clamp), which is resistant to complete denaturation under the conditions used. In analysis of community DNA, a set of bands will be generated representing variations in the sequence of the target gene. A typical analysis of an environmental sample will generate more than 50 bands and many different samples can be compared on a single gel. This provides a rapid and efficient method of community fingerprinting, i.e. monitoring qualitative changes in community composition at different locations or under different conditions. Sometimes, the results of DGGE community analysis are interpreted in a semi-quantitative fashion using the assumption that the intensity of bands indicates the relative abundance of the species producing them. However, great caution is needed in drawing such conclusions because the techniques used for DNA extraction, the presence of inhibitors, the nature of the primers and the PCR reaction conditions can all have dramatic effects on the results obtained. Temperature gradient gel electrophoresis (TGGE) works using similar principles, but the concentration of denaturing chemicals remains the same whilst the temperature of the gel is increased gradually and uniformly.

In terminal restriction fragment length polymorphism (TRFLP) analysis, PCR is performed on DNA extracted from the mixed community with one of the primers labeled with a fluorescent probe. After PCR, the products are cut using specific restriction enzymes. Each PCR product produces a fluorescent molecule with the probe at the end. The size of the fragments is determined by differences in the presence or absence of restriction sites in a particular region of the molecule or by deletions and insertions in the region. Like DGGE, TRFLP can provide a quick semiquantitative picture of community diversity, but results must again be interpreted carefully.

2.6.7 Limitations of environmental analysis of nucleic acids

As previously noted, various protocols and kits are available for nucleic acid extraction, but a considerable amount of prior investigation is needed to ensure efficient extraction and amplification. Design of appropriate primers for PCR is the most critical step in the process. Primer design depends on the quality of relevant information available in sequence databases. So-called universal primers present the least problems. These primers contain consensus sequences designed to amplify DNA from broad phylogenetic groups rather than specific lineages. However, such primers can hybridize more favorably with some templates. For example, DNA templates from thermophilic organisms with a high content of G+C base pairs may denature less rapidly than those from mesophiles. Addition of reagents to improve template denaturation can sometimes overcome these problems. A further problem arises from the production of chimeric PCR products. These are artifacts caused by the joining together of the amplification products of two separate sequences and, if not detected, such sequences could suggest the presence of a novel nonexistent organism. This is a particular pitfall when multitemplate PCR is carried out using environmental samples of DNA. Various techniques are available to limit the formation of chimeras and software to detect chimeric sequences is now available. Analysis of rRNA databases has revealed a number of chimeric sequences which could be potentially confusing in phylogenetic studies. A basic tenet in the use of SSU rRNA in determining phylogenetic relationships is the assumption that rRNA genes are not subject to horizontal gene transfer, because of the stringent constraints on the function of rRNA in the essential process of protein synthesis. However, it must be noted that there is a growing body of evidence that single organisms may contain heterogeneous SSU rRNA sequences. Several protist groups have been shown to contain two types of 18S rRNA genes with significant nucleotide differences (over 5%), and similar evidence of multiple 16S rRNA operons has been found in halophilic *Archaea* and in *Actinobacteria*. Since this phenomenon occurs in all three domains of life, it suggests that horizontal transfer of SSU rRNA genes may be more widespread than previously thought and this may make the interpretation of rRNA-based phylogenetic data more difficult.

2.6.8 Genomic fingerprinting and molecular markers

The methods described above are used mainly for the analysis of microbial communities of unknown composition in environmental samples. Investigation of diversity among closely related individual species (or strains within a species) can be achieved by a variety of genomic fingerprinting techniques. The most widely used methods make use of differences in genetic sequences revealed by the action of restriction enzymes, which cut DNA at specific sites determined by short sequences of nucleotide bases. All of these fingerprinting techniques produce band patterns that lend themselves to computer-assisted pattern analysis, leading to the generation of databases and phylogenetic trees useful for studying diversity among cultured isolates of marine microbes.

Ribotyping is a technique used for bacterial identification that uses the same principles described above for 16S rDNA analysis. However, in this case the PCR products from rRNA gene amplification are cut with restriction enzymes and separated by electrophoresis. The resulting band pattern gives a rapid indication of similarities and differences between strains, but obviously lacks the resolution of sequence analysis.

Restriction fragment length polymorphism (RFLP) analysis involves cutting DNA with one or more restriction endonucleases and separating the fragments according to size by gel electrophoresis. Fragments containing a specific base sequence can be identified by Southern blotting, i.e. transfer to a sheet of nitrocellulose and addition of a radioactively labeled complementary DNA probe. Hybridization of the probe is detected using autoradiography or with an enzyme-linked color reaction. This enables identification of the restriction fragment with a complementary sequence to that of the DNA probe. Fragments of different length caused by mutations in duplicate copies of the gene can be detected and used for strain differentiation.

The RFLP method can be combined with PCR amplification of specific genes (e.g. SSU rRNA genes) followed by restriction endonuclease digestion and electrophoresis. This avoids the need for blotting and radiography but is not always successful in providing useful markers for strain differentiation. The RAPD (random amplified polymorphic DNA) technique is a more commonly used PCR marker technique for strain differentiation. It uses a single short primer in a PCR reaction leading to a fingerprint of multiple bands generated by single nucleotide differences between individuals that prevent or allow primer binding. The method is quick and easy to perform but does not always give reliable results for detecting strain differences. AFLP (amplified fragment length polymorphism) is now a widely used technique with better resolution than either RFLP or RAPD analysis. Genomic DNA is digested using a pair of restriction endonucleases, one of which cuts at common sites and one which cuts at rare sites. Adapter sequences are ligated to the resulting fragments and a PCR performed with primers homologous to the adapters plus selected additional bases. A subset of the restriction fragments is amplified, resulting in a large but distinct set of bands on a polyacrylamide gel. Band patterns are compared using imaging systems and similarities and differences between strains can be computed to generate phylogenetic trees. Despite its advantage, AFLP analysis is a technique requiring very highly purified DNA and its use is generally restricted to laboratories specializing in taxonomy.

Microsatellite markers are repeat regions of short nucleotide sequences that can be analyzed using PCR amplification of unique flanking sequences and separation of the resulting bands by gel electrophoresis. Microsatellite markers are used mainly in the population analysis of eukaryotic organisms such as microalgae.

The genomic fingerprinting method for bacteria known as rep-PCR has been introduced recently as a simple and reproducible method for distinguishing closely related strains and deducing phylogenetic relationships. The method is based on the fact that bacterial DNA naturally contains interspersed repetitive elements, such as the REP, ERIC and BOX elements. Primers against these sequences are used in the PCR. The method requires very simple preparation of DNA samples without the need for extensive purification and is especially useful for the initial screening of cultures collected from marine samples.

Pulsed-field gel electrophoresis (PFGE) is a method in which the orientation and duration of an electric field is periodically changed allowing large-molecular-weight fragments of DNA (up to 1000 kb) to migrate through the gel. Bacteria are embedded in the gel and lysed *in situ* into large fragments, using a restriction enzyme that recognizes rare restriction sites. PFGE typing detects small changes in the genome resulting from the insertion or deletion of gene sequences or mutations that alter the restriction enzyme sites. PFGE is also very useful for analysis of virus diversity.

2.6.9 Genomics

Since the mid 1990s, the whole genome shotgun sequencing method pioneered by Craig Venter and colleagues at the Institute for Genomic Research (TIGR) has been applied to an increasing number of microbes. Automated sequencing methods and robotic handling of samples mean that it is now possible to determine the complete sequence of viral and bacterial genomes within a few months. Nevertheless, genome mapping is expensive and usually relies on collaborative efforts between different laboratories. The method involves construction of a library of DNA fragments in *E. coli,* random sequencing and alignment of contiguous nucleotide sequences. The genomes of several marine *Bacteria* and *Archaea* have now been mapped and some of the more important examples are shown in *Table 2.3*. Initial attention was focused on culturable hyperthermophilic and psychrophilic organisms of biotechnological potential, but the sequencing of other marine bacteria is now gaining pace. Three strains of *Prochlorococcus* (*Section 5.4.4*) have now been sequenced and projects to sequence the ubiquitous ocean bacterium SAR11 (see *Section 5.2.2*) and the bioluminescent symbiont *Vibrio fischeri* (see *Box 10.3*) are now underway. There is intense activity to establish many other marine

Table 2.3 Some representative species of *Archaea* and *Bacteria* found in marine and coastal habitats for which genome sequences are published or almost complete[1]

Archaea	*Bacteria*
Aeropyrum pernix	*Aquifex aeolicus*
Archaeoglobus fulgidus	*Caulobacter crescentus*
Halobacterium salinarum	*Magnetospirillum magnetotacticum*
Methanocladococcus jannaschii	*Nostoc* sp.
Methanococcus maripaludis	*Oceanobacillus iheyensis*
Methanopyrus kandleri	*Prochlorococcus marinus*
Pyrobaculum aerophilum	*Synechocystis* sp.
Pyrococcus abyssi	*Thermotoga maritima*
Pyrococcus horikoshii	*Vibrio cholerae*
Pyrococcus furiosus	*Vibrio parahaemolyticus*
	Vibrio vulnificus

[1]Data from The Center for the Advancement of Genomics, March 2003.

genome sequencing initiatives. These projects will undoubtedly lead to many insights into processes such as nutrient cycling and symbiotic interactions.

Once sequences are complete and gaps are closed, annotation of the genome sequence begins. This involves predicting the sequence regions coding for proteins (open reading frames, ORFs) and the likely homology of proteins between different organisms, giving insights into the genetic organization and likely metabolic properties of the organism. Often, many of the predicted ORFs cannot be reliably linked to a function using current knowledge of bioinformatics. Therefore, the next stages in molecular analysis are functional genomics and proteomics. These involve inactivation of specific regions of the genome and high-resolution separation and identification of proteins. Such postgenomic analyses of marine microbes will undoubtedly reveal many properties and activities that are currently completely unknown.

One of the major methods used in functional genomics is DNA microarray technology. DNA sequences identified from microbial genomes are attached (or synthesized directly) on the surface of a silicon chip in a highly ordered array to produce a device that can act as a probe for thousands of genes. For some bacteria, chips with probes for every expressed gene in the genome have been designed. The target nucleic acids (mRNA or cDNA) are labeled with fluorescent reporter groups and lasers detect hybridization of complementary sequences to the oligonucleotide on the microarray chip. Microarrays are particularly useful for determining the effects of environmental conditions on patterns of gene expression and they will undoubtedly find widespread application in marine microbiology.

To date, no organism's genome has been sequenced without an intervening culture step, but direct environmental genome sequencing is a distinct possibility. Even without full genome sequencing, it is possible to gain valuable information on large segments of DNA obtained directly from environmental sources. With careful handling, fragments up to 300 kilobases can be cloned directly into the large F plasmids of *E. coli* to create bacterial artificial chromosomes (BACs). These have the additional advantage that expression of some of the insert genes may occur in the *E. coli* host harboring the vector. It is possible to sequence such relatively small amounts of DNA by a shotgun cloning approach, in which a series of individual partial sequencing steps is used to build up the overall sequence by assembling overlapping sequences. Alternatively, the technique of 'chromosome walking' can be used. This entails the initial isolation of genomic clones in a library that overlap with one of the markers that flank the gene of interest. More clones are then isolated which overlap these initial clones and this continues progressively until a clone containing the gene of interest is isolated.

2.6.10 Fluorescent *in situ* hybridization (FISH)

The acronym FISH is no accident. This technique permits an investigator to 'fish' for a specific nucleic acid sequence in a 'pool' of unrelated sequences. Originally pioneered by Rudolf Amann of the Max-Planck Institute for Marine Microbiology, it has proved to be one of the most useful culture-independent techniques for identification of particular organisms or groups of organisms in the marine environment. As noted above, oligonucleotide probes can be designed to recognize and hybridize to complementary sequences that are unique to specific microbial groups. In FISH, the oligonucleotide probe is covalently linked to a fluorescent dye such as fluorescein, Texas Red or indocarbocyanines. Hybridization with the target within a morphologically intact microbial cell can therefore be visualized by fluorescence or confocal microscopy (see micrograph in *Box 6.1*). The probe signal that is produced after hybridization and washing away excess unbound probe is proportional to the amount of target nucleic acid present in the sample. This allows an investigator to visualize and quantify the microbes in marine samples and determine cell morphology and spatial distributions *in situ*. FISH has been particularly useful for examining microbes attached to particles, spatial localization of microbes in biofilms, in symbiotic associations and in infected tissues. The widespread use of SSU rRNA molecules for diversity studies means that there are extensive databases of sequence information that can be used to design probes as phylogenetic stains. Cells usually contain thousands of copies of rRNA molecules and they are relatively stable. However, cells which are slow-growing, dormant or in a nonculturable state (see *Section 4.5*) are harder to detect and can give variable results because they may have low numbers of ribosomes, or have membrane modifications which lower the accessibility of the probe. This places some limits on the sensitivity of FISH in investigation of oligotrophic ocean environments. Sometimes, a PCR amplification step is added to overcome this problem. Multiple probes, using different sequences and different fluorochromes can be applied to single samples in order to improve sensitivity and to assess genetic diversity. The FISH technique can be extended to search for a wide range of specific gene sequences characteristic of particular structures or metabolic functions, assuming that sufficient copies of the nucleic acid are present in the sample.

2.6.11 GC ratios and DNA–DNA hybridization in bacterial taxonomy

Determination of the ratio of nucleotide base pairs is often used in bacterial taxonomy. The principle of this method is that two organisms with very similar DNA sequences are likely to have the same GC ratio, i.e., the ratio of guanine (G)+cytosine (C) to the total bases. DNA is extracted from the bacterium and the concentration of the bases is determined by high performance liquid chromatography (HPLC) or by determining the 'melting point' (T_m) of DNA. This is the temperature at which double-stranded DNA dissociates; because the hydrogen bonds connecting G–C pairs are stronger than those connecting adenine (A) and thymine (T), the T_m increases with higher ratios of G+C. Closely related organisms have very similar GC ratios, and this can be used to compare species within a genus. However, bacteria possess a wide range of GC ratios and completely unrelated organisms can share a similar GC ratio.

A more useful measure of relatedness for the definition of species is DNA– DNA hybridization. Purified DNA from two organisms to be compared is denatured by melting and mixing their DNA. When cooled, DNA with a large number of homologous sequences will reanneal to form duplexes. The amount of hybridization can be measured if the DNA from one of the organisms is labeled by prior incorporation of bases containing ^{14}C or ^{32}P. The amount of radioactivity in the reannealed DNA collected on a membrane filter is measured and the percentage hybridization calculated by comparison with suitable controls. Taxonomists use the benchmark of 70% or greater hybridization (under carefully standardized conditions) as the definition of a bacterial species.

2.7 Detecting microbial activities in the marine environment

2.7.1 The microenvironment

The aim of microbial ecology is the study of the diversity and activities of microbes *in situ*. The preceding sections have largely been concerned with the question 'Who's there?', whereas this section is concerned with methods to address the other major question, that is 'What are they doing?' Activity measurements are largely concerned with providing a picture of the overall activities within a microbial community. However, when combined with modern knowledge and tools to investigate diversity, we can begin to build a picture of the activities of individual community members. The importance of scale cannot be overemphasized – *micro*organisms carry out their activities in a *micro*environment. Measuring techniques must be designed to cause the minimum disturbance of the microenvironment that they are probing. Microbial communities may exist in a delicate three-dimensional structured organization, which is easily disrupted. At the micrometer scale, physicochemical gradients are very steep. For example, anaerobic, microaerophilic and aerobic conditions may occur in a sediment particle or a microbial mat just a few hundred micrometers thick. Other conditions such as boundary layer effects, diffusion and flow patterns will affect the microenvironment in the vicinity of a microbe. Most important of all, the microbes carry out biochemical reactions that alter the physical and chemical conditions in their immediate vicinity. Bulk measurements of substrates, metabolites, pH or oxygen in marine samples may therefore not reflect the heterogeneity of the natural situation.

2.7.2 Microelectrodes and biosensors

Special electrochemical microelectrodes constructed of glass and metal with tip diameters as small as 5 μm can be inserted using micromanipulators into certain habitats, in order to measure changes in pH, O_2, CO_2, H_2 or H_2S. Microsensors to detect small changes in light and temperature can also be constructed. These probes can be left in place for long periods and when linked to recorders they provide continuous data about environmental changes and the zonation and rates of microbial and chemical processes. For example, development and use of O_2, H_2S and pH microsensors made it possible to analyze the gradients of these factors in sediments and biofilms. This was used to show how sulfur bacteria live in opposing gradients of sulfide and O_2 with only very small concentrations of both chemical species being present in the oxidation zone. Microsensor techniques are used to investigate oxygenic photosynthesis at a spatial resolution of less than 0.1 mm in biofilms and sediments. Recent innovations have included the development of microscale biosensors suitable for use in seawater and marine sediments. Here, biological components such as enzymes or antibodies are coupled to a signal converter so that the results of a reaction are amplified by production of an electronic reading. The development of microscale biosensors for various inorganic nitrogen species has been a major advance permitting detailed study of the processes of nitrification and denitrification. Biosensors for CH_4 and volatile fatty acids have recently been developed to probe microbial transformations in anaerobic sediments and planktonic aggregates such as marine snow particles. Studies using O_2 microelectrodes have shown that even the tiniest aggregates contain anaerobic niches, but it is not yet known if anaerobes such as methanogens grow in these ephemeral habitats.

2.7.3 Isotope methods

The use of substrates containing radioactive isotopes provides a valuable method for determining microbial transformations in the marine environment. One of the earliest applications in marine microbial ecology was the introduction of $^{14}CO_2$ into bottle experiments in order to measure rates of photosynthesis by picoplankton. Pioneers in the development of isotope incorporation methods in the late 1970s include David Karl and Jed Fuhrman, then working at the Scripps Institute of Oceanography. Fixation of the labeled precursor is detected by measuring incorporation of radioactivity into organic material in a scintillation counter, or by autoradiography. Bacterial production in seawater samples or pure cultures is measured by the incorporation of [^{3}H]-adenine or [^{3}H]-thymidine into RNA or DNA, or by incorporation of labeled amino acids (e.g. [^{3}H]-leucine) into proteins. The flux of organic material in seawater can be measured using [^{14}C]-glucose. Many other microbial processes can be measured using appropriate isotopes. Appropriate correction factors for the discrimination of enzyme processes against the heavier radioisotopes must be included in calculations.

Stable isotopes are different forms of an element that are not radioactive but are acted on differentially during biological and chemical processes. For example, carbon in nature exists predominantly as ^{12}C, but a small amount exists as ^{13}C. Similarly, sulfur exists mainly as ^{32}S, with a small proportion as ^{34}S. Enzymic reactions tend to preferentially utilize the lighter isotope so, for example, cells which have fixed CO_2 tend to have a higher ratio of ^{12}C:^{13}C than exists in nonbiological processes. Similarly, H_2S produced by sulfate-reducing bacteria has a higher proportion of the lighter isotope ^{32}S. Measuring the ratios of stable isotopes is thus a valuable way of distinguishing microbial (enzymic) processes from purely geochemical ones.

2.7.4 Measurement of specific cell constituents

Mention has already been made of the use of membrane lipids and fatty acids as biomarkers of microbial activity. Measurement of ATP has been widely used as an estimate of the size and activity of total microbial communities, for example in the vicinity of hydrothermal vents and in the analysis of microbial contamination of ballast water in ships. ATP occurs universally in all living organisms and its presence is an indicator of active metabolism, since it is lost rapidly from dead cells. ATP is easily extracted from cells using boiling TRIS buffer or nitric acid, but to produce reliable results, care is needed in the processing of samples by filtration. A sensitive chemiluminescent assay for ATP depends on use of the enzyme luciferase, which catalyzes the reaction

$$\text{luciferin} + \text{ATP} \longrightarrow \text{oxyluciferin} + \text{AMP} + \text{PP}_i + \text{CO}_2 + \text{light}$$

Luciferase was originally obtained from fireflies, but is now available as a recombinant product made in *E. coli*. Light production is measured in sensitive photometers and these are available as hand-held instruments for rapid monitoring of ATP levels.

Lipolysaccharide (LPS), a unique component of Gram-negative bacterial cell walls (*Section 3.9.5*), can be detected by a sensitive assay using extracts from the horseshoe crab (*Limulus*), which causes an increase in optical density (due to gel formation) upon the addition of picogram amounts of LPS. The *Limulus* amoebocyte lysate (LAL) assay has been widely used to estimate bacterial biomass in marine samples and to demonstrate the presence of Gram-negative bacteria in the tissues of the hosts of symbionts and pathogens. Great care is needed to prevent contamination during sample preparation. LPS and ATP measurements were commonly used in the 1980s, but have become less 'fashionable' since the introduction of nucleic-acid-based methods.

The diversity, distribution and concentration of photosynthetic pigments in the water column can be used to investigate the structure of phytoplankton. Chlorophylls, bacteriophylls, carotenes and other pigments are analyzed by HPLC.

A number of methods are available for estimating biomass of fungi, for example in ecosystems such as salt marshes and mangrove swamps. Ergosterol is a characteristic component of many fungal membranes (especially ascomycetes) and can be extracted by chromatography and measured by UV spectrophotometry. The rate of synthesis of ergosterol can be measured using a radiolabeled precursor, such as acetate.

2.7.5 Remote sensing

At the opposite extreme from techniques to measure microenvironmental changes, the introduction of remote sensing has provided valuable information about the activities of planktonic organisms in the surface layers of the oceans. Whilst the ocean appears blue to the human eye, sensitive instruments detect subtle changes in ocean color due to the presence of plankton, suspended sediments, and dissolved organic chemicals.

The first such satellite was the Coastal Zone Color Scanner (CZCS) launched in 1978, which collected light reflected from the sea surface at different wavelengths, including the absorption maximum for chlorophyll. This therefore provided valuable information about the distribution of photosynthetic plankton (both microbial and macroalgal). The CZCS ceased operations in 1986 and scanning is now carried out using the Sea-viewing Wide Field-of-View Sensor (SeaWiFS) system operated by the Orbital Sciences Corporation satellite, which makes data available to research agencies worldwide. The results of the application of this technology in the monitoring of phytoplankton productivity is shown in *Table 9.1* and *Section 12.7* considers its use in monitoring algal blooms. *Box 7.1* includes an example of a SeaWiFs satellite image.

References and further reading

Amann, R.I., Ludwig, W., and Schleifer, K.H. (1995) Phylogenetic identification and in-situ detection of individual microbial-cells without cultivation. *Microbiol Rev* **59**: 143–169.

Beja, O., Aravind, L., Koonin, E.V., *et al.* (2000) Bacterial rhodopsin: evidence for a new type of phototrophy in the sea. *Science* **289**: 1902–1906.

Beja, O., Spudich, E.N., Spudich, J.L., Leclerc, M., and DeLong, E.F. (2001) Proteorhodopsin phototrophy in the ocean. *Nature* **411**: 786–789.

Beja, O., Suzuki, M.T., Heidelberg, J.F., *et al.* (2002) Unsuspected diversity among marine aerobic anoxygenic phototrophs. *Nature* **415**: 630–633.

Button, D.K., Schut, F., Quang, P., Martin, R., and Robertson, B.R. (1993) Viability and isolation of marine bacteria by dilution culture: theory, procedures and initial results. *Appl Environ Microbiol* **59**: 881–891.

Connon, S.A., and Giovannoni, S.J. (2002) High-throughput methods for culturing microorganisms in very low-nutrient media yield diverse new marine isolates. *Appl Environ Microbiol* **68**: 3878–3885.

DeLong, E.F. (1997) Marine microbial diversity: the tip of the iceberg. *Trends Biotechnol* **15**: 203–207.

DeLong, E.F. (2001) Microbial seascapes revisited. *Curr Opin Microbiol* **4**: 290–295.

DeLong, E.F. (2002) Microbial population genomics and ecology. *Curr Opin Microbiol* **5**: 520–524.

DeLong, E.F., Taylor, L.T., Marsh, T.L., and Preston, C.M. (1999) Visualization and enumeration of marine planktonic archaea and bacteria by using polyribonucleotide probes and fluorescent in situ hybridization. *Appl Environ Microbiol* **65**: 5554–5563.

Fuhrman, J.A. (2002) Community structure and function in prokaryotic marine plankton. *Ant van Leeuw Int J Gen Molec Microbiol* **81**: 521–527.

Fuhrman, J.A., Griffith, J.F., and Schwalbach, M.S. (2002) Prokaryotic and viral diversity patterns in marine plankton. *Ecol Res* **17**: 183–194.

Gaasterland, T. (1999) Archaeal genomics. *Curr Opin Microbiol* **2**: 542–547.

Gingeras, T.R., and Rosenow, C. (2000) Studying microbial genomes with high-density oligonucleotide arrays. *ASM News* **66**: 463–469.

Giovannoni, S., and Cary, S.C. (1993) Probing marine systems with ribosomal RNAs. *Oceanography* **6**: 95–103.

Kaeberlein, T., Lewis, K., and Epstein, S.S. (2002) Isolating 'uncultivable' microorganisms in pure culture in a simulated natural environment. *Science* **296**: 1127–1129.

Karl, D.M. (2002) Microbiological oceanography – hidden in a sea of microbes. *Nature* **415**: 590–591.

Karner, M.B., DeLong, E.F., and Karl, D.M. (2001) Archaeal dominance in the mesopelagic zone of the Pacific Ocean. *Nature* **409**: 507–510.

Kolber, Z.S., Van Dover, C.L., Niederman, R.A., and Falkowski, P.G. (2000) Bacterial photosynthesis in surface waters of the open ocean. *Nature* **407**: 177–179.

Kolber, Z.S., Plumley, F.G., Lang, A.S., *et al.* (2001) Contribution of aerobic photoheterotrophic bacteria to the carbon cycle in the ocean. *Science* **292**: 2492–2495.

Koonin, E.V., and Galperin, M.Y. (1997) Prokaryotic genomes: the emerging paradigm of genome-based microbiology. *Curr Opin Genet Dev* **7**: 757–763.

Lanoil, B.D., Carlson, C.A., and Giovannoni, S.J. (2000) Bacterial chromosomal painting for in situ monitoring of cultured marine bacteria. *Environ Microbiol* **2**: 654–665.

Paul, J. (2001) Marine Microbiology. *Methods in Microbiology* (Vol. 30). Academic Press, San Diego.

Rappé, M.S., Connon, S.A., Vergin, K.L., and Giovannoni, S.J. (2002) Cultivation of the ubiquitous SAR11 marine bacterioplankton clade. *Nature* **418**: 630–632.

Ribosome Database Project. Centre for Microbial Ecology, Michigan State University. http://rdp.cme.msu.edu (accessed March 31 2003).

Schut, F., de Vries, E.J., Gottschal, J.C., Robertson, B.R., Harder, W., and Prins, R.A. (1993) Isolation of typical marine bacteria by dilution culture: growth, maintenance and characteristics under laboratory conditions. *Appl Environ Microbiol* **59**: 2150–2160.

Schut, F., Gottschal, J.C., and Prins, R.A. (1997) Isolation and characterisation of the marine ultramicrobacterium *Sphingomonas* sp. strain RB2256. *FEMS Microbiol Rev* **20**: 363–369.

Schut, F., Prins, R.A., and Gottschal, J.C. (1997) Oligotrophy and pelagic marine bacteria: facts and fiction. *Aquat Microb Ecol* **12**: 177–202.

The Center for the Advancement of Genomics. http://gnn.tigr.org/main.shtml (accessed March 31 2003).

Turner, P.C., White, M.R.H., McLennan, A.G., and Bates, A.D. (2000) *Instant Notes in Molecular Biology*. Bios Scientific Publishers Ltd., Oxford.

Zengler, K., Toledo, G., Rappé, M.S., Elkins, J., Mathur, E.J., Short, J.M., and Keller, M. (2002) Cultivating the uncultured. *Proc Natl Acad Sci USA* **99**: 15681–15686.

Structure of marine prokaryotes

3

3.1 Overview of the *Bacteria* and *Archaea*

The aim of the next two chapters is to review the main structural features of prokaryotic cells and to provide a broad overview of the nutrition, growth and metabolic activities of the *Bacteria* and *Archaea*. This will provide the background necessary for a more detailed consideration of the major marine representatives of these domains in *Chapters 5* and *6*. As discussed in *Chapter 1*, this book follows the 'three-domain' classification of the living world based on molecular phylogeny, so why are the *Bacteria* and *Archaea* considered together in these chapters, perpetuating the old prokaryote–eukaryote dichotomy? It is important that the student appreciates the modern view that the *Bacteria* and *Archaea* are very different groups, separated by long evolutionary history, but because *Bacteria* and *Archaea* share many common features of cellular structure and physiology, it is still convenient to consider these aspects together. Where major differences are known, these will be emphasized.

3.2 Cell morphology and structure

Under the microscope, various shapes of prokaryotic cells are apparent, and examples of all these types occur in marine habitats, as shown in *Figure 3.1*. The basic cell forms are spherical or ovoid cells (cocci), rods (bacilli) and spiral cells (spirilla). Some groups of marine prokaryotes have distinctive morphologies such as tightly coiled spirals (spirochaetes), cells with buds, stalks or hyphae (e.g. *Planctomycetes* and *Caulobacter*) or filamentous forms (actinomycetes). As noted previously, the 'typical' prokaryotic cell considered in most textbooks will be stated to have a size range in the 1–5 μm range, but marine prokaryotes show enormous diversity in cell size, ranging from about 0.1–750 μm in diameter (see *Table 1.2*).

A schematic diagram showing the features of prokaryotic cells is shown in *Figure 3.2*. It is important to recognize that certain features occur in all prokaryotic cells, whilst others are found only in specific types. The cytoplasmic membrane (CM) is the boundary layer of all cells and damage to the membrane results in the loss of cellular integrity. The presence of the various types of internal and surface structures depends very much on the physiological and ecological properties of the microbial group in question.

3.3 Cytoplasmic and internal membranes

The CM is a selectively permeable structure, typically about 8–10 nm thick, which allows the cell to take up specific gases and solutes from its surroundings and to excrete waste products and extracellular compounds. Water- and lipid-soluble compounds can freely cross the membrane, but polar (charged) molecules and ions cannot. The cell membrane is usually a highly dynamic structure (fluid mosaic model) composed of a lipid bilayer containing up to 200 proteins. The membrane is stabilized by Mg^{2+} and Ca^{2+} and by sterol-like compounds known as hopanoids derived from hopane. Hopanoids are quite resistant to breakdown and large quantities have accumulated in the environment over the many millennia of bacterial evolution. The total global mass of carbon deposited in hopanoids is estimated at about 10^{12} tons, approximately the

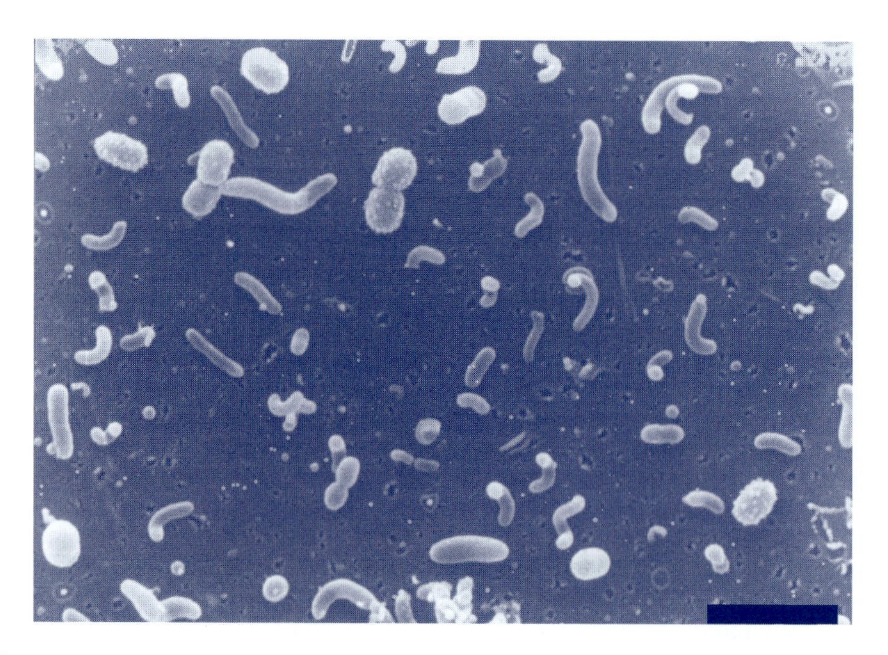

Figure 3.1

Scanning electron micrograph of picoplankton, showing various cell morphologies of marine bacteria. Bar represents ~1 μm. Image courtesy of Ed DeLong, Monterey Bay Aquarium Research Institute, California.

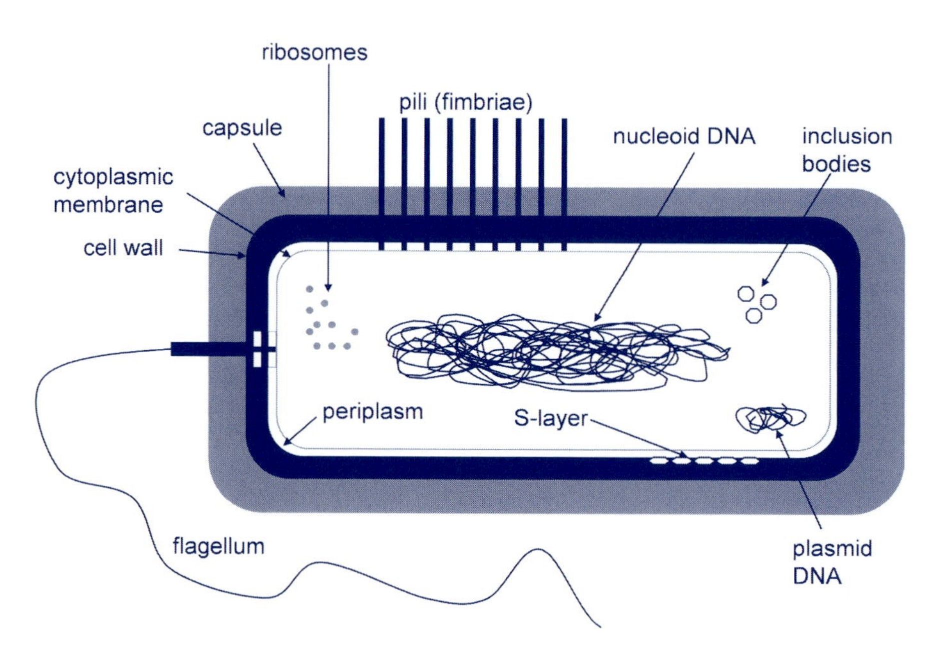

Figure 3.2

Schematic diagram showing structure of a prokaryotic cell. For clarity, only a small number of ribosomes, inclusions and surface proteins are represented. There is considerable variation in the nature of the envelope and surface structures.

same as the mass of organic carbon in all organisms living today. A very large fraction of fossil fuels such as petroleum and coal are composed of hopanoids, much of which has accumulated from the settlement of bacteria in the oceans. Because different bacteria produce characteristic hopanoid structures, these compounds are frequently used as markers for the analysis of microbial communities in marine sediments.

A major difference exists between the CM of *Bacteria* and *Archaea*. The fatty acids of *Bacteria* are linked to the glycerol moieties via ester linkages. Most *Archaea* contain hydrocarbons with repeating 5-carbon units (isoprenoids) rather than fatty acids, and these are linked to glycerol via ether linkages. Many *Archaea,* especially those from high-temperature environments such as hydrothermal vents, have monolayer membranes, consisting of a single layer of diglyceride tetraethers.

As well as controlling the entry/exit of substances to/from the cell by its selective permeability and its specific transport proteins, the CM is the most important site of energy generation and conservation. During metabolism, electron carriers associated with the membrane bring about a charge separation across the membrane. H^+ ions (protons) become concentrated on the exterior surface of the membrane and OH^- ions concentrate on the inner surface. This charge separation creates an electrochemical potential that can be converted to chemical energy. Prokaryotes vary enormously in the oxidation–reduction reactions used to produce this gradient, but the common 'currency' of energy exchange is always the molecule ATP, used for synthesis and mechanical work by the cell. When protons cross the membrane through a specific port, the contained energy is captured in the conversion of ADP to ATP.

A defining feature of prokaryotic cells is that they do not contain complex organelles such as mitochondria or chloroplasts. However, extensive invaginations or aggregates of membrane vesicles are often present in some marine prokaryotes, especially phototrophs and chemolithotrophs with very high respiratory activity. Phototrophs grown in low light intensity will produce more internal membranes and pigments than those grown under high illumination, in order to harvest available light with the greatest efficiency. The densely packed membranes in *Cyanobacteria* are called thylakoids and are analogous to the organelles of the same name in algae and plants. Cyanobacterial thylakoids appear as stacked or radial lamellae and hold the components of photosystems I and II, which contain chlorophyll *a* together with accessory pigments for harvesting of light (*Section 5.3*). Chemolithotrophic bacteria such as *Nitrobacter* and *Nitrococcus* (*Section 5.5*) produce elaborate intracellular membrane systems of densely packed lamellae or tubules, as do methylotrophs when they oxidize CH_4 (*Section 5.8*). The intracellular membranes are much less developed when other 1-carbon compounds are utilized.

Gas vesicles are especially important in the ecology of marine phototrophs such as *Cyanobacteria* (*Section 5.4*) and halophilic and methanogenic *Archaea* (*Section 6.2.4*). Gas vesicles have a 2-nm-thick single-layered gas-permeable wall composed of two hydrophobic proteins, surrounding a central space. These form a very strong structure into which gas diffuses from the cytoplasm, resulting in increased buoyancy of the cell. This enables phototrophic bacteria to maintain themselves in the desired zone of light intensity in the water column. Organisms from deep habitats have narrow vesicles as they are more resistant to hydrostatic pressure.

3.4 Inclusion bodies

Prokaryotes often contain small structures that are visible in the light microscope, especially with the use of phase contrast or special stains. Inclusion bodies are composed of granules of organic or inorganic material used as a store of energy or structural components. Sometimes they occur as simple concentrations of material free within the cytoplasm, but they are often enclosed in a thin single-layered lipid membrane. For example, a wide range of marine *Bacteria* and *Archaea* produce polyphosphate granules and carbon and energy storage polymers such as polyhydroxyalkanoates. Large inclusions of elemental liquid sulfur up to 1.0 μm in diameter occur in sulfur-oxidizing bacteria and can constitute up to 30% of the cell weight (see *Figure 5.6* and *Box 10.2*). Sulfur globules can accumulate freely in the periplasmic space or within the cytoplasm surrounded by a thin membrane. *Cyanobacteria* also contain granules composed of large amounts of α-1,4-linked glucan which is the polymerized carbohydrate product of photosynthesis. They also contain granules composed of polypeptides rich in the amino acids arginine and aspartic acid, which appear to be produced as a reserve of nitrogen.

Many types of *Cyanobacteria* and other autotrophic prokaryotes contain crystal-like regular hexagons known as carboxysomes. These are about 120 nm diameter and surrounded by a thin envelope. They act as stores of the enzyme RuBP carboxylase, the key enzyme in fixation of CO_2 (*Section 4.4.3*). *Cyanobacteria* and some other phototrophs always seem to contain them, but they have not been found in all chemolithotrophs. Packaging of RuBP carboxylase appears to enhance the efficiency of the enzyme in CO_2 fixation. If genetic mutations are generated resulting in structural changes or total loss of the carboxysomes, cells require much higher concentrations of CO_2 than normal for growth.

3.5 Prokaryotic genomes

In prokaryotes, the genetic material is not contained within a membrane-bound nucleus and is termed a nucleoid or nuclear body. The bulk of the genome of marine *Bacteria* and *Archaea* usually occurs as a single circular molecule of DNA containing the principal genes involved in cell structure and 'housekeeping'. This is termed the chromosome because it behaves as a single linkage group in genetic experiment, although it is not a true chromosome like those of eukaryotes. The nucleoid is attached to the CM and replication of DNA occurs at this point, with growth of the membrane leading to separation of daughter nucleoids during cell division. Rapidly growing bacteria may contain more than one copy of the chromosome. The circumference of the DNA molecule may be over 1 mm, which has to be packaged into a cell that is 1 μm or less in diameter. This is achieved by DNA supercoiling and stabilization of secondary structures by proteins. Like eukaryotes, *Archaea* possess histones (proteins that stabilize the DNA structure), but *Bacteria* do not (although they appear to possess basic proteins with a similar function). Various enzymes are involved in the supercoiling and unwinding of DNA enabling regions to be exposed for transcription of the DNA to mRNA and the process of protein synthesis by the ribosomes.

Genes for more specialized functions are often contained in smaller units of DNA termed plasmids, which have been found in many marine *Bacteria* and *Archaea*. Plasmids have their own system for replication. Usually, this is closely linked to replication of the chromosome, so that the plasmid is passed on to both daughter cells. However, if replication of the chromosome and plasmid is not synchronized, one of the daughter cells may fail to inherit the plasmid genes. This process of plasmid 'curing' occurs spontaneously, but can be enhanced by treatment with chemicals that interfere with DNA replication or by shifts of temperature. Induction of curing is an important technique in elucidating the function of plasmid-borne genes. For example, this technique led to the discovery of the role of a plasmid in the virulence of the marine fish pathogen *Vibrio anguillarum* (*Section 14.3.2*). In general, plasmids carry between 1 and 30 genes that are not essential for normal growth and reproduction of the cell, but confer a selective advantage on bacteria under certain conditions. Virulence (the ability to colonize, multiply within a host and produce disease) is one important example and, as well as *V. anguillarum*, several of the marine pathogens described in *Chapters 11, 13* and *14* have plasmid-encoded virulence mechanisms. Other important examples of the functions of plasmids include (a) resistance to antibiotics such as those used in aquaculture (*Section 14.5.2*), (b) resistance to heavy metals such as nickel and mercury in contaminated coastal waters (*Section 11.4*) and (c) the biodegradation of hydrocarbons in oil pollution (*Section 16.3.1*). Plasmids often contain genes that enable the whole plasmid to be mobilized and transferred to another bacterium via sex pili. This process of conjugation allows the spread of genes within a population and sometimes results in transfer to other species.

When the genome sequence of the marine-associated human pathogen *Vibrio cholerae* (*Section 11.2.1*) was elucidated in 2001, it was realized that the genetic information is divided between two distinct chromosomes. The larger chromosome (about 2.9 million bases) contains most of the 'housekeeping' genes and major pathogenicity factors, whilst the smaller chromosome (about 1.1 million bases) contains many genes of unknown function and some that are involved in gene capture and host colonization. The second chromosome represents more than a quarter of the total genome and is clearly more than a plasmid. Several pieces of evidence suggest that an ancestor of *V. cholerae* acquired it at some time during its evolution and this horizontal gene transfer may be the critical factor in explaining how a bacterium normally resident in oceanic and estuarine water developed into a human pathogen.

Besides conjugation, there are two other important mechanisms of transfer of genetic material from one bacterium to another. Transformation can occur when DNA is released into the environment by cell disruption. Large molecules like DNA cannot usually cross the CM, but in some bacteria, this ability (competence) can occur at certain stages of cell division, or it can be induced by treatment with $CaCl_2$. This is an important process in recombinant DNA experiments. Bacterial genes can also be transferred by bacteriophages in a process known as generalized transduction. Bacterial DNA is occasionally packaged into a viral particle, which then adsorbs to another bacterium and transfers the DNA. Assuming that the transferred DNA has some homology with that in the new host, it can be incorporated into the chromosome. Transduction of specific genes can occur when lysogenic bacteriophages become integrated into the bacterial chromosome and pick up host DNA during excision (*Figure 8.4*). As outlined above and in the relevant sections in other chapters, these processes of horizontal gene transfer have significant implications. There is growing evidence that such genetic exchange is a major factor affecting the diversity and evolution of marine prokaryotes and their interactions with other forms of life. Its significance will be increasingly understood as the sequencing of genomes of marine organisms progresses.

3.6 Ribosomes and protein synthesis

As shown in *Figure 2.4*, the 70S prokaryotic ribosome is a complex structure composed of two subunits (50S + 30S), each of which comprises a number of polypeptides and ribosomal RNA (rRNA) molecules. The ribosome is responsible for translating genetic information contained in the nucleotide base sequence of messenger RNA (mRNA) into the amino acid sequence of polypeptides. Although the ribosomes of *Bacteria* and *Archaea* are the same size and the mechanisms of transcription and translation are basically similar, there are some very important distinctions, supporting the concept that the two domains separated at a very early stage in evolution. In *Archaea* and eukaryotes, the initiator codon leads to incorporation of methionine, but in *Bacteria* this is modified to N-formyl methionine. The enzyme responsible for transcription of mRNA from DNA (DNA-dependent RNA polymerase) contains four polypeptide subunits, whereas that in *Archaea* contains eight subunits, a pattern more similar to that found in eukaryotes (*Table 1.1*). Protein synthesis in *Bacteria* is inhibited by several antibiotics such as rifamycin and chloramphenicol, which do not affect the process in *Archaea* and the structure of the elongation factors is significantly different in *Bacteria* and *Archaea*.

Upon completion of translation, the protein folds into a tertiary structure. Protein folding is determined by its primary amino acid sequence, but various proteins called chaperonins help to ensure that improper folding and aggregation do not occur. All prokaryotes produce chaperonins, but their synthesis is increased dramatically when cells are subject to sudden stress such as elevated temperature (the heat-shock response). In thermophilic prokaryotes such as those isolated from hydrothermal vents, very high concentrations of chaperonins are produced to protect newly formed proteins from heat damage.

3.7 Cell walls

All known marine prokaryotes possess some type of cell wall external to the CM. The cell wall is responsible for the shape of the cell and provides protection in an unfavorable osmotic environment. There are important variations in cell wall structure and composition among different prokaryote groups.

The key component of the cell wall in *Bacteria* is peptidoglycan, a mixed polymer composed of alternating residues of two amino sugars (N-acetyl glucosamine and N-acetyl muramic acid) with a tetrapeptide sidechain composed of a small number of amino acids. The great mechanical strength of peptidoglycan lies in the formation of crosslinks between the peptide sidechains, forming a mesh-like molecule. Although variation in the types of amino acids and the nature of crosslinking occurs, peptidoglycan with essentially the same structure occurs in all marine *Bacteria* with the exception of the *Planctomycetes* (see below). By contrast, the cell wall of *Archaea* is much more variable in composition and never contains true peptidoglycan, although some methanogens contain an analogous compound called pseudopeptidoglycan. Some halophiles have a wall composed of a complex heterosaccharide.

S-layers are arrays of a single protein or glycoprotein arranged in a tetragonal or hexagonal lattice-like crystalline structure (*Figure 14.2*). Interestingly, S-layers are found in almost all phylogenetic groups of both *Bacteria* and *Archaea* and appear to represent one of the simplest biological membranes, which probably developed very early in evolution. In most *Archaea,* they appear in place of a cell wall. Some methanogens can have up to three different S-layers, the outermost of which encloses groups of cells in a sheath. However, in *Bacteria* they tend to occur as an additional surface layer external to the peptidoglycan-containing cell wall. An important exception with significance in the marine environment is the genus *Planctomyces,* which appears to have only an S-layer. In the fish pathogen *Aeromonas salmonicida,* the S-layer contributes significantly to the ability of the bacterium to overcome host defense mechanisms (*Section 14.3.4*). The unusual properties of S-layers have important biotechnological applications (*Section 16.9*).

Within the *Bacteria* two broad divisions have been recognized for over a century – the so-called Gram-positive and Gram-negative types. This basic distinction, based on a simple differential staining procedure remains a remarkably useful first stage in the examination of unknown bacteria in culture. As shown in *Figure 3.3,* Gram-positive bacteria have a relatively thick, simple wall composed of peptidoglycan, together with ribitol or phosphate polymers known as teichoic acids. Gram-negative bacteria have a thinner but more complex wall with a very thin peptidoglycan layer that lacks teichoic or teichuronic acids, together with an additional outer membrane (OM) as shown in *Figure 14.2*. The Gram stain can be applied to *Archaea* but it is of little use as a diagnostic test. *Archaea* with pseudopeptidoglycan or other complex walls stain Gram-positive because they retain the crystal violet stain, but those with S-layers are usually Gram-negative.

The OM of cells of Gram-negative *Bacteria* is a lipid–protein bilayer complex, but is very different from the CM. The OM is anchored tightly to the peptidoglycan layer via a lipoprotein. The OM also contains lipolysaccharide (LPS), a compound unique to Gram-negative *Bacteria* and with some very important properties. LPS is composed of lipid A covalently bound to a core polysaccharide and strain-specific polysaccharide sidechains. These are immunogenic and are known as the O-antigen. Serological typing based on differences in the sequence of sugars in the O-antigen is important in the identification of closely related bacteria and strain-to-strain variation in immunological specificity is important in the epidemiology of many bacterial diseases. LPS stimulates the activation of complement and related host responses, which is very important in pathogenesis and development of vaccines, for example in septicemic diseases of fish caused by *Vibrio* spp. and *A. salmonicida* (*Section 14.3*). LPS is also known as endotoxin because the lipid A molecule is responsible for hemorrhage, fever, shock and other symptoms of diseases caused by Gram-negative pathogens. The proteins of the OM also have very distinctive properties. The major proteins, called porins, are trimeric structures that act as channels for the diffusion of low-molecular-weight solutes across the OM. Other

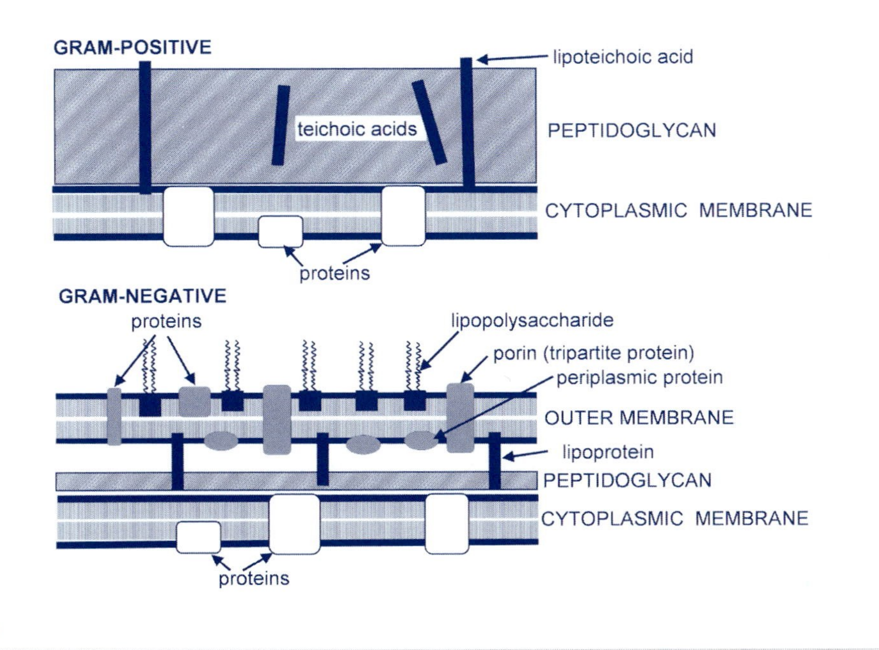

Figure 3.3

Comparison of the envelopes of Gram-positive and Gram-negative bacteria.

proteins act as specific receptors for iron and other key nutrients and their transport into the cell. An important feature of many of these proteins is that they are not produced constitutively. Their synthesis can either be repressed or induced, according to the concentration of the specific substances. An especially important example of this in marine bacteria is in the acquisition of the essential nutrient iron, which may be imported into the cell via secreted compounds known as siderophores (*Section 4.4.7*). Expression of genes encoding siderophores and OM receptors for their entry into the cell (as well as many other genes) may only occur when iron is in short supply. In consequence, bacteria grown in laboratory culture, which provides excess iron unless special precautions are taken, may have very different OM structure and physiological properties from those in natural environments, which are usually very iron-depleted. Other OM proteins are responsible for the detection of environmental changes such as pressure, temperature and pH and communication of this information via signaling systems. In later chapters, numerous examples are given of the importance of the variability of the OM and its role in monitoring the physical and chemical status of the cell's immediate environment. Because the OM is an additional permeability barrier exterior to the CM, but with different selectivity, some substances cross one membrane, but not the other. This means that the periplasmic space, between the membranes, is an important feature of Gram-negative bacteria. Many binding and transport proteins are located here and proteins such as enzymes and toxins may not be released until the cell is lysed, unless there are specific excretion mechanisms. Vesicles of OM containing periplasmic contents may also be released from the cell surface (*Figure 14.2*).

3.8 Capsules and the glycocalyx

In addition to the cell wall, many marine prokaryotes secrete a slimy or sticky extracellular matrix. It is usually composed of polysaccharide, in which case it may form a network extending into the environment and is termed the glycocalyx. In other cases, there may be a more distinct rigid layer (capsule) which may contain protein. The main importance of the glycocalyx in marine systems is in the attachment of bacteria to plant, animal and inanimate surfaces leading to the formation of biofilms (*Section 4.5.4*). The bacterial glycocalyx is critical for the settlement of algal spores and invertebrate larvae and the significance of these processes in biofouling of ships and marine structures is considered in *Chapter 16*. The release of these sticky exopolymers is also very important in the aggregation of bacteria and organic detritus in the formation of marine snow particles (*Section 1.7.1*). Furthermore, a large proportion of the dissolved organic carbon in ocean waters probably derives from the glycocalyx of bacteria. Slime layers can act as a protective layer, preventing the attachment of bacteriophages and penetration of some toxic chemicals. In pathogens, the presence of a capsule can inhibit engulfment by host phagocytes and in free-living forms it is commonly assumed to inhibit ingestion by protists. However, many grazing flagellates appear to feed voraciously on capsulated planktonic bacteria, probably because the capsule contains additional nutrients.

3.9 Motility of marine bacteria

3.9.1 Flagella

Many marine species of *Bacteria* and *Archaea* can swim due to the presence of one or more flagella. Prokaryotic flagella are long helical filaments attached to the cell by means of a hook-like structure and basal body embedded in the membrane. In marine bacteria, the most common arrangement is attachment of a single flagellum to one end of the cell (polar flagellation). Some groups have peritrichous flagella, in which the flagella are arranged all over the cell surface. As flagella are so thin, they can only be visualized in the light microscope by use of special stains, but they are readily visible by TEM or SEM (*Figure 3.4*). The location and number of flagella can be useful in bacterial identification, although some marine bacteria such as *Vibrio* spp. can change the organization of their flagella when in contact with surfaces and this is important in colonization and biofilm formation, as discussed below. In *Bacteria* the flagellum is about 20 nm in diameter and is made of many subunits of a single protein (flagellin) synthesized in the cytoplasm and passed up the central channel of the filament to be added at the tip. Less is known about the structure and synthesis of flagella in *Archaea*. Although the basic structure is similar to that of *Bacteria,* the flagellum is made of several different proteins and is much thinner (about 13 nm in diameter), possibly too thin for the subunits to pass up the central channel. The basal

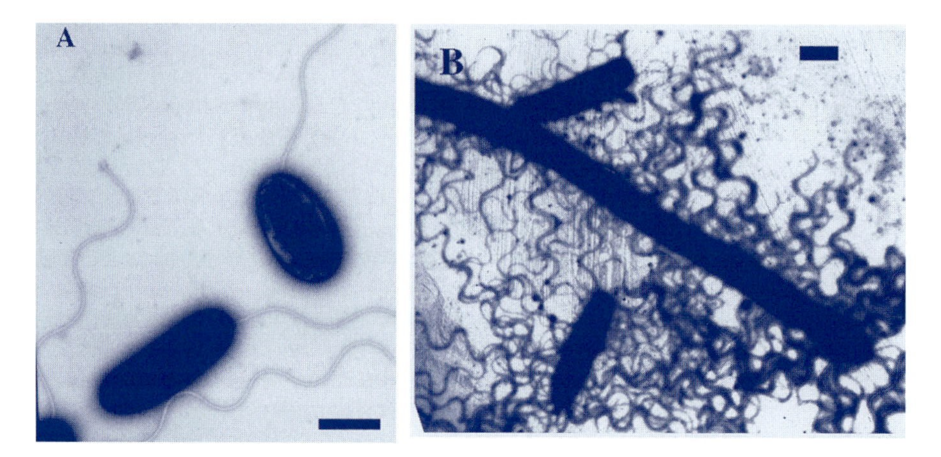

Figure 3.4

Electron micrographs of *Vibrio parahaemolyticus*. (a) Grown in liquid showing the swimmer cells with single, sheathed polar flagellum. (b) Grown on a surface showing the elongated swarmer cell with numerous lateral flagella. Bars represent ~1 μm. Image courtesy of Linda McCarter, University of Iowa.

body of the flagellum consists of a central rod and a series of rings embedded in the cell wall and membrane, composed of about 20 different proteins. In *Archaea,* the basal body resembles that of Gram-positive *Bacteria,* although the proteins are different.

Movement is caused by the rotation of the basal body acting as a tiny motor, which couples the flow of ions to the generation of force. The flagellum is rigid and behaves like a propeller. The remarkable mechanics of the motor and the process of self-assembly of the fine flagella tubules have attracted the attention of engineers for possible exploitation in nanotechnology. The mechanism has been extensively studied in *E. coli,* where rotation is caused by the movement of H^+ ions from the outside to the inside of the cell through protein subunits (Mot) surrounding the ring in the membrane. However, in *Vibrio* spp. (and probably most other marine bacteria) Na^+ ions are used instead, which has an advantage in alkaline (pH 8) seawater. Sodium-driven flagellar motors are exceptionally fast and vibrios can rotate their flagella at 1700 revolutions per second and swim at up 400 μm per second. In fact, recent evidence suggests that most marine bacteria may possess both H^+- and Na^+-driven flagella motors (*Box 3.1*). When *Vibrio parahaemolyticus* or *V. alginolyticus* are grown on solid surfaces or in viscous environments, they undergo a remarkable change in morphology. The cells stop dividing normally and elongate up to about 30 μm. At the same time, they start synthesizing large numbers of lateral or peritrichous flagella as well as the normal polar flagellum (*Figure 3.4*). The polar flagellum is effective when the cell is in the free-living, planktonic state whilst the lateral flagella enable cells to move more easily through viscous environments. This can be observed in the laboratory by the phenomenon of swarming on the surface of agar plates. In the natural environment, differentiation to the swarmer form enables vibrios to adopt a sessile lifestyle, which is significant in the formation of biofilms on surfaces or the colonization of host tissues. The polar flagellum acts as a sensor and increased viscosity slows down its rotation, leading to induction of the lateral phenotype. Experimentally induced interference of the Na^+ ion flux with chemical agents also induces differentiation to the swarmer state. Genetic analysis of the Na^+ motor is leading to an understanding of this mechanism. However, little is currently known about how the signal from the flagellum induces changes in the expression of genes for altered cell division and synthesis of different flagella. In vibrios, the polar (Na^+-driven) flagellum is enclosed in a sheath formed from the OM. It is not known whether the flagellum rotates within the sheath or whether the filament and sheath rotate as a single unit. The lateral (H^+) flagella do not appear to possess a sheath.

3.9.2 Chemotaxis and related behaviors

When motile bacteria like *E. coli* are observed in the microscope, they move in an apparently random fashion. They swim on a relatively straight path ('run') before executing a brief 'tumble' and swimming off in another direction. The tumble is caused by a change in direction of rotation of the flagellum and reorientation of the cell due to bombardment by water molecules (Brownian motion). By observation of the three-dimensional movement of a single bacterium with digital imaging systems, it can be seen that movement in a neutral environment is completely random. However, if an attractant or repellent substance is present, the frequency of tumbling is modified and the 'random walk' becomes biased. Cells tumble less frequently and therefore spend longer swimming *up* a gradient of increasing concentration of an attractant. Consequently, they show a net movement towards the source of the substance (positive chemotaxis). The bacterium senses minute changes in concentration with time, through a series of chemoreceptors on the cell surface and a chemical signaling system (involving changes in the methylation state of proteins) that relays information to a tumble generator at the base of the flagellum. When swimming *down* a concentration gradient of repellents, decreases in concentration promote fewer tumbles and bacteria will therefore make longer runs. This complex process involves many different proteins encoded by numerous genes, which have been extensively studied in *E. coli* and some other enteric bacteria, but much less so in marine bacteria. For marine bacteria associated with surfaces in high-nutrient environments, the advantages of motility and chemotaxis by this 'biased random walk'

Box 3.1 RESEARCH FOCUS

Fast laps and quick lane turns

How marine bacteria swim and respond to nutrient and physical gradients

The ability to swim towards high concentrations of favorable nutrients (chemotaxis) is an important property of many marine bacteria (*Section 3.9*). Study of this phenomenon relies on the fusion of direct observation, experimental manipulation, theoretical biophysics and mathematical modeling. One of the leading investigators, Jim Mitchell of Flinders University, Australia, has concluded that the swimming behavior of most marine bacteria is very different to the 'classical' system of enteric bacteria (Mitchell 2002). The movement of bacteria occurs over tiny, micrometer distances whereas energy flow into the ocean through convection, wind and planetary rotation creates turbulence on a scale many orders of magnitude greater. However, the effects of turbulence translate down to powerful micrometer-scale shear forces. Furthermore, because most free-living marine bacteria are very small (as little as 0.2 μm diameter), the effects of Brownian motion (random bombardment of the bacterium by molecules in the suspending fluid) are very great. Therefore, marine bacteria need to swim very fast and show very rapid responses to altered conditions if motility and chemotaxis are to be effective. There is growing evidence that marine bacteria can indeed swim at staggering speeds, up to 400 μm per second (Mitchell *et al.* 1996). This is the equivalent of several hundred body-lengths per second. How are such speeds possible? Studies of marine vibrios show that the Na^+-driven motor enables rotation speeds hundreds of times faster than the H^+-driven system described in enteric bacteria such as *E. coli.* Analysis of the genes encoding components of the polar flagellar motor of vibrios reveals that the Na^+ driven motor has additional components that are responsible for detecting the speed of rotation or the flux of Na^+ ions (Asai *et al.* 1997). Research by Linda McCarter of the University of Iowa, USA has shown that vibrios can switch to an additional H^+-driven flagellar system motor when swimming in viscous fluid (McCarter 1999), accompanied by dramatic changes in cell morphology and behavior (see *Figure 3.5*). It is possible to use selective inhibitors of the Na^+ and H^+ ion transport systems and this approach has been applied to other marine bacterial isolates and communities. Mitchell and Barbara (1999) showed that, when used individually, the respective inhibitors slowed the movement of several isolated marine bacteria and those in a mixed community. However, the movement did not stop completely. Analysis of this effect leads to the conclusion that most high-speed bacterial community members use both H^+ and Na^+ -driven ion motors simultaneously. The Na^+ motors are responsible for about 60% of the speed of marine bacteria.

Investigators have inferred that most marine bacteria are capable of motility, but that they use it intermittently. For example, Grossart *et al.* (2001) conducted long-term field and mesocosm studies and concluded that the proportion of the natural population in seawater showing motility varies considerably. Motility is highest in the spring and summer and during the release of nutrients from phytoplankton blooms. Also, Mitchell *et al.* (1995) and Johansen *et al.* (2002) have shown that marine bacteria do not swim continuously or at constant speeds. The most important factor is the ability of bacteria to respond, through motility, to the presence of localized concentrations of organic carbon. Certainly, clusters of bacteria can be observed around microscopic nutrient sources (Mitchell *et al.* 1996).

Due to the limitations of small size and shear effects mentioned earlier, the 'run and tumble' strategy of chemotaxis does not seem suitable to explain the behavior of marine bacteria (Mitchell 2002). Very high speeds enable small marine bacteria to swim in a relatively straight line before stopping and reversing direction. The reversal of flagellar rotation is almost instantaneous, so that the bacterium is not 'knocked off course' by shear or Brownian motion. Presumably, as in enteric bacteria, receptors on the bacterial surface sense a temporal gradient in nutrient concentration resulting in a signal for change in direction of rotation of the motor. Little is known about this mechanism in marine bacteria, apart from the fact that it must occur much more rapidly than in bacteria like *E. coli*. In a generally low-nutrient environment such as the open ocean, it is energetically impossible for bacteria to seek out nutrients by biased random movement. Instead, they must rely on turbulent flow to bring them within a patch of increased nutrient concentration and then use the 'run and reverse' movement strategy to maximize the likelihood that they will remain within it. Observation of the motility characterisitics (speed and angle of change of direction) of a large collection of phylogenetically different bacteria supports the idea that reversal of swimming direction is the main method of chemotaxis in most marine bacteria (Johansen *et al.* 2002).

Although most marine prokaryotes are very small, there are also some exceptionally large organisms. For example, *Thiovulum majus* is large enough (about 16 μm diameter cocci) to show directional swimming up to 600 μm per second in a nutrient or oxygen gradient (Fenchel 1994). Like protozoa, cells of this bacterium can rotate and swim in a helical path. Because they live at a microaerophilic interface between sediments and water, when they encounter increasing concentrations of oxygen they increase their swimming speed and follow a U-shaped path that returns them to the preferred location. The U-turns are thought to occur because of sudden acceleration or deceleration of flagellar rotation.

In summary, cell size and physical factors have enormous effects on the ability of marine bacteria to adapt and respond to their environment through movement and chemotaxis. There is growing evidence that bacterial motility in the open ocean has great ecological and geochemical significance because of the heterogeneous distribution of nutrients. It is an intriguing concept that study of such microscale processes may improve our understanding of global-scale processes like CO_2 flux and climate change.

mechanism are clear. Swimming brings the bacteria into close contact with surfaces and may help to overcome electrostatic repulsion. Chemotaxis towards host surfaces is an important factor in colonization by several pathogenic and symbiotic marine bacteria. However, as discussed in *Box 3.1,* recent research suggests that marine bacteria in the open ocean use a 'back and forth' or 'run and reverse' movement, rather than a 'biased random walk', to enable them to localize within a zone of favorable nutrient concentration.

Movement towards light (phototaxis) is common in phototrophic marine bacteria, which detect changes in the concentration of visible light of different wavelengths and swim towards higher illumination. The halophilic archeaon *Halobacterium salinarum* possesses rhodopsin molecules (*Section 6.2.4*), which act as light sensors influencing rotation of the flagella.

Other tactic movements include those towards or away from high concentrations of O_2 (aerotaxis) and ionic substances (osmotaxis). Magnetotaxis is a specialized response seen in certain species of bacteria found in marine and freshwater muds which contain intracellular inclusion bodies called magnetosomes. These bacteria orient themselves in the Earth's magnetic field although, as discussed in *Section 5.14.2,* their movements are primarily in response to gradients of chemicals and O_2.

3.9.3 Gliding motility

Gliding motility does not involve flagella, and only occurs when bacteria are in contact with a solid surface. It is especially important in the responses of *Cyanobacteria* and members of the *Cytophaga* and *Flavobacteria* genera to gradients of oxygen and nutrients in microbial mats. Little is known about the mechanisms of gliding motility. The most common mechanism appears to be a kind of jet propulsion through the exclusion of slime via minute pores in the cell surface. However, in *Flavobacterium* it is thought that rotatory motors, similar to flagellar motors, transmit force to special OM proteins, which results in a ratchet-like movement of the cell envelope analogous to the movement of a tank on caterpillar tracks.

3.10 Pili (fimbriae)

Pili (also known as fimbriae) are fine hair-like protein filaments on the surface of many bacteria. Pili are typically about 3–5 nm in diameter and 1 μm long. They are composed of a single protein, although a cell may have different types of fimbriae and the amino acid sequence of each type can show significant strain-to-strain variation. The most common pili are those involved in the attachment of bacteria to surfaces; some microbiologists restrict the use of the term fimbriae to these adhesive structures. Their function has been particularly well studied in the interaction of pathogenic bacteria with mucous membranes of animals. There is often a specific receptor recognition site on the pili, which can explain why certain strains of bacteria attach specifically to particular hosts or tissues. Host immune responses to bacterial attachment are often important in resisting infection. Adhesive pili are often encoded by plasmids or bacteriophages and may be exchanged by conjugation or transduction. Again, the human pathogen *V. cholerae* provides a good example; here, pili play a crucial role in infection and the acquisition of pilus genes and their expression are important factors in explaining the transition of this bacterium from its estuarine or marine habitat to the human gut (*Figure 11.3*). Pili are also likely to be involved in colonization of the squid light organ by *V. fischeri* (*Box 10.3*). Another example is the attachment of *Roseobacter* spp. to invertebrate larvae, where pili develop into a form of holdfast. Pili play a similar role in the attachment of bacteria such as *Caulobacter* and *Hyphomicrobium* to inanimate surfaces (*Section 5.15*).

As well as the more common adhesive pili, some cells possess structures known as type IV pili. These are responsible for a type of cell movement on surfaces known as twitching motility. This involves reversible extension and retraction of the pili and can result in sudden jerky movement of a few μm. Twitching motility is thought to be important in the formation of microcolonies during the initial formation of biofilms. It is likely that pili are involved in many other processes involving marine bacteria and this area warrants further investigation.

Sex pili are quite different structures, often several μm long, formed only by donor ('male') bacteria involved in the process of conjugation. As noted above, sex pili and the machinery needed for replication and transfer of DNA from donor to recipient cells are encoded by conjugative plasmids. Sex pili are important in the transfer of genes, such as those encoding antibiotic resistance or degradative ability, between cells of *Bacteria* in the marine environment. In addition, sex pili are the receptors for certain bacteriophages and an application of this is the use of F^+ RNA bacteriophages as an indicator of fecal pollution in seawater (*Section 11.3.7*). Conjugative plasmids have recently been found in several species of *Archaea* (especially thermophiles) and there is currently considerable interest in determining the extent of horizontal gene transfer and its application for genetic recombination systems in this group.

References and further reading

Adams, D.G. (2001) How do cyanobacteria glide? *Microbiol Today* **28**: 131–133.

Asai, Y., Kojima, S., Kato, H., Nishioka, N., Kawagishi, I., and Homma, M. (1997) Putative channel components for the fast-rotating sodium driven flagellar motor of a marine bacterium. *J Bacteriol* **179**: 5104–5110.

Berg, H.C. (2000) Motile behavior of bacteria. *Physics Today on the Web*. http://www.aip.org/pt/jan00/berg.htm (accessed March 31 2003).

Fenchel, T. (1994) Motility and chemosensory behavior of the sulfur bacterium *Thiovulum majus*. *Microbiology* **140**: 3109–3116.

Grossart, H.-P., Riemann, L., and Azam, F. (2001) Bacterial motility and its ecological implications. *Aquat Microb Ecol* **25**: 247–258.

Johansen, J.E., Pinhassi, J., Blackburn, N., Zweifel, U.L., and Hastrom, A. (2002) Variability in motility characteristics among marine bacteria. *Aquat Microb Ecol* **28**: 229–237.

Madigan, M.T., Martinko, J.M., and Parker, J. (2003) *Brock Biology of Microorganisms*. Prentice-Hall, NJ.

McCarter, L. (1999) The multiple identities of *Vibrio parahaemolyticus*. *J Mol Microbiol Biotechnol* **1**: 51–57.

Mitchell, J.G. (2002) The energetics and scaling of search strategies in bacteria. *Am Nat* **160**: 727–740.

Mitchell, J.G., and Barbara, G.M. (1999) High speed marine bacteria use sodium-ion and proton driven motors. *Aquat Microb Ecol* **18**: 227–233.

Mitchell, J.G., Pearson, L., Dillon, S., and Kantalis, K. (1995) Natural assemblages of marine bacteria exhibiting high-speed motility and large accelerations. *Appl Environ Microbiol* **61**: 4436–4440.

Mitchell, J.G., Pearson, L., and Dillon, S. (1996) Clustering of marine bacteria in seawater enrichments. *Appl Environ Microbiol* **62**: 3716–3721.

Neidhardt, F.C., Ingraham, J.L., and Schaechter, M. (1990) *Physiology of the Bacterial Cell: A Molecular Approach*. Sinauer Associates Inc., Sunderland, MA.

Paustian, T. (2001) *Structure of Procaryotes*. University of Wisconsin-Madison. http://www.bact.wisc.edu/microtextbook (accessed March 31 2003).

Sleytr, U.B., Sára, M., and Pum, D. (2000) Crystalline bacterial cell surface layers (S-layers): a versatile self-assembly system. In: **Ciferri A.** (ed.) *Supramolecular Polymerization*, pp. 177–213. Marcel Dekker, NY.

Physiology of marine prokaryotes

4.1 Metabolic diversity and the importance of microbial communities

Bacteria and *Archaea* have been evolving for between three and four billion years. As a result, there is a huge diversity of metabolic types. Different species have evolved to obtain energy from light or various inorganic substances or by breaking down different organic compounds to their basic constituents. Some species obtain energy and key elements directly from simple sources, whilst others require 'ready-made' compounds. The selective advantage of the ability to utilize particular nutrients under a particular set of physical conditions has led to the evolution of the enormous metabolic diversity that we see today. It is important to emphasize the role of microbial interactions in marine microbial ecology. Most microbiologists follow the pure culture methods, in which single clones of bacteria are studied in isolation, developed by the pioneers of microbiology in the nineteenth century. However, pure culture approaches cannot provide all the answers because only a tiny fraction of all the known types of marine prokaryotes can be cultured. Perhaps even more important is the fact that pure culture technique overlooks the importance of microbial *communities*. In the marine environment, different microbial species interact and their metabolic activities are interdependent. In subsequent sections, numerous examples of the activities of such communities in marine habitats will be apparent.

4.2 Modes of nutrition in marine prokaryotes

All cells need to obtain energy and conserve it in the compound ATP. It is convenient to classify bacteria in terms of their energy source as either chemotrophs (which produce ATP by the oxidation of inorganic or organic compounds) or phototrophs (which use light as a source of energy for the production of ATP). Chemotrophs may be further divided into those that can obtain energy solely from inorganic compounds (chemolithotrophs) and those that use organic compounds (chemoorganotrophs). Chemolithotrophy is unique to the prokaryotes and occurs in many species of marine *Bacteria* and *Archaea*. All cells need carbon as the major component of cellular material. Those organisms that can use CO_2 as the sole source of carbon are termed autotrophs, whilst those that require organic compounds as a source of carbon are termed heterotrophs. Autotrophs are primary producers that fix CO_2 into cellular organic compounds, which are then used by heterotrophs. Most chemolithotrophs and phototrophs are also autotrophic, but some phototrophs (e.g. purple nonsulfur bacteria) can switch between heterotrophic and autotrophic metabolism. Some chemotrophs are also mixotrophic and are capable of growing heterotrophically or autotrophically (e.g. some species of hydrogen bacteria and *Thiobacillus*). *Table 4.1* shows a summary of these nutritional categories.

4.3 Energy-yielding processes

4.3.1 Methods of ATP generation

All energy-yielding metabolic processes depend on coupled reduction–oxidation (redox) reactions, with conservation of the energy in the molecule ATP using the reaction

$$ADP + P_i + \text{energy} \rightleftharpoons ATP + H_2O$$

Table 4.1 Nutritional categories of microorganisms

Nutritional type	Energy source	Carbon source	Hydrogen or electron source	Representative examples
Photolithotroph (photoautotroph)	Light	CO_2	Inorganic compounds	*Cyanobacteria*, purple sulfur bacteria
Photoorganotroph (photoheterotroph)	Light	Organic compounds (or CO_2)	Organic compounds (or H_2)	Purple nonsulfur bacteria
Chemolithotroph (chemoautotroph)	Inorganic chemicals	CO_2	Inorganic chemicals	Sulfur bacteria, hydrogen bacteria, nitrifying bacteria
Chemoorganotroph (chemoheterotroph)	Organic compounds	Organic compounds	Organic compounds	Wide range of *Bacteria* and *Archaea*

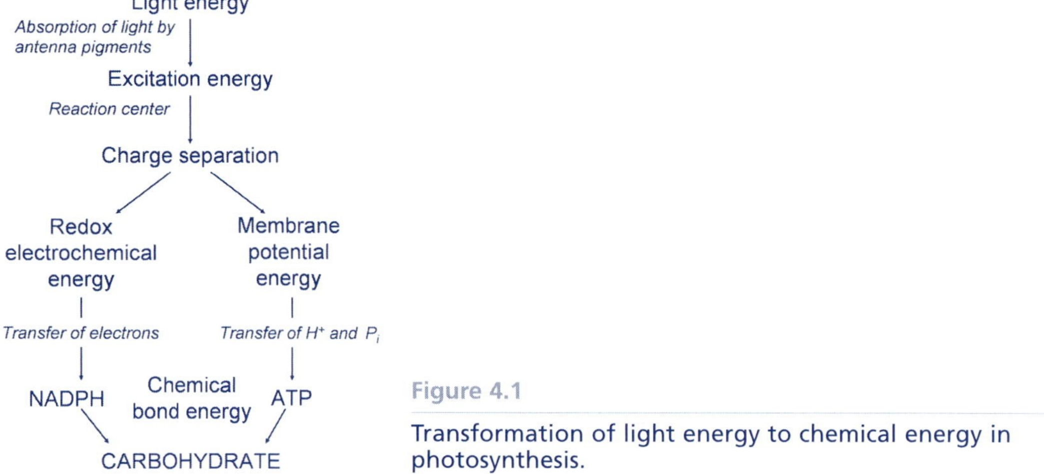

Figure 4.1

Transformation of light energy to chemical energy in photosynthesis.

ATP transfers energy from energy-yielding reactions to energy-requiring reactions. *Bacteria* and *Archaea* synthesize ATP by one of three mechanisms:

(a) Photophosphorylation, in which light energy creates a charge separation across membranes
(b) Fermentation, in which high-energy phosphate is transferred directly from a phosphorylated organic compound via ADP to ATP (substrate-level phosphorylation)
(c) Respiration, in which reduced inorganic or organic compounds are oxidized with the concomitant reduction of an electron acceptor (electron transport phosphorylation).

4.3.2 Phototrophy and primary productivity

In photophosphorylation, light energy is used to remove electrons from H_2O, H_2S or certain other reduced substances. The excited electrons are used to generate ATP and most phototrophs use this for the fixation of CO_2 into cellular material (photoautotrophy). Virtually all life on Earth depends directly or indirectly on solar energy via these reactions and it is likely that simple mechanisms for harvesting light energy developed very early in the evolution of life. Photophosphorylation 'traps' light energy from the excited electrons into a chemically stable molecule (ATP) and also generates NADPH, which is used in the reduction of CO_2 into organic compounds (*Figure 4.1*). Various photochemical systems exist in microorganisms, but

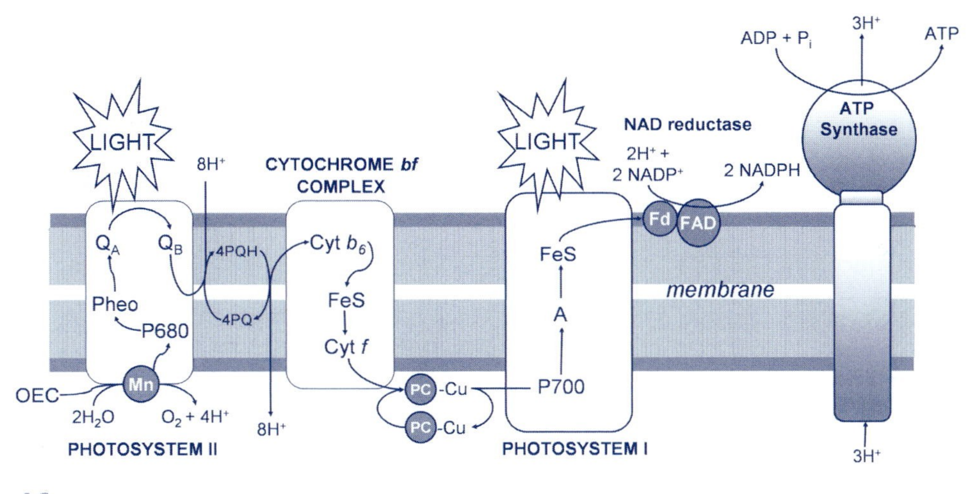

Figure 4.2

Electron flow in oxygenic photosynthesis. The light-driven electron flow generates an electrochemical gradient across the membrane, leading to ATP synthesis. Electrons are provided by water and the oxygen-evolving complex (OEC) generates oxygen.

true photosynthesis occurs only in those *Bacteria* that possess the magnesium-containing tetrapyrrole pigments known as chlorophylls and bacteriochlorophylls. These form complex structures containing accessory pigments, which ensure the effective harvesting of light at different wavelengths and transfer of the energy to chlorophyll.

Several phyla of the *Bacteria* contain phototrophic genera or consist exclusively of them. By contrast, photosynthesis is unknown in the *Archaea,* although some members (extreme halophiles) contain the carotenoid pigment bacteriorhodopsin, which appears as purple patches in the CM of cells grown under low O_2 concentration and high light intensity. Light energy is used to form a proton pump, which generates ATP, and the main function of this appears to be in maintenance of the Na^+/K^+ balance in the cell (*Section 4.6.6*). However, light-mediated ATP production does seem to permit slow growth when other energy-yielding reactions are not possible. The recent surprising discovery in oceanic *Proteobacteria* of a similar light-mediated ATP synthesis system based on rhodopsin-like molecules is discussed in *Box 2.2.* The ecological significance of this process is currently unclear.

Oxygenic photosynthesis (in which H_2O is the electron donor leading to production of O_2) is the most important contributor to primary productivity and the oxygenic *Cyanobacteria* play a major role in ocean processes (*Section 5.4*). Oxygenic photosynthesis is characterized by the presence of two coupled photosystems operating in series. In *Cyanobacteria,* chlorophylls (of which there are two main types, *a* and *b*) and the accessory pigments (phycobiliproteins) are assembled into arrays called antennas, which occupy a large surface area to trap the maximum amount of light. As shown in *Figure 4.2,* photosystem I absorbs light at wavelengths >680 nm and transfers the energy to a specialized reaction center chlorophyll (P700). Photosystem II traps light at lower wavelengths and transfers it to chlorophyll P680. Absorption of light energy by photosystem I leads to a very excited state, which then donates a high-energy electron via chlorophyll *a* and iron–sulfur proteins to ferredoxin. The electron can then pass cyclically through an electron transport chain, returning to P700 and leading to ATP synthesis (cyclic photophosphorylation). Electrons can also pass in a noncyclic route involving both photosystems, leading to ATP formation and reduction of $NADP^+$ to NADPH. These processes are known as the light reaction of photosynthesis. In the dark reaction, ATP and NADPH molecules are used to reduce CO_2 leading to its fixation into carbohydrate. The reaction may be represented as

$$CO_2 + 3ATP + 2NADPH + 2H^+ + H_2O \longrightarrow (CH_2O) + 3ADP + 3\ P_i + 2NADP^+$$

There are some important differences between the photosynthetic machinery of *Prochlorococcus* and other *Cyanobacteria*, which are discussed in *Section 5.4.4*.

Green and purple phototrophs (*Section 5.3*) lack photosystem II and cannot use water as electron donor in noncyclic electron transport. The light reaction generates insufficient reducing power to produce NADPH and these bacteria therefore use H_2S, H_2 or S^0 because they are better electron donors. For this reason, they do not generate O_2 during photosynthesis and most are strict anaerobes found in shallow sediments and microbial mats. This mode of photosynthesis was almost certainly the first to evolve.

4.3.3 Fermentation

Fermentation occurs only under anaerobic conditions and is the process by which substrates are transformed (dismutated) by sequential redox reactions without the involvement of an external electron acceptor. Fermentative prokaryotes can be either strict or facultative anaerobes; the latter can often grow using respiration when O_2 is present. Fermentation yields much less ATP than aerobic respiration for each molecule of substrate and growth of facultative anaerobes is better in the presence of O_2. A very wide range of fermentation pathways exists in marine *Bacteria* and *Archaea*, especially those associated with degradation of organic material in sediments, animal guts and microbial mats. Carbohydrates, amino acids, purines and pyrimidines can all serve as substrates. Some individual species are adapted to utilize only a limited range of substrates, whilst others can carry out fermentations with various starting materials. Characterization of fermentation end products is one of the main criteria in the identification of bacteria using biochemical tests.

4.3.4 Respiration

In respiration, electrons are transferred via a sequence of redox reactions through an electron transport chain located in the CM, which leads to the transfer of protons to the exterior of the membrane and generates a proton-motive force. Electron transport chains in most organisms contain flavoproteins, iron–sulfur proteins and quinones. In obligate and facultatively aerobic *Bacteria,* the iron-containing cytochrome proteins are key components, but they are not present in all aerobic *Archaea*. The terminal electron acceptor in aerobic respirations is O_2. In chemoorganotrophs, respiration occurs via oxidation of organic compounds as substrates. Again, the great diversity of *Bacteria* and *Archaea* that employ respiratory metabolism is based on specialization in the use of particular substrates. Aerobic chemolithotrophs oxidize inorganic compounds such as NH_4^+, H_2S and Fe^{2+}. Important marine examples include the nitrifying bacteria, sulfide-oxidizing bacteria and methanotrophs. Anaerobic respiration is a process in which terminal electron acceptors other than O_2 are used. Reduction of sulfate is especially important in marine sediments (*Section 5.17* and *Box 5.3*). Sulfate-reducing bacteria (SRB) often form metabolic consortia with other organisms, which supply simple organic compounds and H_2 as substrates. SRB have also recently been found to form consortia with *Archaea* in the anaerobic oxidation of methane (*Box 6.1*).

4.3.5 Methanogenesis

Methanogenesis is a special type of metabolism, which occurs as the final step in anaerobic degradation of organic material and is carried out only by particular groups of *Archaea*. The reactions in methanogenesis and their great importance in carbon cycling in marine sediments is discussed in *Section 6.2.1.*

4.4 Nutrients needed for growth

4.4.1 Macronutrients, micronutrients and trace elements

A small number of macronutrients make up over 95% of the dry weight of prokaryotic cells. Carbon, oxygen, hydrogen, nitrogen, sulfur and phosphorus are the major elements in the cell's most important macromolecules, namely nucleic acids, proteins, carbohydrates and lipids. A number of cations (potassium, calcium, magnesium and iron) are also required in significant amounts to maintain the structure and function of various cellular components such as enzymes and coenzymes. In addition to these major nutrients, most bacteria require small quantities of a range of trace elements, such as manganese, cobalt, zinc, molybdenum, copper and nickel. Some heterotrophs also require low concentrations of preformed growth factors or micronutrients, such as amino acids, pyrimidines, purines and vitamins, because they lack the biochemical pathways for the synthesis of key intermediates.

A particular feature of marine *Bacteria* and *Archaea* is their requirement for sodium. Some microbiologists define true marine prokaryotes as those having an absolute requirement for Na^+ ions (usually in the range 0.5–5.0%) and which fail to grow when K^+ is substituted in the cul-ture medium. This distinguishes them from closely related terrestrial and freshwater species. As might be expected, most grow optimally at a concentration of NaCl similar to that found in seawater (about 3.0–3.5% NaCl), although they differ greatly in the lower and upper limits for growth. The majority of marine prokaryotes are moderate halophiles and can grow in media with NaCl up to about 15%. Extreme halophiles are discussed in *Section 4.6.6.*

4.4.2 Carbon

The carbon source for marine microorganisms ranges from simple molecules such as CO_2, CO and CH_4 to complex organic molecules. Heterotrophs require preformed organic compounds and these include carbohydrates, amino acids, peptides and organic acids. The origin of these compounds in the marine environment (dissolved organic carbon, DOC) is discussed in *Chapter 9.* Generally, macromolecules cannot be assimilated by prokaryotic cells, so many marine bacteria produce extracellular enzymes such as chitinases, amylases and proteases for their degradation into monomers of complex macromolecules. Chitinase is especially significant in marine bacteria, since chitin (a polymer of N-acetyl glucosamine, NAG) is such an abundant compound in the sea. Chitin is a major structural component of the exoskeleton of many invertebrates (especially *Crustacea*), and the cell walls of fungi and algae. Many marine bacteria possess surface-associated chitinases and the released NAG is an excellent source of both carbon and nitrogen. As in all cases where bacteria degrade macromolecules in their environment, mechanisms must exist for transporting the monomers produced into the cell. We know little about this, although it is likely that specific transport systems exist. In a natural bacterial assemblage, degradation of complex macromolecules by one type of microbe may make nutrients available to many others. Such processes are particularly important in particles of marine snow and may lead to a plume of DOM as the particle falls through the water column, because colonizing bacteria produce DOM more quickly than they can use it (*Section 1.7.1*). Bacteria inhabiting this plume probably grow at much faster rates than those in bulk seawater.

4.4.3 Carbon dioxide fixation in autotrophs

The great majority of photoautotrophic and chemolithoautotrophic *Archaea* and *Bacteria* incorporate CO_2 into cellular material via the Calvin cycle. The key enzyme in this pathway is

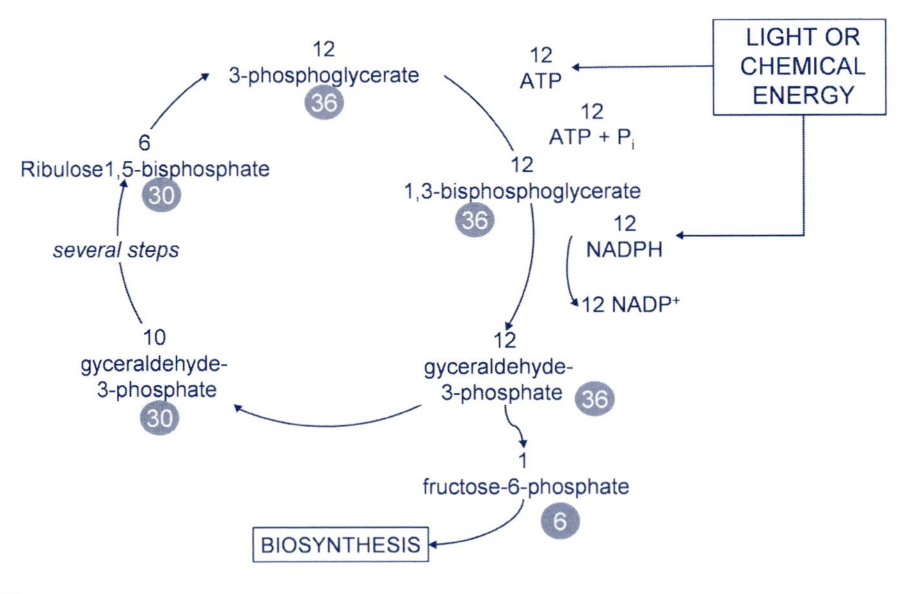

Figure 4.3

The Calvin cycle for CO_2 fixation in autotrophic organisms. One molecule of hexose sugar is produced from each six molecules of CO_2 incorporated; figures in circles show the total number of carbon atoms involved.

ribulose bisphosphate carboxylase (RubisCO), which is unique to autotrophs. Each cycle incorporates one molecule of CO_2 by carboxylation of ribulose bisphosphate (RUBP). The intermediate molecule is unstable and immediately dissociates to two molecules of 3-phosphoglycerate. This is then phosphorylated to form glyceraldehyde-3-phosphate, which is a key intermediate of the glycolytic pathway. Therefore, monomers of hexose sugars are synthesized by a reversal of glycolysis. As shown in *Figure 4.3,* the overall reaction mechanism leads to the incorporation of six molecules of CO_2 to generate one molecule of hexose sugar. It requires 12 molecules of NADPH and 18 molecules of ATP. Hexose monomers are converted by various reactions to other metabolites and the building blocks of macromolecules. If energy and reducing power are in excess supply, hexoses are converted to storage polymers such as starch, glycogen or polyhydroxybutyrate, and deposited as cellular inclusions. The final step in the Calvin cycle is the regeneration of one molecule of RUBP by another unique enzyme, phosphoribulokinase.

Some hyperthermophilic *Archaea* and *Bacteria* in marine sediments and other anaerobic habitats can fix CO_2 via two other possible routes. The reverse (reductive) TCA cycle utilizes ferredoxin-linked enzymes leading to the formation of acetate and was first described in *Aquifex* and *Chlorobium*. By using sequence analysis of microbial communities from hydrothermal vents, it has recently been found that genes encoding the key enzyme of this pathway (ATP citrate lyase) appear to be more common than genes for RubisCO. This suggests that the reverse TCA cycle may be the main mechanism of CO_2 fixation in chemolithoautotrophs in these habitats. Another route, the hydroxypropionate pathway, occurs in *Chloroflexus* in terrestrial hot springs, but it is not known if this pathway also occurs in marine prokaryotes.

4.4.4 Nitrification and denitrification

Marine prokaryotes assimilate nitrogen in the form of NH_4^+, NO_3^- or via fixation of N_2 from the atmosphere. Heterotrophs can also assimilate preformed amino acids, purines and pyrimidines. Inorganic nitrogen is most readily assimilated in the form of NH_4^+, using two main routes:

(a) $NH_4^+ + HCO_3^- + 2ATP \quad NH_2\text{-}COO\text{-}PO_3^{2-} + 2ADP + P_i + 2H^+$ (reactions catalyzed by carbamoyl-phosphate synthetase)
(b) $2NH_4^+ + \alpha$-ketoglutarate + NADPH + 2ATP glutamine + $NADP^+ + 2P_i$ (reactions catalyzed by glutamate dehydrogenase and glutamine synthase)

The amino group of the products is transferred by various reactions to form the numerous amino acids, pyrimidines and purines in the cell.

Whilst almost all prokaryotes can utilize NH_4^+ as a source of nitrogen, assimilation of NO_3^- occurs only in certain groups, when NH_4^+ is absent. These organisms use an assimilatory nitrate reductase system to convert NO_3^{2-} to ammonia via a number of intermediate steps. The dissimilatory pathway of NO_3^- (denitrification) is very important in the global nitrogen cycle, because it results in loss of fixed nitrogen to the atmosphere. Denitrification occurs in heterotrophs and chemoautotrophs and is widespread among marine genera of *Bacteria* and *Archaea*. Denitrification only occurs when organisms carry out nitrate reduction in sediments rich in organic matter. The synthesis of the enzymes involved in this denitrification (nitrate reductase and nitrite reductase) is repressed by molecular O_2 and induced by high levels of NO_3^-. Recently, it has been found that some prokaryotes can carry out denitrification under aerobic conditions, but the significance of this in marine systems is not known.

4.4.5 Nitrogen fixation

The direct incorporation of atmospheric nitrogen into cellular material is very restricted in its distribution. It occurs only in certain species of *Bacteria* and *Archaea*, including photolithotrophs, chemolithotrophs and chemoorganotrophs. It depends on the nitrogenase complex, consisting of the enzymes dinitrogenase (which contains Fe) and dinitrogenase reductase (which contains Fe and Mo). Dinitrogenase is very sensitive to O_2. This poses no problem under anaerobic conditions, but aerobic nitrogen-fixers must employ special strategies for protecting the enzyme from O_2 (*Section 5.4.3*). The reduction of N_2 to NH_4^+ is extremely energy demanding, because N_2 is a very unreactive molecule. *Figure 4.4* shows the overall reactions in nitrogen fixation and further consideration of the process is given in *Sections 5.4.3* and *5.10*. Ultimately, all life on Earth depends on the activity of these prokaryotes incorporating atmospheric N_2 into cellular material, which is then recycled by other organisms. The aerobic *Cyanobacteria* (especially *Trichodesmium*) and the anaerobic archaeal methanogens play a particularly significant role in nitrogen cycling in the oceans, as discussed in *Section 9.4.2*.

4.4.6 Sulfur and phosphorus

Most prokaryotes can assimilate sulfur under aerobic conditions in the form of SO_4^{2-}, which is usually abundant in seawater. Reduction of SO_4^{2-} proceeds through a series of reactions involving thioredoxin, ATP and NADPH. The first product is adenosine-5′-phosphosulfate. In assimilative sulfate reduction, another P is added from ATP to form phosphoadenosine-5′-phosphosulfate which is reduced to SO_3^{2-} and further converted to HS^- by the enzyme sulfite reductase. Sulfide is used for the synthesis of the amino acids cysteine and methionine, co-enzyme A plus other cell factors that depend on S for their function. Under anaerobic conditions, HS^- can be incorporated directly without the need for reduction. Assimilation of sulfur must be distinguished from the use of sulfur compounds as electron donors or acceptors for energy generation. The ability to use SO_4^{2-} as an electron acceptor is restricted to the δ-*Proteobacteria* known as the sulfate-reducing bacteria (*Section 5.17*). H_2S is used as an electron donor by the purple sulfur bacteria (phototrophy) and the sulfide-oxidizing bacteria such as *Thiobacillus*, *Thioplaca* and *Thiomargarita* (*Section 5.6*).

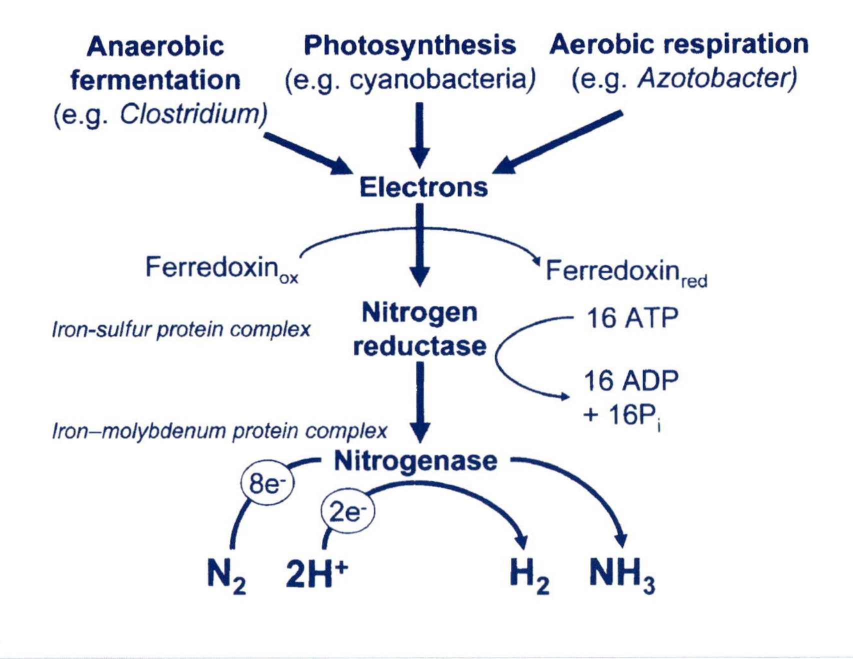

Figure 4.4

Reactions in nitrogen fixation.

Phosphorus is an essential constituent of nucleic acids, ATP and phospholipid membranes. It is usually assimilated in the form of PO_4^{2-} directly (without change in valence) from seawater. Phosphate availability can have a marked effect on planktonic growth and this is important in coastal regions subject to terrestrial runoff and in regions that experience nutrient upwelling.

4.4.7 Iron

Iron is an essential constituent of cytochromes and iron–sulfur proteins, which have crucial roles in electron transport processes of energy generation. Although iron is one of the most common elements in the Earth's crust, the amount of free iron in the open ocean can be virtually undetectable (picograms per liter) without ultraclean assay techniques.

In oxygenated seawater it occurs in the ferric Fe(III) state and forms highly insoluble compounds so it is a major challenge for marine microbes to obtain sufficient iron for growth. Many iron-containing proteins are highly conserved across diverse phylogenetic lineages and it is thought that the early evolution of life occurred under anoxic conditions (before the development of oxygenic photosynthesis), in which circumstances iron was readily available to life in the soluble Fe(III) form. As the atmosphere and oceans became oxidizing, iron precipitated in large quantities and the amount of iron in seawater today is extremely low. Aquatic microbes (including the cyanobacterial ancestors that caused the change in oxygen content of the atmosphere) were faced with an iron-shortage crisis and this led to the development of efficient mechanisms for scavenging the increasingly low levels in the environment. Many bacteria, both autotrophs and heterotrophs, secrete chelating agents known as siderophores that bind to iron, preventing its oxidation and allowing transport into the cell via specific receptors on the cell surface. In culture, these compounds are produced under conditions of iron restriction and some structures have been characterized. These include phenolics and derivatives of amino acids or hydroxamic acid. The first siderophores to be characterized from oceanic bacteria are structurally quite different from those previously described and behave as self-assembling amphiphilic molecules. Aquachelins from *Halomonas aquamarina* and marinobactins from

Marinobacter both have a water-insoluble fatty acid region and a water-soluble region of amino acids. In laboratory studies in the absence of iron, these compounds form clusters of molecules attached together by their fatty-acid tails. Upon binding to Fe(III), the micelles come together to form vesicles. It is not yet known whether this phenomenon of aggregation occurs in the natural environment, but it might be important for bacteria associated with particles containing local high concentrations of organic matter. The mechanism by which these siderophores are taken up by the cell is also unknown. Iron deprivation is also a problem for bacteria growing in the tissues of vertebrate animal hosts, which produce iron-binding proteins as a defense mechanism. For example, *Vibrio vulnificus* and *V. anguillarum* produce siderophores, known as vulnibactin and anguibactin respectively, and these are important factors in virulence (*Section 14.3.2*).

4.5 Growth and the effects of nutrient concentration

4.5.1 The bacterial growth cycle

During growth, a typical prokaryotic cell must synthesize additional cell membranes, cell wall components, nearly 400 different RNA molecules, 2000 proteins and a complete copy of its genome. The coordination of these activities is subject to complex regulation of gene expression. With a few exceptions (such as the budding and stalked bacteria), almost all prokaryotes reproduce by binary fission into two equal-sized cells. When growing under optimal or near-optimal conditions, the cell doubles in mass before division; regulation of DNA replication and cell septation are largely controlled by the rate of increase in cell mass. The growth cycle of culturable heterotrophs in the laboratory is well known. Upon introduction to the growth medium, there is a lag phase during which unbalanced growth (without cell division) occurs. During this period, the cell synthesizes metabolites, enzymes, coenzymes, ribosomes and transport proteins needed for growth under the prevailing conditions.

Under ideal conditions, bacteria grow exponentially (logarithmic phase). Different species vary enormously in the time taken for the population to double (generation time). Environmental factors such as temperature, pH and O_2 availability (for aerobes) have a major influence on the rate of growth. Growth rate increases with temperature; most marine bacteria isolated from water at 10°C grow with generation times of about 7–9 h whilst those isolated at 25°C typically double every 0.7–1.5 h. All organisms have a minimum, optimum and maximum temperature for growth and this reflects their evolution in a particular habitat. Above the minimum temperature, metabolic reactions increase in rate as temperature increases and so the generation time reduces. The maximum temperature is usually only a few degrees above the optimum; this is determined by the thermostability of enzymes and membrane systems. Some marine prokaryotes have very low generation times. For example, *Vibrio parahaemolyticus* can divide every 10–12 min at 37°C (although this temperature is not encountered in its normal estuarine habitat) and the hyperthermophile *Pyrococcus furiosus* doubles every 37 min at 100°C. At the other extreme, some psychrophiles growing just above 0°C may have generation times of several days.

Bacteria continue exponential growth until they enter the stationary phase because (a) a limiting nutrient is exhausted, (b) toxic by-products of growth accumulate to become self-limiting or (c) some density-dependent signal limits growth. Cells entering the stationary phase do not simply 'shut down', but undergo a number of genetically programed changes that involve the synthesis of many new starvation-specific proteins induced via global regulatory systems. When cells of laboratory-cultured marine bacteria are starved, they become smaller and spherical. Morita coined the term 'ultramicrobacteria' to describe the very small cells (0.3 μm diameter or less) which develop in this way. To achieve reduction in cell size, some proteins and ribosomes are degraded in a controlled fashion, but 30–50 new proteins are induced in response to starvation. These include proteins that are essential for survival with limited sources of energy and nutrients as well as proteins that enable cells to acquire scarce substrates at very low concentrations. Cell surfaces become more hydrophobic, which increases the adhesion of bacteria to surfaces. Surface association in biofilms appears to be a major adaptation to nutrient limitation. Changes in the fatty acid content of membranes promote better nutrient uptake and provide increased stability whilst cell walls become more resistant to autolysis. Starved cells have a much-reduced content of ATP, but maintain a proton-motive force across the membrane. Starved cells are more resistant to a variety of stress conditions and mutations in any of the genes in the survival induction pathway may lead to death in the stationary phase.

4.5.2 Cell viability

Studies by Morita and other workers showed that the majority of bacteria in the open sea are also very small and have similar morphology to the ultramicrobacteria formed by starvation of cultured marine isolates in the laboratory. Therefore, the paradigm developed that marine bacteria are 'normal' cells that have undergone reduction in size and are in a state of perpetual near-starvation. This led to the view that bacterioplankton are virtually inactive in the natural environment. In fact, since the 1980s, evidence has accumulated to show that the naturally occurring small cells found in the sea are *more* active than larger bacteria on a per volume basis and a large fraction of supposedly inactive cells can be induced to high metabolic activity by addition of substrates. However, determination of *in situ* growth rates of marine heterotrophic prokaryotes remains one of the most difficult aspects. Recent techniques have relied on the use of different fluorochromes in FCM (see *Section 2.3*) to determine cell biomass. Some protocols attempt to distinguish dead and live bacteria by employing fluorochromes directed against nucleic acids, whilst others use fluorochromes that detect respiratory activity. One method combines nucleic acid staining with a permeable dye and an impermeable dye both attached to gene probes to measure membrane integrity and DNA content. Cell viability can also be measured by use of fluorescent electron acceptors such as CTC, to monitor the presence of an active electron transport chain. Although many studies have produced conflicting results, the most likely hypothesis is that many planktonic prokaryotes in the open sea are metabolically active and their small size is an evolutionary (genotypic) adaptation to ensure efficient utilization of scarce nutrients in their oligotrophic environment.

The concept of a 'viable but nonculturable' (VBNC) state is used to explain the discrepancy between the number of bacterial cells that can be visualized by direct observation (e.g. epifluorescence) and those which can be enumerated using viable plate counts. The terms VBNC and somnicells ('sleeping' cells) were originally introduced in the 1980s by Rita Colwell and co-workers, who observed that *Vibrio cholerae* do not die off when introduced into the environment (*Section 11.2*). Since then, the VBNC state has been studied extensively and shown to occur in a wide range of Gram-negative (and a few species of Gram-positive) bacteria, including pathogens of humans, fish and invertebrates. The phenomenon is thus especially important in studies of the ecology of indigenous pathogenic marine bacteria and the survival characteristics of introduced pathogens and indicator organisms. The VBNC state could constitute a dangerous reservoir of pathogens, which cannot easily be detected and is also relevant to evaluation of the survival of introduced genetically modified organisms into the environment. A variety of factors, especially nutrient deprivation and changes in pH, temperature, pressure or salinity can initiate the cascade of cellular events leading to the VBNC state. These events are generally believed to be an inducible response to promote survival under adverse conditions, a property of particular importance for bacteria whose natural habitat is the aquatic environment.

Several methods can be used to detect VBNC cells. For example, radiolabeling techniques can be used to show that VBNC cells are metabolically active (i.e. they incorporate a suitable substrate) and demonstrating the reduction of reduced tetrazolium salts can be used to indicate an active electron transport chain. One method, which has been widely used to distinguish metabolically active from inactive bacteria, is the 'direct viable count', in which cell growth is stimulated by the addition of yeast extract whilst cell division is inhibited by the antibiotic nalidixic acid. This results in greatly enlarged cells that can be easily seen in the microscope. Another widely used method is the use of nucleic-acid binding fluorochromes in conjunction with epifluorescence microscopy or flow cytometry. Some dyes have differential ability to penetrate dead and 'alive' (i.e. metabolically active) cells.

It is of fundamental importance to distinguish the transition shown by bacteria that can be grown in laboratory culture to a VBNC state from the 'normal' state of most 'unculturable' marine prokaryotes. As discussed in *Box 2.1,* these are better described as 'not yet cultured', since although they cannot be cultured using conventional methods there has been considerable progress using dilution methods. The VBNC state remains a matter of considerable controversy among microbiologists and some current thinking on this important concept is summarized in *Box 4.1.*

4.5.3 Effects of nutrient concentration

In laboratory cultures, most nutrients are usually provided at concentrations well in excess of that required for balanced growth. Therefore, the rate of growth is independent of the nutrient concentration, although the final growth yield (number of cells or cell mass) does increase with higher concentrations of nutrient. Bacterial cells possess very efficient mechanisms for transporting nutrients into the cell. However, the growth rate will reduce at very low threshold concentrations of the limiting nutrient.

Prokaryotes synthesize special proteins that span the membrane to form a channel through which nutrients (and certain other substances) pass into the cell. Often, these transport proteins are specific to particular substrates. The proton motive force drives simple transport across the membrane. Facilitated diffusion can occur when the external concentration of nutrients is higher than in the cell and active transport enables the cell to acquire nutrients even when the external concentration is very low, as will occur in most aquatic environments. As the substrate is transported across the membrane, the conformation of the transport protein changes so that the substrate is released on the inner side of the membrane. This process requires energy in the form of ATP hydrolysis or the proton motive force. In the latter case, H^+ must be transported into the cell at the same time (symport). In another method of transport, the substrate is chemically modified during entry to the cell. Many bacteria accomplish this using the phosphotransferase system, a group of enzymes that phosphorylates molecules such as sugars. Finally, in many Gram-negative bacteria, amino acids, peptides, sugars and amino acids may cross the outer membrane through porins into the periplasmic space, where they bind to specific binding proteins. These proteins have extremely high affinities for their substrate, which means that the cell can acquire nutrients when present at micromolar or nanomolar concentrations. The periplasmic binding proteins then transfer the nutrient to a membrane-spanning protein for active transport across the membrane. Oligotrophic bacteria must possess very efficient transport systems in order to ensure that nutrients can enter the cell against a very steep concentration gradient and it is likely that ATP-driven periplasmic binding protein systems are used most commonly. To date, little research has been conducted on these processes in marine bacteria.

Growth efficiency can be defined as the ratio of yield of biomass produced (grams) to amount of substrate utilized (moles). Efficiency can also be expressed in terms of the amount of energy in the form of ATP (y_{ATP}) produced per mole of substrate. In studies of bacterial culture under constant conditions (chemostat culture), the value of y_{ATP} is always much lower (usually less than 50%) than that predicted by theoretical calculations predicted from the energetics of biochemical pathways. Organisms require a certain amount of nutrient for cell maintenance such as transport, maintenance of internal pH and a proton gradient across the membrane, DNA repair and other essential cell functions. These functions take priority over biosynthesis for

Box 4.1 RESEARCH FOCUS

Just resting, or suicidal?

The continuing controversy over the VBNC phenomenon

The interpretation of experiments to investigate the viable but nonculturable (VBNC) state has provoked much discussion and controversy in the scientific literature. What do the terms living, viable or culturable mean when we are discussing bacteria? Is there such a thing as the 'normal state' of bacteria; if so, what is it? Many factors can induce a state of nonculturability (i.e. reduced plate counts on normal media). The most consistent of these is nutrient deprivation, but salinity, temperature, light and other factors all have an effect depending on the organism in question. A complicating factor is the fact that many bacteria entering the stationary phase of growth will undergo a programed 'starvation-survival' adaptation to the onset of nutrient deprivation. This includes alterations in membrane permeability, reduction in cell size and metabolic changes. Adaptation involves complex gene regulation with synthesis of a repertoire of new proteins. These cells remain culturable, and can be recovered on agar plates. Cells that enter the VBNC state also undergo reproducible changes in cellular composition, cell morphology, cell wall and membrane structure. For example, there is a 60% reduction of the major fatty acids in *V. vulnificus* (Linder & Oliver 1989) and *Aeromonas salmonicida* shows a loss of DNA and protein content (Effendi & Austin 1995). Although related changes occur, this is not the same as the VBNC state. For example, if *V. vulnificus* is subjected to nutrient deprivation at 20–25°C, it will initiate the starvation-survival response, but at 10°C it will enter the VBNC state (Paludan-Miller *et al.* 1996).

Can VBNC cells be resuscitated, that is induced to revert to a form that can be cultured in the laboratory? Many studies have found that addition of nutrients or temperature changes can cause VBNC bacteria to revert, provided (usually) that the cells have not been in the VBNC state for too long. Others have refuted these findings and have found that the effect is due to the regrowth of a few remaining culturable cells in the population (e.g. Weichart *et al.* 1992). Bacteria may by injured sublethally but not be in a true VBNC state. One explanation for the inconsistency of results in this area is the possibility that remaining viable cells within the population produce some kind of 'wake up' signal molecule which allows VBNC cells to resuscitate (Mukalamova *et al.* 1998).

A key question is whether VBNC cells of pathogens retain their virulence. An observation with major epidemiological implications is that of Colwell *et al.* (1996), who showed that human volunteers developed clinical symptoms of cholera after being fed VBNC *V. cholerae*, and that the culturable form of the pathogen was excreted from the gut. However, cells which had been in the VBNC state for long periods had reduced infectivity.

Bloomfield *et al.* (1998) have proposed an important new hypothesis, namely that bacteria possess a kind of 'self-destruct' mechanism. This is said to occur because, when exponentially-growing bacteria are faced with a sudden insult such as the deprivation of nutrients, growth is decoupled from metabolism and the cells suffer a burst of oxidative metabolism, resulting in lethal free radicals. Bacteria can avoid this suicidal mechanism if they can induce changes to protect DNA, proteins and cell membranes from these effects (as they do when they enter stationary phase 'normally'). The shock of sudden transfer of cells which are in the process of adaptation to life in the aquatic environment (or to changed temperature or other conditions) to a rich medium could result in rapid death. Thus, evidence of metabolism does not indicate viability, nor does failure to grow in a laboratory culture medium constitute death, because the very isolation conditions may trigger lethal changes. So-called VBNC cells may be in transition to a truly dormant state, in which they can survive for long periods in the environment.

Kell and Young (2000) propose a redefinition of terms which renders the term VBNC an obsolete oxymoron. They equate 'viable' with 'culturable', that is capable of division and able to proliferate in culture media (provided suitable conditions are provided). Dormancy is defined as a state of low metabolic activity, in which bacteria will be unable to divide or form colonies on agar plates without a preceding resuscitation phase. Investigation of the VBNC state and vigorous debate will clearly continue for the practical reasons mentioned above and, perhaps, because it 'strikes at the heart' of microbiologists (most of us are only really comfortable when we can see a colony on an agar plate!). See also *Box 2.1.*

growth. At low growth rates (due to nutrient limitation), the cell will use proportionately more of the substrate for maintenance energy. Below a certain threshold, cells will divert all the substrate to just 'staying alive' rather than growing. Uncoupling of catabolic and anabolic processes often occurs at low growth rates when growth is limited by a substrate other than the energy source. Maintaining the maximum possible energy state of the cell enables it to resume growth rapidly when conditions change. We might conclude that bacterioplankton in the open sea are in a constant low-nutrient environment and that the efficiency with which the nutrients are converted to biomass is inherently low. These effects of nutrient levels and growth efficiency are important concepts in assessing the role of bacteria in carbon flux and productivity in marine systems. Because nutrient concentrations in the sea are generally low, growth rates are necessarily slow and a large proportion of energy intake is used for maintenance energy, especially active transport and synthesis of extracellular enzymes for acquiring nutrients. Therefore, net bacterial growth might be low, despite considerable turnover of organic material. However, as noted in *Section 1.7.1,* there is an increasing recognition that the ocean environment is very heterogeneous with respect to nutrients and that many marine prokaryotes are associated with local concentrations of organic matter.

4.5.4 Growth on surfaces – microbial interactions and biofilm communities

The physiological properties of bacteria in the sea are greatly affected by population density and interactions with other microbes, and significant differences exist between free planktonic cells and those that are associated as biofilms on surfaces and with particles such as marine snow. The importance of biofilms, i.e. microbial communities associated with surfaces, was introduced in *Section 1.7.3.* Although biofilm formation has been recognized for over a century, it is only in recent years that significant advances in the study of biofilm physiology have been achieved, largely as a result of the application of CLSM and FISH techniques (*Section 2.2.3*). Bacteria and diatoms have both been shown to initiate biofilm formation due to environmental cues, particularly nutrient availability. They undergo considerable developmental changes during transition from the suspended, planktonic form to the sessile, attached form. The events occurring in biofilm formation by single species of bacteria have been studied extensively in laboratory investigations and it can be regarded as a form of cellular development towards a multicellular lifestyle (some researchers liken it to a tissue). Organic molecules (largely polysaccharides and proteins) will coat any surface with a conditioning film within minutes of placing it into seawater. The initial stage in colonization by bacteria usually involves motility towards the surface, and changes in microviscosity as the cell approaches the surface may cause motility to slow. This causes a transient, reversible adsorption to the surface, mediated by electrostatic attraction and van der Waals forces. Genetic studies involving the creation of biofilm-deficient mutants have shown the importance of motility due to flagella and type IV pili (*Sections 3.9.1, 3.10*) in the movement of bacteria across the surface in order to contact other cells to form microcolonies. As noted in *Section 3.9.1,* bacteria may undergo dramatic changes in cell morphology and flagellar synthesis during contact with surfaces, involving significant changes in regulation of gene expression. Flagellar synthesis may be repressed completely once the bacteria are firmly attached. The expression of specific genes depends very much on the nature of the substrate. In experimental studies, bacteria will quickly colonize inert surfaces such as pieces of plastic, glass or stainless steel immersed in seawater, but the genes expressed are quite different from those expressed when they colonize organic substrates such as chitin. Some marine heterotrophic bacteria express chitin binding and chitinase genes selectively when they encounter chitin, so that they can begin to utilize it as a nutrient source. This is especially important in the colonization of the surface of crustacean animals.

The most important developmental change following attachment is the expression of copious quantities of exopolymeric substances (EPS). These provide a strong and sticky framework that cements the cells together. The chemical and physical properties of EPS are very variable and the nature and amount produced depends on the bacterial species, concentration of specific substrates and environmental conditions. The majority are polyanionic due to the presence of acids (e.g. D-glucuronic, D-galacturonic or D-mannuronic acids), ketal-linked pyruvate or the presence of phosphate or sulfate residues. The EPS often form a complex network of long interlinked strands surrounding the cell (glycocalyx). Both rigid and flexible properties can be conferred on the biofilm, depending on the secondary structure of the EPS and interactions with other molecules such as proteins and lipids may produce a gel-like structure. The mature biofilm often has a complex architecture composed of pillars and channels. Stalked bacteria and diatoms may increase the length of their stalk when growing in dense biofilms. Once established, the dense packing of bacteria and the diversion of metabolism to the production of EPS may mean that the cells are metabolically active but divide at very slow rates.

As noted above, the activity of extracellular enzymes leads to the dissolution of organic material from marine snow particles as they sink through the water column, and much of this DOM becomes available to other members of the plankton. Extracellular enzyme activity is likely to provide little 'return' for free-living bacteria at low density. Composition of the bacterial community inhabiting the particle and the plume of DOM that follows it could therefore have significant effects on the release and subsequent utilization of nutrients. We know that there are extensive antagonistic interactions between microbes in dense communities inhabiting the soil or the gut of animals, by the production of antibiotics and bacteriocins that inhibit growth of organisms unrelated to the producer strain, but there have been few such investigations of antagonistic interactions between marine bacteria. However, some recent studies suggest that a large proportion of marine particle-associated bacterial isolates possess antibiotic activity against other pelagic bacteria. Such antibiotic interactions are likely to be widespread in sediments, microbial mats and biofilms on the surface of plants and animals.

Quorum sensing (QS) is an intercellular communication system enabling bacteria to detect the concentration of cells in a population and regulate gene expression accordingly. It is described in detail in *Section 5.12.3* and *Box 16.1*. Study of this phenomenon is leading to new insights into the physiology and ecology of attached marine microbial communities. Experimental studies of single species biofilms have shown that the production of acyl homoserine lactones (AHLs) is important in determining the three-dimensional structure of mature biofilms. AHL mutants produce densely packed biofilms that are more easily dislodged from the surface. Little is known about QS in multispecies marine biofilms, but this is an area of active research. Bacteria may utilize signaling molecules produced by other species, or actively inhibit or degrade them by production of specific enzymes. Bacteria, algae, protozoa and viruses will all interact in the mature biofilm and there is growing evidence that horizontal gene transfer between different microbial species is greatly enhanced within biofilms; a fact of great significance in the evolution of organisms with altered characteristics. Marine-snow-associated bacteria have recently been shown to produce AHLs, suggesting that QS is important in the production of extracellular enzymes and antagonistic interactions in these densely populated habitats.

4.6 Extreme environmental conditions

4.6.1 Low temperature

The temperature of over 90% of the ocean is 5°C or colder. The temperature of seawater in the deep sea and in polar regions ranges from −1 to 4°C whilst internal fluids in sea ice can be as low as −35°C in winter. With the exception of sea ice, these temperatures are very stable and little affected by seasonal changes. The overwhelming majority of marine microorganisms are adapted to this cold environment and conditions can only be considered 'extreme' from a human perspective. Psychrophilic (cold-loving) prokaryotes are defined as those with an optimum growth temperature of less than 15°C, a maximum growth temperature of 20°C and a minimum growth temperature of 0°C or less. In fact, many deep sea and polar bacteria have quite a narrow temperature growth range and may lose viability after brief exposure to typical laboratory temperatures. Therefore, special precautions are needed in their collection, transport and culture. Psychrotolerant bacteria are those which can grow at temperatures as low as 0°C, but have optima of 20–35°C; many organisms from shallow seawater or coastal temperate regions fall into this category. Apart from understanding their considerable importance in nutrient cycling and ocean processes, there has also been recent interest by astrobiologists in psychrophiles because of the planned future exploration and search for possible life on Europa, the ice-covered moon of Jupiter. Biotechnologists also study psychrophiles because of the industrial potential of their fatty acids and extracellular polymer-degrading enzymes such as chitinase, chitobiase and xylanase (*Section 16.8*). Proteins from psychrophiles are more flexible at low temperatures because they have greater amounts of α-helix and lesser amounts of β-sheet than those from other organisms. Also, specific amino acids are often located at particular regions of the active site of the enzymes, ensuring better access of the substrate. Study of several enzymes has been facilitated by using recombinant DNA technology for their expression and manipulation in *E. coli*. *Colwellia psychroerythrea* (domain *Bacteria*), an obligate psychrophile that is widely distributed in Antarctic and Arctic regions, recently became the first cold-adapted organism for which the genome sequence has been elucidated. Comparison of the genome with that of mesophiles confirms the importance of amino acid composition and the prevalence of certain residues for protein function. Since the discovery of the abundance of *Archaea* (especially *Crenarchaeota*) in deep ocean waters (*Section 6.3.3*), it is appropriate that there are increasing studies of proteins from psychrophilic *Archaea*. For example, the elongation factor 2 of psychrophilic methanogens is overproduced and has higher affinity for GTP compared with that of mesophilic relatives, permitting efficient protein synthesis at low temperatures. Recently, homologs of genes for cold-shock proteins from *Bacteria* have been discovered in pelagic *Crenarchaeota*, suggesting gene transfer between the two domains driven by the environmental influence of low temperature. Active transport of substances across membranes must also be efficient at low temperatures. This is achieved by the incorporation of large amounts of unsaturated fatty acids into the membrane, which helps to maintain membrane fluidity and transport processes. Omega-3-polyunsaturated fatty acids (PUFAs), once thought to be nonexistent in bacteria, have been found in Antarctic and deep-sea isolates and have considerable biotechnological potential (*Section 16.8*).

4.6.2 High temperature

Only prokaryotes can grow at temperatures above 60°C. Such thermophilic organisms are found in the marine environment in areas of geothermal activity (*Section 1.7.5*) including shallow submarine hydrothermal systems, abyssal hot vents (black smokers) and active volcanic

Table 4.2 Growth conditions of hyperthermophilic marine *Archaea* and *Bacteria*

Species	Growth temperature °C			Nutrition[1]	Aerobic/ Anaerobic
	Minimum	Optimum	Maximum		
Archaea					
Archeoglobus fulgidus	60	83	95	CL	An
Ferroglobus placidus	65	85	95	CL	An
Igniococcus sp.	65	90	103	CL	An
Methanocaldococcus jannaschii	46	86	91	CL	An
Methanococcus igneus	45	88	91	CL	An
Methanopyrus kandleri	84	98	110	CL	An
Pyrobaculum aerophilum	75	100	104	CL	Ae/An
Pyrococcus furiosus	70	100	105	CO	An
Pyrodictium occultum	82	105	110	CL	An
Pyrolobus fumarii	90	106	113	CL	An
Staphylothermus marinus	65	92	98	CO	An
Thermococcus celer	75	87	93	CO	An
Bacteria					
Aquifex pyrophilus	67	85	95	CL	Ae
Thermotoga maritima	55	80	90	CO	An

[1] CL =chemolithotrophic; CO =chemoorganotrophic. Data from Huber & Stetter (1998).

seamounts. In deep-sea vent systems, the temperature of seawater can exceed 350°C. As this superheated water mixes with cold seawater, a temperature gradient is established and diverse communities of thermophilic organisms with different temperature optima occur. Those organisms that can grow above temperatures of 80°C are termed hyperthermophiles. *Table 4.2* shows the temperature growth ranges of representative marine species. The majority of hyperthermophiles described to date belong to two major groups of *Archaea* (*Sections 6.2* and *6.3*). About 70 species of hyperthermophilic *Archaea* have been described and the full genome sequences of several of these have now been elucidated. Only two major genera of *Bacteria* (*Aquifex* and *Thermotoga*) are hyperthermophilic. In both domains, hyperthermophiles occupy very deep branches of the phylogenetic tree and the significance of this is discussed in *Section 6.2.2*.

As shown in *Table 4.2,* hyperthermophiles show a range of physio-logical types and can be aerobic or anaerobic, chemoorganotrophic or chemolithotrophic. The enzymes and structural proteins are adapted to show optimum activity and stability at high temperatures. The overall structure of proteins from hyperthermophiles often shows relatively little difference from that of homologous proteins in mesophiles. However, variation in a small number of amino acids at critical locations in the protein appears to affect the three-dimensional conformation, permitting greater stability and function of the active site of enzymes. Intracellular proteins from hyperthermophiles also contain a high proportion of hydrophobic regions and disulfide bonds, which improve thermostability. There is great interest in the use of enzymes from hyperthermophiles in high-temperature industrial processes (*Section 16.6*). Adaptations of the CM also occur to ensure stability and effective nutrient transport at high temperatures. As noted in *Section 3.4,* the CM of *Archaea* contains ether-linked isoprene units and hyperthermophiles usually possess monolayer membranes, which appear to be more stable at high temperatures.

4.6.3 **High pressure**

Microorganisms that inhabit the deep ocean must withstand a very high hydrostatic pressure, which increases by one atmosphere (atm = 0.103 MPa) for every 10 m depth. Over 75% of the ocean's volume is more than 1000 m deep. As we now know that *Bacteria* and *Archaea* are distributed in great numbers throughout the water column (see *Figure 6.2*), as well as in marine sediments many hundreds of metres deep, growth under conditions of very high pressure is the normal state of affairs. Zobell and Morita pioneered the study of deep-sea bacteria in the 1970s and 1980s. Recent advances in sampling and cultivation methods together with the application of molecular techniques to the study of diversity and physiology are leading to some significant new insights. Those *Bacteria* and *Archaea* that have been isolated and cultured in the laboratory are usually found to be barotolerant, i.e. they can grow over a wide range of pressures from 1–400 atm. For such organisms, high pressure usually results in lower growth rates and metabolic activity. Many organisms isolated from coastal environments can only tolerate pressures up to about 200 atm. By contrast, some prokaryotes can only grow at pressures greater than 400 atm; these are known as obligate or extreme barophiles (the alternative term piezophiles is also used). Many of these organisms die if brought to the surface and temperature or exposure to light exacerbate this lethal effect. By the use of special isolation techniques involving collection in pressurized chambers and cultivation in solid silica gel media, an increasing number of species of obligate barophiles has been cultured in recent years, including some from the deepest habitats such as the Marianis Trench in the Pacific Ocean (10 500 atm). Genetic studies of extreme barophiles indicate that many have a close resemblance to common barotolerant or barosensitive species (e.g. the *γ-Proteobacteria Shewanella, Photobacterium, Colwellia,* and *Moritella*) although some unique taxa have also been discovered. At abyssal depths (>4000 m), barophiles appear to be ecologically dominant over bacteria from shallow waters carried there by sinking organic matter. The most abundant sources of obligate barophiles are nutrient-rich niches such as decaying animal carcasses or the gut of deep-sea animals. However, oligotrophic barophiles adapted to very low nutrient concentrations also occur in seawater. Some obligately barophilic chemolithotrophic *Archaea* have been found near hydrothermal vents, but little is known about their physiology.

A range of adaptations appears to be present in deep-sea bacteria. It is important to note that very low temperatures (apart from hydrothermal vents) and very low nutrient conditions (apart from the localized occurrence of concentrations of organic matter) characterize the deep sea, as well as high pressure. We must therefore consider adaptation of deep-sea microorganisms as a response to the combined effects of these factors. High pressure decreases the binding of substrates to enzymes, which explains why the metabolism of shallow-water (or terrestrial) barotolerant organisms is much slower when incubated in the laboratory under pressure. A remarkable demonstration of this effect is the '*Alvin* sandwiches' story. In 1968, the submersible vessel *Alvin* sunk accidentally to a depth of 1500 m. The crew had inadvertently left a lunchbox containing meat sandwiches on board the vessel. These were found in almost perfect condition nearly a year later, when *Alvin* was recovered. At atmospheric pressure at 2°C, growth of contaminating bacteria would spoil the food in a few days. Protection of enzymes from the effects of pressure seems to be due mainly due to changes in the conformation of proteins. Proteins in barophiles seem less flexible and less subject to compression under pressure, due to a decreased content of the amino acids proline and glycine. Cells grown under high pressure also contain high levels of osmotically active substances, which are thought to protect proteins from hydration effects of high pressure. The most well studied adaptation to high pressure and low temperature is a change in membrane composition. Membranes of barophiles contain a higher proportion of PUFAs and have a more tightly packed distribution of fatty acyl chains. Pressure also affects DNA secondary structure.

Application of molecular genetics to the study of two barophilic bacteria, *Photobacterium profundum* and *Shewanella* sp., has recently revealed some insight into the mechanisms of regulation of the pressure response. When *P. profundum* is shifted from atmospheric to high pressure, the relative abundance of two OM proteins (OmpH and OmpL) is altered. These proteins act as porins for the transport of substances across the OM. An increased production of OmpH probably provides a larger channel, suggesting that the pressure response enables the bacteria to take up scarce nutrients more easily (as would occur in the deep-sea environment). A pair of cytoplasmic membrane proteins regulates transcription of the *ompL* and *ompH* genes. Interestingly, these have a high sequence homology to the ToxR and ToxS proteins, which were first discovered in *V. cholerae,* where they are responsible for detecting changes in temperature, pH, salinity and other conditions during the transition from the aquatic environment to the host, leading to the expression of virulence factors (see *Figure 11.2*). The homology suggests some common ancestry of this environmental sensing system that has evolved to perform different functions according to habitat. Some other genes have also been shown to be important in the response to pressure in a deep-sea *Shewanella* sp. and these may be grouped into pressure-regulated operons. Use of gene probes directed against these genes could aid in the identification of new species of barophiles that we cannot currently culture.

4.6.4 Toxic effects of oxygen

Prokaryotes vary in their requirements for O_2 or tolerance of its presence and can be classified as aerobic or anaerobic. Obligate aerobes always require the presence of O_2 and use it as the terminal electron acceptor in aerobic respiration. Facultative aerobes can carry out anaerobic respiration or fermentation in the absence of O_2 or aerobic respiration in its presence. Even though O_2 is not required, the growth of facultative organisms is better in its presence due to the greater yield of ATP from aerobic respiration. Microaerophiles carry out aerobic respiration but require an O_2 level lower than that found in the atmosphere. Obligate anaerobes carry out fermentation or anaerobic respiration and many are killed by exposure to O_2, although some are aerotolerant and survive (but do not grow) in its presence. Examples of all categories occur in marine *Bacteria* and *Archaea*.

Apart from the gas O_2, oxygen can exist in various forms which are highly reactive and toxic to all cells unless they possess mechanisms to destroy them. Singlet oxygen ($^1O_2$) is a high-energy state that causes spontaneous oxidation of cellular materials. In particular, singlet oxygen forms during photochemical reactions and phototrophs usually contain carotenoid pigments, which convert singlet oxygen to harmless forms (quenching). For this reason, non-phototrophic marine organisms exposed to bright light (such as those inhabiting clear surface waters) are also often pigmented. Various other toxic oxygen species form during the oxidation of O_2 to water during respiration (*Figure 4.5*). Superoxide and hydroxyl radicals are particularly

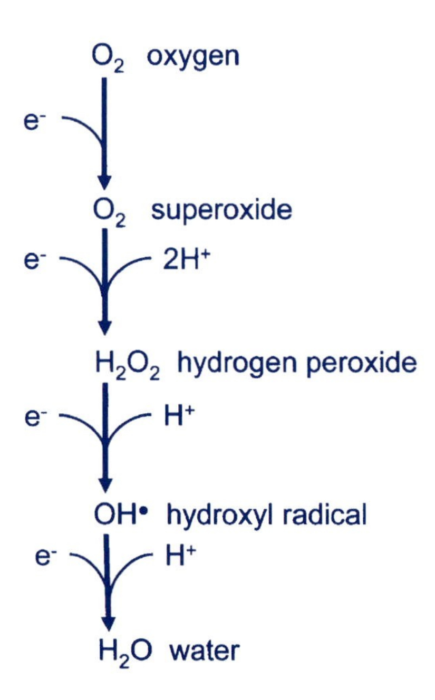

Figure 4.5

Formation of toxic intermediates during reduction of oxygen to water.

destructive and react rapidly with cellular compounds. The evolution of mechanisms for the removal of toxic oxygen species was a major step in the transition of the biosphere from anaerobic to aerobic following the development of oxygen-evolving photosynthesis. Organisms capable of aerobic growth usually contain the enzymes catalase ($2H_2O_2 \rightarrow H_2O + O_2$), superoxide dismutase ($2O_2^- + 2H^+ \rightarrow H_2O_2 + O_2$) and peroxidase ($H_2O_2 + NADH + H^+ \rightarrow 2H_2O + NAD^+$). Superoxide reductase is an enzyme originally found in *Pyrococcus furiosus* and thought to be unique to the *Archaea,* but genome sequence analysis has shown that it may be widely distributed in obligate anaerobes in place of superoxide dismutase. This enzyme reduces superoxide to H_2O_2 without the formation of O_2 ($O_2^- + 2H^+ + \text{cyt } c_{reduced} \rightarrow H_2O_2 + \text{cyt } c_{oxidized}$).

4.6.5 Ultraviolet (UV) irradiation

Research into the effects of UV radiation on marine microbes is needed because of growing evidence that UV radiation is increasing at certain locations on Earth, due particularly to ozone depletion in the upper atmosphere and the formation of an ozone 'hole' over Antarctica and the Southern Ocean. The lethal and mutagenic effects of UV radiation result from damage to DNA. UV-B causes direct damage to DNA through the formation of pyrimidine dimers, whilst the main effects of UV-A are due to formation of toxic oxygen and hydroxyl radicals. Various mechanisms for the repair of UV-induced damage exist, including nucleotide excision repair and light-activated enzyme repair (photoreactivation, see *Box 5.2*). Studies of DNA damage in bacteria in surface waters show that there is a pronounced effect over the course of the day, with maximal damage evident in the late afternoon and repair occurring during the night. We do not yet fully understand the ecological significance of these processes. Bacteria produce a range of UV-screening products such as mycosporine-like amino acids and scytonemin, a complex aromatic compound formed in the sheath of some *Cyanobacteria.* Some bacteria isolated from corals in very clear surface waters show extreme resistance to the effects of UV radiation by enhancing the activity of NAD(P)H quinine oxidoreductase, a powerful antioxidative enzyme. These mechanisms could have significant biotechnological potential in human health, as products for skin protection treatments and overcoming the effects of oxidative stress during aging.

4.6.6 High salt concentrations

Several genera of *Archaea* are extreme halophiles and can grow at very high NaCl concentrations (15–35%) found in salterns, submarine brine pools and brine pockets within sea ice (*Section 6.2.4*). Extreme halophilicity is rare in the *Bacteria,* but *Salinibacter rubrum* is an exception. In order to protect themselves from dehydration due to loss of water from the cell to the external environment, marine prokaryotes must maintain the concentration of intracellular solutes at a high level. One way of achieving this is by accumulating noninhibitory substances known as compatible solutes or osmoprotectants. Usually, these are types of sugars, alcohols or amino acids that are extremely soluble in water. For example, many marine Gram-negative *Bacteria* can synthesize glycinebetaine or glutamate. Many marine *Cyanobacteria* make this substance as well as α-glucosylglycerol. These substances are released when cells lyse and some bacteria accumulate glycinebetaine from the environment rather than synthesizing it themselves. Most Gram-positive *Bacteria* accumulate the amino acid proline as an osmoprotectant. In algae, dimethylsulfide propionate (DMSP) is the main osmoprotectant (see *Section 9.7*).

The extremely halophilic *Archaea* use a different method to prevent water loss. They have an active mechanism for pumping K^+ ion into the cell until the internal concentration balances the high concentration of Na^+ outside. In some species, a large proportion of the proton motive force for the ion pump derives from a light-mediated generation of ATP via the pigment bacteriorhodopsin (*Section 6.2.4*). Extreme halophiles also have other adaptations for growth at high NaCl concentrations. Their enzymes and structural proteins have a high proportion of acidic amino acids, which protects the conformation from disruption by high salt concentrations. Internal cellular components, such as the ribosomes and DNA-replication enzymes, require high K^+ concentrations for their integrity and activity. By contrast, those exposed to the environment (such as the S-layer) depend on high Na^+ levels.

References and further reading

Azam, F., and Long, R.A. (2001) Oceanography – sea snow microcosms. *Nature* **414**: 495–498.

Bartlett, D. (1999) Microbial adaptations to the psychrosphere/piezosphere. *J Molec Microbiol Biotechnol* **1**: 93–100.

Bloomfield, S.F., Stewart, G.S.A.B., Dodd, C.E.R., Booth, I.E.R., and Power, E.G.M. (1998) The viable but non-culturable phenomenon explained? *Microbiology* **144**: 1–3.

Choi, J.W., Sherr, B.F., and Sherr, E.B. (1999) Dead or alive? A large fraction of ETS-inactive marine bacterioplankton cells, as assessed by reduction of CTC, can become ETS-active with incubation and substrate addition. *Aquat Microb Ecol* **18**: 105–115.

Colwell, R.R., and Grimes, J. (eds) (2000) *Nonculturable Microorganisms in the Environment.* ASM Press, Washington.

del Giorgio, P.A., and Cole, J.J. (1998) Bacterial growth efficiency in natural aquatic systems. *Ann Rev Ecol Syst* **29**: 503–541.

Deming, J.W. (1998) Deep ocean environmental biotechnology. *Curr Opin Biotechnol* **9**: 283–287.

Deming, J.W. (2002) Psychrophiles and polar regions. *Curr Opin Microbiol* **5**: 301–309.

Effendi, I., and Austin, B. (1995) Dormant/unculturable cells of the fish pathogen *Aeromonas salmonicida. Microb Ecol* **30**: 183–192.

Gram, L., Grossart, H.-P., Schlingloff, A., and Kiørboe, T. (2002) Possible quorum sensing in marine snow bacteria: production of acylated homoserine lactones by *Roseobacter* strains isolated from marine snow. *Appl Environ Microbiol* **68**: 4111–4116.

Huber, H., and Stetter, K.O. (1998) Hyperthermophiles and their possible potential in biotechnology. *J Biotechnol* **64**: 39–52.

Hutchins, D.A., Witter, A.E., Butler, A., and Luther, G.W. (1999) Competition among marine phytoplankton for different chelated iron species. *Nature* **400**: 858–861.

Kato, C., and Bartlett, D.H. (1997) The molecular biology of barophilic bacteria. *Extremophiles* **1**: 111–116.

Kato, C., and Qureshi, M.H. (1999) Pressure response in deep-sea piezophilic bacteria. *J Molec Microbiol Biotechnol* **1**: 87–92.

Kell, D.B., and Young, M.Y. (2000) Bacterial dormancy and culturability: the role of autocrine growth factors. *Curr Opin Microbiol* **3**: 238–243.

Lebaron, P., Bernard, L., Baudart, J., and Courties, C. (1999) *The Ecological Role of VBNC Cells in the Marine Environment.* Proceedings of the 8th International Symposium of Microbial Ecology. Atlantic Canada Society for Microbial Ecology, Halifax, Canada. http://plato.acadiau.ca/isme/Symposium23/lebaron.PDF (accessed March 31 2003).

Linder, K., and Oliver, J.D. (1989) Membrane fatty acid changes in the viable but nonculturable state of *Vibrio vulnificus. App Environ Microbiol* **55**: 2387–2842.

Martinez, J.S., Zhang, G.P., Holt, P.D., Jung, H.T., Carrano, C.J., Haygood, M.G., and Butler, A. (2000) Self-assembling amphiphilic siderophores from marine bacteria. *Science* **287**: 1245–1247.

Morita, R.K. (1997) *Bacteria in Oligotrophic Environments: Starvation-Survival Lifestyle.* Chapman & Hall, New York.

Mukamalova, G.V., Yanoplaskaya, N.D., Kell, D.B., and Kaprelyants, A.S. (1998) On resuscitation from the dormant state of *Micrococcus luteus. Anton van Leeuw Int J Microbiol* **73**: 237–243.

Neidhardt, F.C., Ingraham, J.L., and Schaechter, M. (1990) *Physiology of the Bacterial Cell: A Molecular Approach.* Sinauer Associates Inc., Sunderland, MA.

Oliver, J.D., Hite, F., McDougald, D., Andon, N.L., and Simpson, L.M. (1995) Entry into, and resuscitation from, the viable but nonculturable state by *Vibrio vulnificus* in an estuarine environment. *App Environ Microbiol* **61**: 2624–2630.

Paludan-Miller, C., Weichart, D., McDougald, D., and Kjelleberg, S. (1996) Analysis of starvation conditions that allow for prolonged culturability of *Vibrio vulnificus. Microbiology* **142**: 1675–1684.

Paustian, T. (2001) Growth and Nutrition of Bacteria. Metabolism. University of Wisconsin-Madison. http://www.bact.wisc.edu/microtextbook (accessed March 31 2003).

Riemann, L., and Azam, F. (2002) Widespread N-acetyl glucosamine uptake among pelagic marine bacteria and its ecological implications. *Appl Environ Microbiol* **68**: 5554–5563.

Roszak, D.B., and Colwell, R.R. (1987) Survival strategies of bacteria in the natural environment. *Microbiol Rev* **51**: 365–379.

Rothschild, L.J., and Mancinelli, R.L. (2001) Life in extreme environments. *Nature* **409**: 1092–1101.

Schut, F., Prins, R.A., and Gottschal, J.C. (1997) Oligotrophy and pelagic marine bacteria: facts and fiction. *Aquat Microb Ecol* **12**: 177–202.

Sherr, B.F., del Giorgio, P., and Sherr, E.B. (1999) Estimating abundance and single-cell characteristics of respiring bacteria via the redox dye CTC. *Aquat Microb Ecol* **18**: 117–131.

Sinha, R.P., Klisch, M., Groniger, A., and Hader, D.P. (2001) Responses of aquatic algae and cyanobacteria to solar UV-B. *Plant Ecol* **154**: 221–236.

Stoodley, P., Sauer, K., Davies, D.G., and Costerton, J.W. (2002) Biofilms as complex differentiated communities. *Ann Rev Microbiol* **56**: 187–209.

Sutherland, I. (2001) Biofilm exopolysaccharides: a strong and sticky framework. *Microbiology* **147**: 3–9.

Wai, S.N., Mizunoe, Y., and Yoshida, S. (1999) How *Vibrio cholerae* survive during starvation. *FEMS Microbiol Lett* **180**: 123–131.

Watnick, P., and Kolter, R. (2000) Biofilm, city of microbes. *J Bacteriol* **182**: 2675–2679.

Weichart, D., Oliver, J.D., and Kjelleberg, S. (1992) Low temperature induced non-culturability and killing of *Vibrio vulnificus*. *FEMS Microbiol Lett* **100**: 205–210.

Marine *Bacteria*

5.1 Approaches to the study of prokaryotic diversity

As discussed in earlier chapters, the application of molecular biological methods, especially the sequencing of 16S rRNA, has revolutionized the study of marine microbes (and microbiology in general). In addition to the classical methods of isolation, culture and determination of morphological or biochemical properties, using a phylogenetic approach has led to major rethinking about the relationships between different prokaryotic groups. Many *Bacteria* and *Archaea* are known only on the basis of genetic evidence and cannot be cultured, whilst others can be grown and studied in the laboratory. Thus, there is a large discrepancy in the extent of our knowledge of the various organisms. For those marine prokaryotes that cannot yet be cultured, we can only infer their likely properties by considering their relationship to well-studied species, their habitats and geochemical evidence relevant to their actvities. Fortunately, advances in gene sequencing mean that we are beginning to be able to predict the metabolic nature of uncultured types through analysis of genes encoding key enzymes. In addition, as discussed in *Box 2.1,* new techniques are enabling the culture of marine bacteria previously regarded as unculturable. Because of the great disparity in knowledge of different types (and limitations of space), discussion in this chapter is necessarily skewed to those organisms whose properties are best known and those whose activities are of particular importance when their ecological role or applications is considered in later chapters. A phylogenetic treatment and taxonomic discussion is beyond the scope of this book and it is impossible to consider other than a selection of the many marine prokaryotes. For further information, the reader is referred to appropriate sources in the *Further Reading* given at the end of this chapter.

Figure 5.1 shows the *Bacteria* divided into 17 well-characterized major divisions; representatives of most of these groups are found in marine habitats. The view presented here is considerably simplified because the phylogenetic classification of *Bacteria* is currently in a state of great flux. When diversity is assessed using 16S rRNA sequences from environmental samples the number of distinct divisions probably exceeds 40, many of which contain no cultured representatives, and the taxonomy of these will not be resolved until more information is available. After discussing our knowledge of bacterial diversity revealed by molecular, culture-independent methods this chapter contains a brief description of some of the major groupings of *Bacteria* that can be studied in culture. Selected examples of important genera and their distinctive properties are given. It is very important to emphasize that the section headings used sometimes represent artificial groupings of bacteria that share significant physiological or morphological properties, rather than phylogenetic groups. Molecular methods often show that organisms currently grouped together because of such shared properties may be quite distantly related.

Our consideration of bacterial diversity is somewhat dominated by the *Proteobacteria,* because this is one of the largest and most physiologically diverse groups of bacteria. All members share the Gram-negative cell structure, but there is a huge diversity in metabolic activities. The group is further divided into five subdivisions, namely the alpha (α), beta (β), gamma (γ), delta (δ) and epsilon (ε) *Proteobacteria,* based on 16S rRNA sequences. There are marine representatives of all of these groups, with the α and γ types being especially important in the bacterioplankton of ocean waters. There are two major phylogenetically related groups (clades) of marine α-*Proteobacteria*, namely the *Roseobacter* and *Sphingomonas* clades. Many species from these groups have been cultured from marine environments. However, culture-independent molecular methods reveal an enormous diversity not evident from culture-based studies and

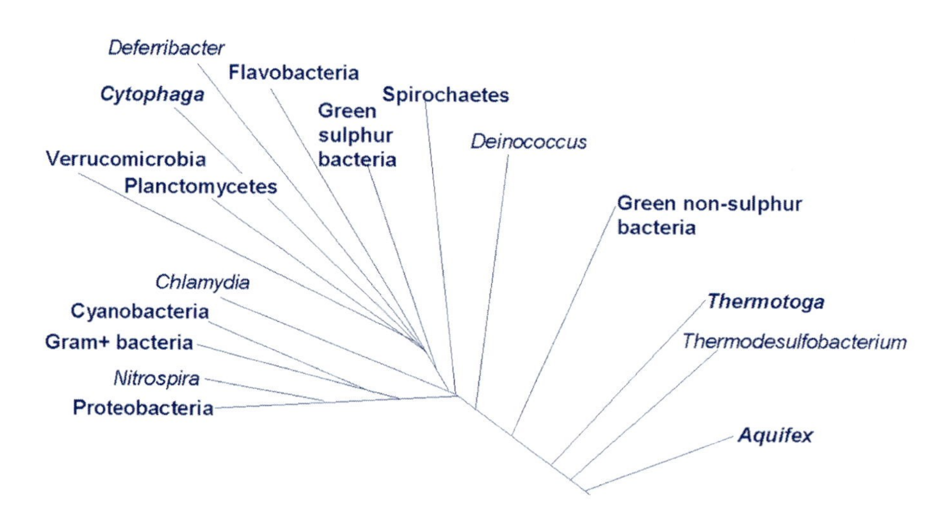

Figure 5.1

Simplified phylogenetic tree of the domain *Bacteria* showing relationships revealed by 16S rRNA sequencing. Representatives of almost all the major divisions are found in marine habitats (bold type). N.B. this tree does not reflect the emerging rRNA-based taxonomic system for the domain. Adapted from Madigan *et al.* (2003).

show that representatives of the α-*Proteobacteria* are among the most dominant bacteria in marine waters. Conversely, members of the γ-*Proteobacteria* are most common when culture methods are used, but are less well represented in databases constructed using molecular methods. Most of the easily cultured marine bacteria fall within this group and, historically, its members were regarded as being the most dominant marine bacteria. Although members of this group are the most easily isolated from waters all over the world because many will form colonies on conventional agar culture media, 16S rRNA studies show that their dominance as free-living heterotrophs has been overestimated in comparison with other groups. However, some of the most well known genera fall within the γ-*Proteobacteria* and are of special significance for discussions later in the book, because of (a) their interactions with other marine organisms, (b) their established role in ecological processes, (c) their pathogenicity for humans, (d) their biotechnological potential, or (e) their value in yielding information about processes of general microbiological importance.

5.2 Prokaryote diversity in marine ecosystems revealed by culture-independent methods

5.2.1 Cloning of 16S rRNA sequences from the environment

One of the most important recent breakthroughs in marine microbiology has been the application of molecular techniques (especially cloning of 16S rRNA sequences) to the direct analysis of environmental samples, without the need to isolate and culture microorganisms (for methods, see *Section 2.6*). Widespread application of this approach in the last decade has led to a complete reevaluation of the importance of marine *Bacteria* and *Archaea*, which are both more abundant and more diverse than we could possibly have imagined before the advent of these techniques. Microbiologists have long realized that there is a large discrepancy between the numbers of organisms counted using direct microscopic observation and those recovered on culture media, even if a range of nutrients and growth conditions are used. It is currently thought that less than 1% of marine prokaryotes have been cultured.

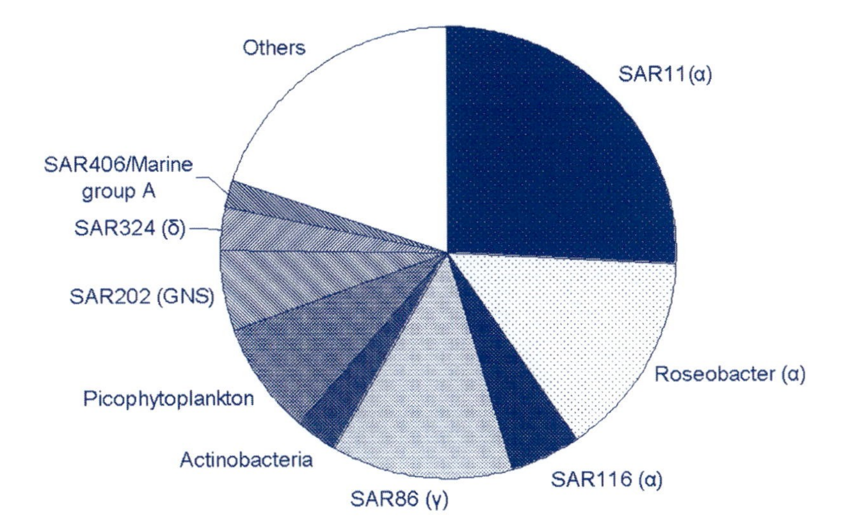

Figure 5.2

Frequency of the most common 16S rRNA gene clusters of *Bacteria* in marine plankton. The 'SAR' prefix indicates the original isolation from samples from the Sargasso Sea. Symbols = group of the *Proteobacteria;* GNS = green nonsulfur bacteria. Based on analysis of 578 clones. Redrawn from data in Giovannoni and Rappé (2000).

Since the first studies in the early 1990s, a number of research groups have conducted studies in diverse marine habitats throughout the world, and established databases containing thousands of sequences from marine prokaryotes. Different investigators, using variations of the basic methods and sampling different geographic regions and depths have repeatedly found similar phylogenetic groups.

5.2.2 The major phylogenetic groups of planktonic *Bacteria*

The most abundant clones of 16S rRNA sequences from *Bacteria* found in ocean plankton do not correspond to cultured species at all. When phylogenetic trees are constructed based on genetic similarity, over three-quarters of the marine *Bacteria* revealed by this approach belong to nine phylogenetic groups with worldwide distribution. These form clusters of related types, rather than single lineages, and genetic variability within these populations is very high. In some clusters, differences in 16S rRNA sequences may be due to genetic variability associated with adaptation to different depths of water. Bacterial diversity in suspended particles (marine snow) and free picoplankton is very different, and varies with depth and environmental conditions. *Figure 5.2* shows the frequency of the most common gene clusters. Of the major clades, only the *Roseobacter* clade in the α-*Proteobacteria* and the marine picophytoplankton clade (*Cyanobacteria* including prochlorophytes) contain cultivable bacteria, although representatives of previously uncultivated clades are now being cultured.

The SAR11 cluster of 16S rRNA sequences has been found in almost every pelagic environment ranging from shallow coastal waters to depths of over 3000 m. It is probably the most abundant microorganism in the sea and on average accounts for nearly a third of the cells present in surface waters and a fifth of the cells in the mesopelagic zone. SAR11 is a deeply branching member of the α-*Proteobacteria* and is phylogenetically distinct from all cultured members of the group. Very recently, members of the SAR11 cluster have been cultured (*Box 2.1*) and this has allowed an initial study of its physiological and morphological properties. In culture, the SAR11 isolates grow with doubling times of about 2 days and reach maximum densities in seawater of 10^5–$10^6\,ml^{-1}$.

The cells are crescent-shaped, 0.4–0.9 μm long and 0.1–0.2 μm in diameter. The cell volume is only about 0.01 μm^3 (less than 1/100th that of familiar bacteria such as *E. coli*) and the high SA:V ratio of the cell is undoubtedly a key factor in the ability of SAR11 bacteria to thrive in low-nutrient conditions of the open ocean. The genome size is about 1.54 Mb (about one-third that of *E. coli*). A project to sequence the SAR11 genome is now underway.

Members of the α-proteobacterial SAR116 cluster are also ubiquitous, although they may be more prevalent in the surface layers of the oceans and shallower coastal waters. As noted previously, members of the *γ-Proteobacteria* dominate collections of marine bacteria obtained by culture, whereas members of the *α-Proteobacteria* are more prevalent in studies of bacterioplankton by 16S rRNA methods. However, the commonly isolated SAR86 gene cluster is a member of the *γ-Proteobacteria,* but is phylogenetically very distant from the culturable members of this group. The SAR86 cluster may have some phylogenetic relationships to the methanotrophs and chemolithotrophs that occur as endosymbionts of marine invertebrates. The *Actinobacteria* clade forms a smaller proportion of sequences and is a branch of the high G+C Gram-positive bacteria of gene clones, only distantly related to culturable representatives. They occur primarily in the upper, photic layers of the ocean.

Three major groups appear to be primarily associated with the aphotic zone of the deep ocean. These are the SAR202 clade (a deeply branching group of the green nonsulfur bacteria), the marine group A clade (probably a unique previously undiscovered branch of the *Bacteria*) and the marine group B clade (a member of the *δ-Proteobacteria*).

In conclusion, the extensive analysis of bacterioplankton communities by 16S rRNA gene cloning conducted in the 1990s has shown a remarkably consistent pattern of diversity in waters from many locations. Some authors have reevaluated sequence data from both culture and culture-independent methods and drawn the conclusion that the overall diversity (at the species level) in marine bacterioplankton is quite low. Other authors believe that there are hundred of thousands of taxa sufficiently different to be called 'species'. It is important to note that the assumed diversity of bacterial species in marine plankton is based very much on 16S rRNA diversity and that this leads to an empirical scheme for classification which has some drawbacks. This controversial topic is discussed further in *Box 5.1*.

5.3 Anoxygenic phototrophic bacteria

5.3.1 Purple sulfur and nonsulfur bacteria

These phototrophic *Proteobacteria* are commonly referred to as the purple bacteria and have been studied for many years. Purple bacteria have vari-able morphology (rods, ovoid cocci and spirals) and representatives of this group occur in the α-, β- and *γ-Proteobacteria*. Unlike *Cyanobacteria*, algae and plants, these bacteria do not evolve O_2 during photosynthesis and the discovery of this group was of major importance in the development of a unifying theory for the mechanism of photosynthesis. Purple phototrophs contain bacteriochlorophylls as the photosynthetic pigment, which together with additional carotenoid pigments give the bacteria their distinctive color. The pigments are located inside multifolded invaginations of the cytoplasmic membrane, which allow the bacteria to make efficient use of available light. Various types of bacteriochlorophyll occur and these absorb light of different wavelengths. This, together with the light-absorbing properties of associated proteins and the source of electrons used to reduce carbon dioxide during photosynthesis, determines the habitat and ecological role of the various species.

The group known as the purple sulfur bacteria utilize H_2S or other reduced sulfur compounds as the source of reductant. The overall reaction for photosynthesis can be represented as

$$CO_2 + H_2S + H_2O \longrightarrow (CH_2O) + S + H_2O$$

(where CH_2O represents the reduced carbon compounds that are the initial products of photosynthesis). Granules of sulfur are deposited inside the cells. Purple sulfur bacteria (all of which

Box 5.1 RESEARCH FOCUS

How many species of prokaryotes are there in the oceans?

Uncertainty surrounds the definition of species and estimates of diversity

Cloning and sequencing of 16S rRNA sequences from seawater, without the need for culture, has been one of the most significant advances in marine microbial ecology. These studies have suggested a picture of far greater bacterial diversity than that envisaged by cultural methods alone. To date, about 6000 species of prokaryotes (from all habitats, not just marine), have been identified and named (Garrity *et al.* 2002). This must surely be a great underestimate, but do the recent molecular studies enable revaluation of this number? Are we able to say how many species there might be? The concept of species in bacteriology has always been a difficult one. In plants and animals, the presence of distinct morphological differences, sexual reproduction and geographic separation can all be used to explain the concept of species as a group of individuals that can produce fertile offspring and are reproductively isolated from other species. This definition is meaningless for prokaryotes, which are haploid and do not have sexual reproduction. Microbiologists use a combination of phenotypic (e.g. morphology and biochemical reactions) and genetic properties to define prokaryotic species.

A generally accepted 'gold standard' is that two prokaryotes belong to the same species if they show more than 70% DNA–DNA cross-hybridization (*Section 2.6.11*). It is found that this threshold corresponds to a 16S rRNA sequence similarity of 97%. Therefore, a 16S rRNA sequence that differs by more than 3% from all known sequences might be regarded as a new species. However, it is important to note that some organisms with very similar 16S rRNA sequences have very different genomes and would not meet the >70% hybridization standard. A further complication is that the same organism can have multiple copies of 16S rRNA, with distinct sequence differences. Fortunately, scientific protocol requires that publication of experimental work using sequence methods is accompanied by deposition of the gene sequences in databases such as the Ribosome Database Project or GenBank, freely accessible via the Internet. Therefore, with the help of powerful software tools, it is possible to analyze sequences from many independent studies. Such a 'meta-analysis' has recently been compiled by Hagström *et al.* (2002). After searching the database and excluding sequences which were not specifically related to bacteria in marine plankton samples, they identified 1117 unique sequences with 97% similarity (the proposed cut-off point for species). Of these, 508 came from cultured bacteria and 609 came from uncultured bacteria.

The chart below shows the assignment of these sequences into known taxonomic groups. Members of the γ-*Proteobacteria* are the most common bacterioplankton when cultured species are considered, whereas the α-*Proteobacteria* dominate the uncultured species. The major difference between the two sets is the presence of the groups *Planctomycetes* and *Verrucomicrobia,* only found in the uncultured bacterioplankton. Cottrell and Kirchman (2000) have suggested that cloning bias results in similar 16S rRNA sequences being retrieved from many sites while other 16S rRNA genes are missed. For example, this could be due to differences in efficiency of PCR amplification with different primers and the existence of multiple copies of 16S rRNA with slight sequence variations. Future studies employing additional techniques will be needed to produce a definitive view of the abundance of different taxonomic groups. Hagström and coworkers also note that the frequency of publication of new sequences from uncultured bacterioplankton has reached a plateau after its peak in 1996. This concurs with the conclusion by Giovannoni and Rappé (2000) that a small number of clades account for the vast majority of marine bacterial 16S rRNA. Analysis of the diversity of *Archaea* in seawater leads to a similar conclusion that a small number of archaeal types are dominant in the plankton (*Section 6.3.3*).

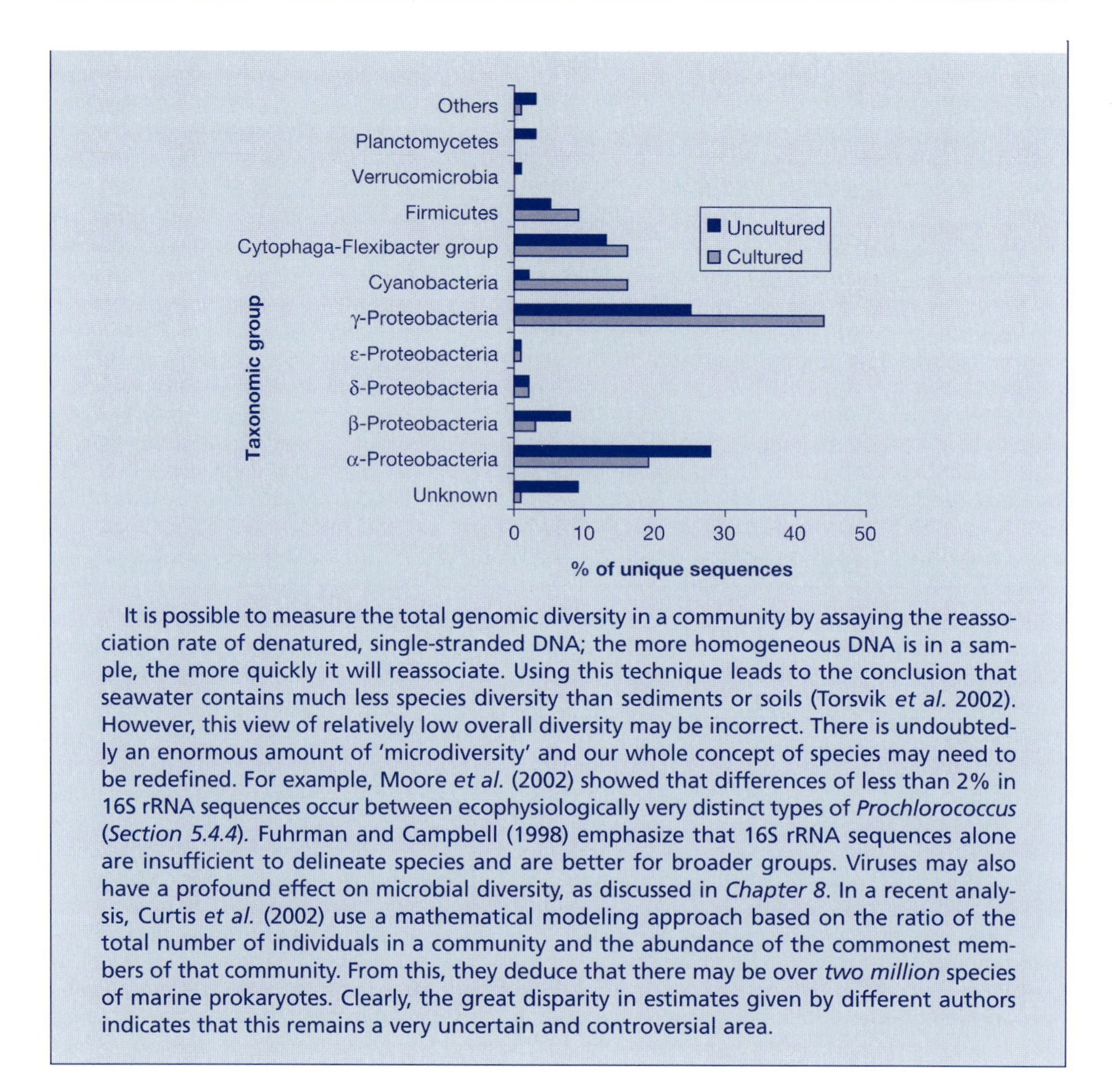

It is possible to measure the total genomic diversity in a community by assaying the reassociation rate of denatured, single-stranded DNA; the more homogeneous DNA is in a sample, the more quickly it will reassociate. Using this technique leads to the conclusion that seawater contains much less species diversity than sediments or soils (Torsvik *et al.* 2002). However, this view of relatively low overall diversity may be incorrect. There is undoubtedly an enormous amount of 'microdiversity' and our whole concept of species may need to be redefined. For example, Moore *et al.* (2002) showed that differences of less than 2% in 16S rRNA sequences occur between ecophysiologically very distinct types of *Prochlorococcus* (*Section 5.4.4*). Fuhrman and Campbell (1998) emphasize that 16S rRNA sequences alone are insufficient to delineate species and are better for broader groups. Viruses may also have a profound effect on microbial diversity, as discussed in *Chapter 8*. In a recent analysis, Curtis *et al.* (2002) use a mathematical modeling approach based on the ratio of the total number of individuals in a community and the abundance of the commonest members of that community. From this, they deduce that there may be over *two million* species of marine prokaryotes. Clearly, the great disparity in estimates given by different authors indicates that this remains a very uncertain and controversial area.

belong to the γ-*Proteobacteria*) are most commonly found in anaerobic sediments of shallow lakes and sulfur springs. However, several types (e.g. *Thiocapsa* and *Ectothiorhodospira*) also occur in shallow marine sediments where anaerobic conditions and high levels of H_2S occur at depths to which sufficient light of the appropriate wavelength can penetrate.

Another group, known as the purple nonsulfur bacteria (all of which are α- or β-*Proteobacteria*) can grow aerobically in the dark using organic compounds or molecular H_2 as electron donor. Many can grow either photoautotrophically using CO_2 and H_2, or photoheterophically using a range of organic compounds as carbon source. Despite their name, some can use low levels of H_2S as a source of electrons. Marine genera include *Rhodospirillum* and *Rhodomicrobium*.

5.3.2 *Roseobacter* and *Erythrobacter*

Aerobic phototrophic bacteria of the α-proteobacterial genera *Erythrobacter* and *Roseobacter* are widely distributed on the surface of marine algae, plants, animals and on suspended particles (marine snow) in coastal and ocean waters. Until recently, it was assumed that anoxygenic bacterial photosynthesis is an anaerobic process. However, the discovery of representatives of a new bacterial group that contain bacteriochlorophyll, yet grow aerobically, has forced a revision of this view (*Box 2.2*). Although they grow aerobically, these types do not produce O_2 from photosynthesis. As noted in *Section 5.2.2*, the *Roseobacter* clade is well-represented in culture-independent surveys. Like the purple bacteria, the aerobic anoxygenic (AAnP) photobacteria contain a range of carotenoid pigments. However, the photosynthetic apparatus is not as well structured and the complex membrane invaginations typical of the anaerobic phototrophs are not seen in the aerobic types. This probably explains why the AAnP bacteria cannot use light as a sole source of energy and rely on various organic compounds as a source of carbon and energy.

It is now known that the *Roseobacter* clade contains both culturable and nonculturable types, as well as phototrophic and nonphototrophic representatives. These are widely distributed in coastal and oceanic plankton and occur in a wide range of associations with other marine organisms. In molecular studies, the *Roseobacter* clade emerges as the second most abundant 16S rRNA gene clone type (over 30% of clones). Their association with blooms of algae and dinoflagellates have been particularly studied, and they may play a role in the formation of dinoflagellate toxins (*Box 12.2*). Despite their obvious ubiquitous presence in ocean water, it is currently difficult to be sure of the role of this group in ecological processes. The diverse metabolic properties of the group undoubtedly play a large part in nutrient cycling. For example, *Roseobacter* has a major role in the breakdown of DMSP leading to the formation of dimethyl sulfide (DMS), which has great significance in the global climate (*Box 8.1*).

5.3.3 Green sulfur bacteria

The group known as green sulfur bacteria forms a separate lineage distinct from the *Proteobacteria,* but resembles the purple sulfur bacteria in metabolism. However, sulfur is produced outside of the cell rather than as intracellular granules. Also, in addition to bacteriochlorophyll *a*, green sulfur bacteria contain bacteriochlorophyll *c, d* or *e*. These pigments are contained in a membrane-bound structure known as the chlorosome. Members of the family *Chromatiaceae* are common on intertidal mudflats and as members of consortia in microbial mats and sediments.

5.4 Oxygenic phototrophs – the *Cyanobacteria*

5.4.1 Nature of the *Cyanobacteria*

The *Cyanobacteria* is a large and diverse group, and members are characterized by their ability to carry out photosynthesis in which O_2 is evolved, although some anoxygenic *Cyanobacteria* have recently been described. *Cyanobacteria* contain chlorophyll *a*, together with accessory photosynthetic pigments called phycobilins. This group was formerly known as the 'blue-green algae' (due to the presence of the blue pigment phycocyanin together with the green chlorophyll) and is still treated as a division of the algae by many marine biologists and phycologists, although *Cyanobacteria* are clearly prokaryotes and form one of the major divisions of the domain *Bacteria*. Furthermore, many marine genera contain phycoerythrins, which give the cells a red-orange, rather than blue-green, color. Fossil evidence, in the form of morphological structures and distinctive biomarkers typical of the group (hopanoids), suggests that organisms resembling *Cyanobacteria* may have evolved about 3 billion years ago and the evolution of O_2 from the photosynthetic activities of *Cyanobacteria* (or their ancestors) was probably responsible for changes in the Earth's early atmosphere. *Cyanobacteria* occupy very diverse habitats in terrestrial and aquatic environments, including extreme temperatures and hypersaline conditions. In the marine environment, habitats include the plankton, sea ice and shallow sediments, as well as microbial mats on the surface of inanimate objects, algae or animal tissue. Some marine isolates require NaCl plus other marine salts for growth in culture, whilst others tolerate a range of salt concentrations.

5.4.2 Morphology and taxonomy

The *Cyanobacteria* are morphologically very diverse, ranging from small undifferentiated rods to large branching filaments showing cellular differentiation. Unicellular *Cyanobacteria* divide by binary fission, whilst some filamentous forms multiply by fragmentation or release of chains of cells. Many types are surrounded by mucilaginous sheaths that bind cells together. The chlorophyll *a* is contained within lamellae called phycobilosomes which are often complex and multi-layered. *Cyanobacteria* show remarkable ability to adapt the arrangement of their photosynthetic membranes and the proportion of phycobilin proteins to maximize their ability to utilize light of different wavelengths. Many *Cyanobacteria* form intracellular gas vesicles (*Section 3.3*), which enable cells to maintain themselves in the photic zone. Gas vesicles, as well as mucilaginous sheaths and pigments, also protect cells from extreme effects of solar radiation (*Section 4.6.5*).

Gliding motility is very important in *Cyanobacteria* that colonize surfaces. Gliding movement, up to 10 μm per second, occurs parallel to the cell's long axis and involves the production of mucilaginous polysaccharide slime. There are two possible mechanisms by which gliding occurs. One is the propagation of waves moving from one end of the filament to the other, created by the contraction of protein fibrils in the cell wall. The other mechanism is secretion of mucus by a row of pores around the septum of the cell. Some types, such as *Nostoc,* are only motile during certain stages of their life cycle, when they produce a gliding dispersal stage known as hormogonia. *Synechococcus* also seems able to swim in liquid media without using flagella.

Until the use of 16S rRNA analysis established *Cyanobacteria* as a group within the *Bacteria,* they were classified by botanists into about 150 genera and 1000 species based on morphological features (*Table 5.1*). Subsequent phylogenetic analysis shows that these groupings are very unreliable and many genera are polyphyletic. Since bacteriologists can now grow many of the *Cyanobacteria* in pure culture, analysis of biochemical characteristics and molecular features are taking over as a basis of classification, and a major revision of the group is in progress. Pure culture studies show that the physiological properties of *Cyanobacteria* are more variable than previously thought. Many are capable of anaerobic growth, some can use H_2S, H_2 or reduced organic compounds as electron donors and some can be photoheterotrophic. However, little is known about the significance of these modes of nutrition in natural marine environments.

5.4.3 Nitrogen fixation

All major groups of marine *Cyanobacteria* contain members which fix atmospheric N_2. *Figure 4.5* shows an outline of the reactions involved in this process. The bond in molecular N_2 is very

Table 5.1 Examples of marine *Cyanobacteria*

Order	Features	Major marine genera
Chroococcales	Unicellular or aggregates of single cells. May be motile	*Prochlorococcus* *Synechococcus* *Synechocystis*
Pleurocapsales	Aggregates of single cells. Reproduce by small spherical gliding cells (baeocytes) formed by multiple fission	*Cyanocystis* *Pleurocapsa*
Oscillatorales	Filamentous cells (trichome), often sheathed. Intercalary binary division at right angles to long axis. Motile	*Trichodesmium* *Lyngba*
Nostocales	Filamentous trichome with heterocysts	*Nostoc*
Stigonematales	Branching clusters, filamentous with heterocysts	Mainly freshwater or terrestrial

stable, and its reduction to ammonia is an extremely energy-demanding process, requiring 16 molecules of ATP for each molecule of N_2 fixed. The key enzyme, nitrogenase, consists of two separate protein components complexed with iron, sulfur and molybdenum. In the marine environment, N_2 fixation is carried out by a wide range of heterotrophic and autotrophic bacteria and is of fundamental significance in primary production in the oceans (*Chapter 9*). Most N_2-fixers are anaerobic, but the *Cyanobacteria* are aerobic and, because the enzyme nitrogenase is highly O_2-sensitive, N_2 fixation is often restricted to the night when no O_2 is generated. Many of the more efficient N_2-fixers contain differentiated cells known as heterocysts within the filament. Because the heterocysts contain no photosytem II, they provide an O_2-free environment which protects the enzyme nitrogenase. However, one of the most prolific marine N_2-fixers is *Trichodesmium*, which does not contain heterocysts. Recently it has been shown that *Trichodesmium* is able to switch the two processes of the O_2-producing photosynthetic system and the O_2-sensitive N_2-fixing system on and off over timescales of a few minutes. There also appears to be a spatial separation of O_2 evolution and N_2 fixation, because N_2 fixation is limited to certain parts of the cell. *Trichodesmium* forms dense filamentous masses which are responsible for large blooms, especially in tropical seas.

5.4.4 ***Prochlorococcus* and *Synechococcus***

Although many types of *Cyanobacteria* are found in marine environments, two genera dominate the picoplankton in large areas of the Earth's oceans, namely *Synechococcus* and *Prochlorococcus*. These organisms are major contributors to the carbon cycle through photosynthetic CO_2 fixation, accounting for between 15 and 40% of carbon input to ocean food webs. *Prochlorococcus* is a very small (about 0.6 μm diameter) cyanobacterium which was not discovered until 1988 (following the use of FCM), despite the fact that it inhabits large parts of the oceans at a density between 10^5 and $10^6\,ml^{-1}$, making it the most abundant photosynthetic organism on Earth. *Prochlorococcus* is most abundant in the region from 40°S to 40°N, temperature range 10°C to 33°C, to a depth of about 200 m. *Prochlorococcus* contains modified forms of chlorophyll (divinyl chlorophylls *a* and *b*), but lacks phycobilins. The photosynthetic apparatus seems to be adapted to allow *Prochlorococcus* to grow at considerable depths, where the amount of light is very low (below 1% that at the surface). The small cell size gives a large SA:V ratio, which helps *Prochlorococcus* obtain scarce nutrients in oligotrophic ocean waters. Other organisms with similar properties include *Prochloron*, which is an intracellular symbiont of certain marine invertebrates. These organisms were originally placed in a group called the prochlorophytes, but phylogenetic analysis shows that this is not a distinct lineage within the *Cyanobacteria*. Prochlorophytes appear to have evolved divinyl chlorophylls *a* and *b*, which allows them to harvest longer wavelengths of blue light, which penetrate deeper waters. Indeed, studies on *Prochlorococcus* cultures and populations from the field have shown that there are two distinct populations of ecotypes, which occupy different light niches. A high-light-adapted ecotype dominates the top 100 m of water, which is characterized by high light flux and very low nutrient concentrations. The second ecotype thrives at depths of 80–200 m, which have low light intensity but higher nutrient concentrations. The two ecotypes have different ratios of chlorophyll a_2 to b_2 and differ in their optimal irradiances for photosynthesis. These significant differences in ecotype are determined by genetic differences of only about 2% in 16S rRNA sequences. They should probably be assigned to separate species, because the two types differ markedly in the number of genes encoding the light-harvesting complex. (See *Box 5.1* for a discussion of the difficulties of species definition based on 16S rRNA studies.)

The discovery of *Prochlorococcus* is also highly significant for evolutionary theory. It has been thought for many years that the chloroplasts present today in algae and plants evolved from *Cyanobacteria* in accordance with the endosymbiosis theory. Phylogenetic analyses suggest that prochlorophytes, despite resembling chloroplasts without phycobilins, are not the immediate ancestral origin of the chloroplast. It is possible that the prochlorophytes and the rest of the

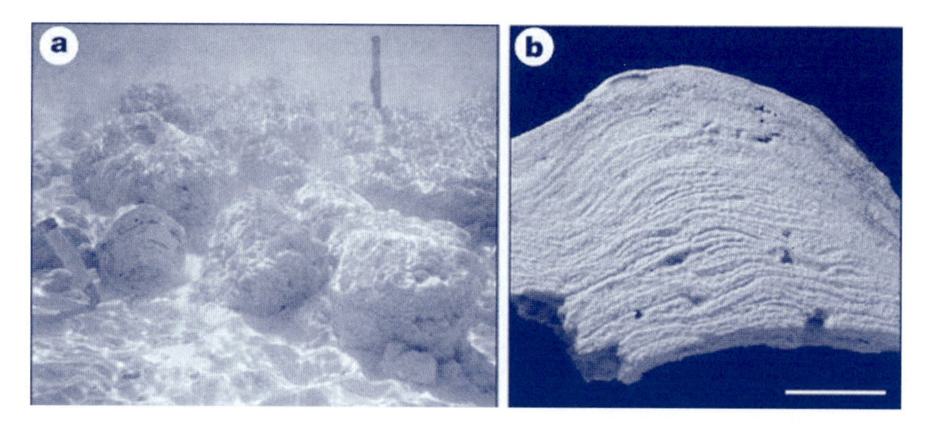

Figure 5.3

Stromatolites. (a) Columnar build-ups in shallow water, Highborne Cay, Bahamas. (b) Vertical section showing lamination; scale bar, 2 cm. From Reid *et al.* (2000), reproduced with permission, Nature-Macmillan Ltd.

Cyanobacteria may have evolved from ancestors that contained phycobilins and more than one type of chlorophyll. Prochlorophytes may have lost their phycobilins and the other *Cyanobacteria* may have lost their chlorophyll *b* during evolution, whilst the eukaryotic chloroplast evolved from the hypothetical ancestor of both groups. We do not know when this divergence occurred.

5.4.5 Microbial mats and stromatolites

Cyanobacteria are especially important in the formation of microbial mats in shallow water. Complex stratified communities of microorganisms develop at interfaces between sediments and the overlying water. Filamentous *Cyanobacteria* such as *Phormidium, Oscillatoria* and *Lyngbya* are often dominant members of the biofilm in association with unicellular types such as *Synechococcus* and *Synechocystis*. Steep concentration gradients of light, O_2, H_2S and other chemicals develop across the biofilm. The mat becomes anoxic at night and H_2S concentrations rise. *Cyanobacteria* (and other motile bacteria in the biofilm) can migrate through the mat to find optimal conditions. Anoxygenic phototrophs as well as aerobic and anaerobic chemoheterotrophs are also present.

Stromatolites are fossilized microbial mats of filamentous prokaryotes and trapped sediment. These ancient structures were widespread in shallow marine seas over three billion years ago. Ancient stromatolites were probably formed by anoxygenic phototrophs, but modern stromatolites are dominated by a mixed community of *Cyanobacteria* and heterotrophic bacteria. Growth of modern marine stromatolites represents a dynamic balance between sedimentation and intermittent lithification of cyanobacterial mats (*Figure 5.3*). Rapid sediment accretion occurs when the stromatolite surfaces are dominated by pioneer communities of gliding filamentous *Cyanobacteria*. During intermittent periods, surface films of exopolymer are decomposed by heterotrophic bacteria, forming thin crusts of microcrystalline calcium carbonate. Other types of *Cyanobacteria* modify the sediment, forming thicker stony plates.

5.5 The nitrifying bacteria

This term describes bacteria that grow using reduced inorganic nitrogen compounds as electron donors. Marine examples, which are present in suspended particles and in the upper layers of sediments, include *Nitrosomonas* and *Nitrosococcus* (which oxidize ammonia to nitrite) and

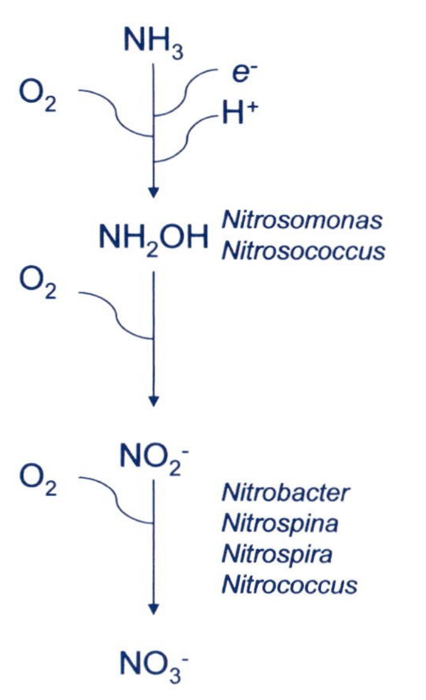

Figure 5.4

Oxidation of ammonia by nitrifying bacteria. Examples of marine genera known to carry out these processes are shown. These reactions are energetically unfavorable and oxidation of 35 ammonia molecules or 15 nitrite molecules is required to produce fixation of one molecule of carbon dioxide.

Nitrosobacter, Nitrobacter and *Nitrococcus* (which oxidize nitrite to nitrate). No organisms that can carry out both reactions are known. The ammonia-oxidizers are obligate chemolithoautotrophs and fix carbon via the Calvin cycle. The nitrite-oxidizers are usually chemolithoautotrophic, but can be mixotrophic using simple organic compounds heterotrophically. Because of these activities, nitrifying bacteria play a major role in nitrogen cycling in the oceans, especially in shallow coastal sediments and beneath upwelling areas such as the Peruvian coast and the Arabian Sea. Previously, nitrifying bacteria were classified mainly on morphological characteristics. However, 16S rRNA analysis shows that they occur in several branches of the *Proteobacteria*, and one type, *Nitrospira*, forms a distinct bacterial phylum. Members of the γ subdivision have been found only in marine environments. Like the phototrophs, nitrifying bacteria have extensive internal structures in order to increase the surface area of the membrane.

It is difficult to obtain estimates of the abundance and community structure of nitrifying bacteria. Although most can be cultivated in the laboratory, the energetics of this mode of chemolithotrophy mean that the bacteria grow slowly and are difficult to work with. Immunofluorescence methods (*Section 2.4*) reveal that *Nitrosococcus oceani* and similar strains are widespread in many marine environments, with worldwide distribution, at concentrations between 10^3 and 10^4 cells ml^{-1}. This organism is thought to be responsible for significant oxidation of ammonia in the open ocean. *Nitrospira* also appears to be distributed worldwide. Study of their activities and contribution to nitrogen cycling is usually carried out using isotopic methods with $^{15}NO_3^-$ or $^{15}NH_4^+$ (*Section 2.7.3*) or by using various inhibitors of nitrification enzymes (e.g. nitrapyrin inhibits ammonia monoxygenase). Nitrification is a strictly aerobic process and sufficient O_2 usually only penetrates a few millimeters into sediments. Activity of burrowing worms can increase O_2 availability to deeper levels of sediments. Nitrification rates are high in waters where plant photosynthesis releases O_2 and the release of nitrate stimulates plant growth. This is of great importance in the productivity of seagrass beds. The overall reactions for oxidation of ammonia to nitrite and its subsequent oxidation to nitrate are shown in *Figure 5.4*.

5.6 Sulfur- and iron-oxidizing chemolithotrophs

5.6.1 *Thiobacillus, Beggiatoa, Thiothrix,* and *Thiovulum*

A wide range of *Proteobacteria* can grow chemolithotrophically using reduced sulfur compounds as a source of electrons, leading to the formation of sulfate. The rod-shaped *Thiobacillus* is the

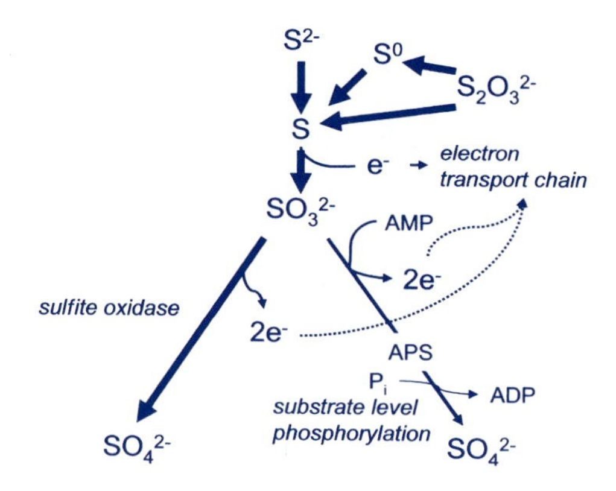

Figure 5.5

Oxidation of reduced sulfur compounds by chemolithotrophs. Sulfite is most commonly oxidized with the enzyme sulfite oxidase (left side of diagram) resulting in the direct generation of ATP via the electron transport chain and a proton motive force generated across the membrane that leads to ATP synthesis by ATPase. Some sulfur-oxidizing bacteria use the enzyme adenosine phosphosulfate (APS) reductase (right side).

best-known genus, using H_2S, elemental sulfur or thiosulfate as electron donors. Filamentous bacteria such as *Beggiatoa, Thiothrix,* and *Thiovulum* are also well represented in the marine environment. These bacteria are usually strict aerobes found in the top few millimeters of marine sediments which are rich in sulfur. They frequently show chemotaxis to seek out the desired gradient of O_2 and sulfur compounds. They are also very prominent at hydrothermal vents and cold seeps, both as free-living forms and as symbionts of animals (see *Section 10.3*), where they form the base of the food chain. *Beggiatoa* and other filamentous forms commonly show gliding motility, and become intertwined to form dense microbial mats, often with a complex community structure containing sulfate-reducers and phototrophs. Although *Beggiatoa* obtains energy from the oxidation of inorganic sulfur compounds, it does not possess the enzymes needed for autotrophic fixation of CO_2 and therefore uses a wide range of organic compounds as carbon source. An outline of the biochemical processes in the oxidation of sulfide is shown in *Figure 5.5.*

5.6.2 *Thioploca* and *Thiomargarita*

These genera are filamentous sulfur-oxidizing chemolithotrophs whose importance in the oxidation of sulfide in anaerobic sediments has only recently been discovered. *Thioploca* spp. are multicellular filamentous bacteria which occur in bundles surrounded by a common sheath and contain granules of elemental sulfur. *Thioploca* is one of the largest bacteria known, with cell diameters from 15 to 40 μm and filaments many cm long, containing thousands of cells. Several species, including *T. chilaee, T. araucae* and *T. marina,* have been described. In the late 1990s huge communities of *Thioploca* spp. were discovered along the Pacific coast of South America where upwelling creates areas of nitrate-rich water, with bottom waters becoming anoxic. Blooms of *Thioploca* can be very dense, up to 1 kg wet weight m^{-2}. Anoxic reduction of H_2S is coupled to the reduction of nitrate. Each cell contains a very thin layer of cytoplasm around the periphery and a liquid vacuole that constitutes 80% of the cell volume. The vacuole stores very high concentrations of nitrate, which is used as an electron acceptor for sulfide oxidation. The bacteria can grow autotrophically or mixotrophically using organic molecules as carbon source. The filaments stretch up into the overlying seawater, from which they take up nitrate, and then glide down 5–15 cm deep into the sediment through their sheaths to oxidize sulfide formed by intensive sulfate reduction. *Thiomargarita namibiensis* was only discovered in 1999 and holds the current

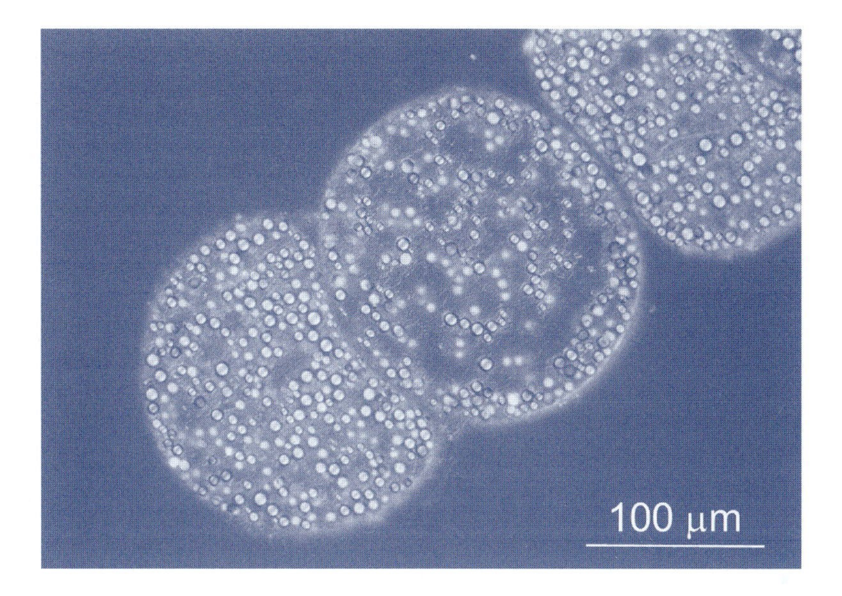

Figure 5.6

Thiomargarita namibiensis. The pearl-like inclusions are globules of sulfur. Image courtesy of Heide Schulz, Max Planck Institute for Marine Microbiology, Bremen, Germany.

record as the largest prokaryote known. The spherical cells are normally 100–300 μm wide, but some reach diameters of 750 μm (*Figure 5.6*). They occur in large numbers in coastal sediments off Namibia, and occur in filaments with a common mucus sheath. Microscopic granules of sulfur reflect incident light and the name derives from their resemblance to a string of pearls. The hydrographic conditions off this coast bring large quantities of nutrients to the surface. Massive phytoplankton growth results in settlement of organic material to the seabed where it is degraded by bacteria, forming large amounts of H_2S. *Thiomargarita* oxidizes sulfide using nitrate and, like *Thioploca*, the interior of the cell is filled with a large vacuole. The nitrate stored in the vacuole and the sulfur stored in the peripheral cytoplasm act as nutrient reserves that allow these bacteria to grow for several months in the absence of external nutrients.

5.7 Hydrogen-oxidizing bacteria

Many bacteria are capable of using molecular H_2 as the electron donor and O_2 as the electron acceptor. Examples found in marine habitats include the *Proteobacteria Alcaligenes, Pseudomonas* and *Ralstonia*. Most of these bacteria can fix CO_2 to grow autotrophically, although they are usually capable of growing heterotrophically with reduced organic compounds. H_2-oxidizing bacteria are typically associated with sediments and suspended particles where a reduced O_2 concentration (less than 10%) provides the optimum conditions for growth.

Species in a major division of the *Bacteria*, the *Aquifex* group, oxidize H_2 chemoautotrophically and are described in *Section 5.22*.

5.8 Aerobic methanotrophs and methylotrophs

These physiological groups are widespread in coastal and oceanic habitats, especially in the top layers of marine sediments, where they utilize methane produced by anaerobic methanogenic *Archaea* (see *Section 6.2.1*). Methylotrophs are able to use various one-carbon (C_1) compounds both as a source of carbon and as an electron donor. Of these, DMSP is the most significant because of its importance in global processes. A very wide range of bacteria in different phylogenetic groups can carry out this process, including common heterotrophs such as *Vibrio* and *Pseudomonas*. However, some bacteria in the α- and γ-*Proteobacteria* are obligate methylotrophs and use only C_1 compounds in their metabolism. A subset known as the methanotrophs can grow only on methane and a few other simple C_1 compounds. The methanotrophs possess a unique copper-complexed enzyme, methane monoxygenase, which leads to the formation of methanol in the reaction

$$CH_4 + O_2 + XH_2 \longrightarrow CH_3OH + H_2O + X$$
(where X is a reduced cytochrome)

Methanol is subsequently converted to formaldehyde by methanol dehydrogenase in the reaction

$$CH_3OH \longrightarrow HCHO + 2e^- + 2H^+$$

The membranes of methanotrophs contain sterols, a feature that is very unusual in prokaryotes and shown in only one other group, the mycoplasmas. Methanotrophs such as *Methylomonas, Methylobacter* and *Methylococcus* (γ-*Proteobacteria*) contain intracytoplasmic vesicles and utilize the unique ribulose monophosphate pathway for carbon fixation, whereas genera such as *Methylosinus* and *Methylocystis* (α-*Proteobacteria*) contain membranes running around the periphery of the cell and utilize the serine pathway for carbon assimilation. Apart from the importance of free-living forms in methane oxidation, methanotrophs also occur as symbionts of mussels found near 'cold seeps' of methane-rich material on the ocean floor, providing the animals with a direct source of nutrition (*Section 10.3.3*). Methanotrophs are also important in bioremediation of low-molecular-weight halogenated compounds, an important process in contaminated marine sediments.

5.9 *Pseudomonas, Alteromonas* and *Shewanella*

This is a heterogeneous group of chemoorganotrophic, aerobic rod-shaped *Proteobacteria*. Whilst *Pseudomonas* (normally found in soils, plant material and as a human pathogen) can be isolated from coastal waters, it is probably not an indigenous marine organism. However, many salt-requiring organisms with some similar properties and relatedness to *Pseudomonas* can be isolated from coastal and ocean seawater, and in association with marine plants and animals. Of these, *Alteromonas* and *Shewanella* are probably the best known. As with other groups, 16S rRNA studies are leading to large-scale reclassification of these genera and they are probably members of the large clade that includes the vibrios and enteric bacteria.

Alteromonas spp. are frequently isolated on marine agar and are often distinguished by brightly colored colonies due to the production of various pigments. Because of their dominance in culture-based surveys, it is assumed that they play a major role in heterotrophic nutrient cycling. However, *Alteromonas* spp. are not as well represented in molecular-based surveys, so it is difficult to determine their ecological importance.

Shewanella spp. are frequently isolated from the surfaces of marine algae, shellfish, fish and marine sediments. Some are extreme barophiles. *Shewanella* spp. show great metabolic versatility and can use a wide range of compounds, including Fe^{2+}, as electron acceptors. Some are important in the spoilage of fish.

5.10 Free-living aerobic nitrogen-fixing bacteria

N_2 fixation by *Azotobacter* (a member of the γ-*Proteobacteria*) is especially important in estuarine, intertidal, sea grass and salt marsh sediments. Apart from the *Cyanobacteria, Azotobacter* is one of the few marine aerobic N_2-fixers. The O_2-sensitive nitrogenase is protected because of the very high respiratory rate of *Azotobacter* and the presence of a slimy capsule. Nitrogenase is also complexed with a protective protein. *Azotobacter* is heterotrophic, and can use a wide range of carbohydrates, alcohols and organic acids as a growth substrate. In soil, *Azotobacter* forms cysts, a resting stage with reduced metabolic activity and resistance to adverse environmental factors, but it is not clear if cysts are formed in marine habitats.

5.11 The Enterobacteriaceae

The Enterobacteriaceae is a large and well-defined family of γ-*Proteobacteria*. They are best known (hence the name) as commensals and pathogens in the gut of warm-blooded animals, and include genera such as *Escherichia, Salmonella, Serratia* and *Enterobacter.* They are fermentative, facultatively anaerobic bacteria and are oxidase-negative and Gram-negative rods, usually motile by means of peritrichous flagella. These properties help to distinguish them from other Gram-negative rods such as *Vibrio, Pseudomonas* or *Alteromonas*. Enterobacteria can be isolated from coastal waters that are polluted from terrestrial sources and they are found in the gut of fish and marine mammals. Apart from these specific cases, enterobacteria would not be regarded as indigenous marine organisms and their main importance for the purpose of discussion in this book is as indicators of fecal pollution (*Section 11.3.3*).

5.12 *Vibrio* and related genera

5.12.1 *Vibrio, Photobacterium, Aeromonas* and related genera

Members of the family *Vibrionaceae* (γ-*Proteobacteria*), of which the principal marine examples are *Vibrio* and *Photobacterium,* have worldwide distribution in coastal and ocean water and sediments. They are oxidase-positive and facultatively anaerobic. Typically, they form curved rods with sheathed polar flagella, although differentiation to lateral flagella may occur during biofilm formation (*Section 3.9.1*). They are typically associated with the surfaces of many marine animals and plants and suspended organic matter. They play a major role in initial colonization of surfaces and biofilm formation and as symbionts and pathogens, as summarized in *Table 5.2*. Taxonomy of this group is frequently revised; currently, nearly 50 species of *Vibrio* are recognized following the application of genomic fingerprinting and DNA hybridization methods. Other genera of predominantly marine types include *Allomonas, Listonella, Enhydrobacter,*

Table 5.2 Principal species of *Vibrio* and *Photobacterium* and their interactions with marine animals

Light organ symbionts (*Chapter 9*)
V. fischeri
P. leiognathi
P. phosphoreum
Human pathogens (*Chapter 11*)
V. cholerae
V. parahaemolyticus
V. vulnificus
V. alginolyticus
V. mimicus
V. cincinnatiensis
V. hollisae
V. fluvialis
V. furnissii
V. harveyi
Coral pathogens (*Chapter 15*)
V. shiloi (=*V. mediterranei*)
V. corallilyticus
V. carchariae (=*V. harveyi?*)
Fish pathogens (*Chapter 14*)
P. damselae
P. damselae subsp. *piscicida*
V. anguillarum
V. ichthyoenteri
V. salmonicida
V. splendidus
V. wodanis
Mollusc pathogens (*Chapter 15*)
V. pectenicida
V. tapetis
V. tubiashii
Crustacean pathogens (*Chapter 15*)
V. harveyi
V. penaeicida
V. proteolyticus
V. parahaemolyticus

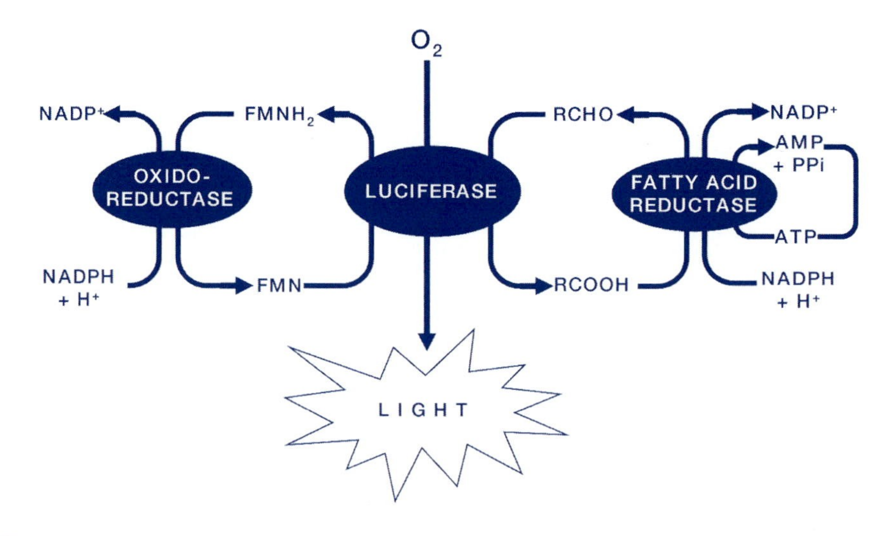

Figure 5.7

Reactions in bacterial bioluminescence. The *luxAB* genes encode the α and β subunits of luciferase. The *luxCDE* genes code for the transferase, synthetase and reductase enzymes used for the formation of the long-chain aldehyde substrate (e.g. tetradecanal) from fatty acids.

Salinivibrio and *Enterovibrio*. The genus *Aeromonas* is usually regarded as a freshwater organism. However, *Aeromonas salmonicida* is a major pathogen of farmed salmon and trout in both freshwater and marine systems (*Section 14.3.4*).

5.12.2 Bioluminescence

Luminous bacteria are very common in the marine environment, and occur as free-living forms in seawater, on organic debris, as commensals in the gut of many marine animals and as light organ symbionts (*Section 10.4*). Some light organ symbionts have never been cultured, but all of the marine bioluminescent bacteria that can be isolated and cultured are members of the family *Vibrionaceae*; the commonest types being *Photobacterium phosphoreum, P. leiognathi, Vibrio fischeri* and *V. harveyii*. (There has been disagreement about the taxonomic position of *V. fischeri,* which some authors designate as *P. fischeri*.)

The reaction mechanism for bacterial bioluminescence is shown in *Figure 5.7*. The enzyme responsible, luciferase, is a mixed-function oxidase that simultaneously catalyzes the oxidation of reduced flavin mononucleotide ($FMNH_2$) and a long chain aliphatic aldehyde (RCHO) such as tetradecanal. Blue-green light with a wavelength of about 490 nm is emitted because of the generation of an intermediate molecule in an electronically excited state. Luciferases from all bioluminescent bacteria are dimers of α (~40 kDa) and β (~35 kDa) subunits, encoded by the *luxA* and *luxB* genes that occur adjacently in the *lux* operon, along with other genes encoding enzymes leading to the synthesis of the aldehyde substrate via fatty acid precursors. The overall process is dependent on ATP and NADPH. The *lux* operon genes from several vibrios have been cloned and appear to have very similar structures. The α- and β-subunits of strains show about 30% identity in amino acid sequences and the β-subunit probably arose by gene duplication. Recombinant *lux* gene technology is widely used as a reporter system for monitoring gene expression, with important biotechnological applications. Inhibition of bioluminescence in *V. fischeri* is also used in a proprietary test for measuring environmental pollution (*Section 16.4*). Why bacteria such as vibrios are bioluminescent has always been difficult to explain, and new ideas about its function are discussed in *Box 5.2*.

Box 5.2 RESEARCH FOCUS

Let there be light ... but why?

Study of mutants leads to new ideas about why bacteria bioluminescence

The evolution of bioluminescence (*Section 5.12.2*) in bacteria is something of an enigma, as it is difficult to explain its biological role. How did this process, which can consume up to 20% of the cell's energy, evolve? What benefits do bacteria derive from emitting light? The discovery of sophisticated mechanisms of regulation of bioluminescence by density-dependent QS adds a further complication to any attempt to explain evolution of the process. It is tempting to say that QS 'makes sense' from the bacterium's point of view, as there is clearly no advantage in single or well-isolated bacterial cells initiating the energetically expensive process of light emission because the amount of light will be too small. In nature, bioluminescence will only occur when the bacteria are numerous enough to be seen, such as in a symbiotic light organ or on a particle. It could be that luminescent bacteria are not truly free-living, as they usually seem to be associated with particulate aggregates or as commensals or symbionts of animals. Bacteria will be shed from animals in feces or exudates and these aggregate to form marine snow particles. Bacteria growing in these particles could cause them to glow and make them more attractive as food items for animals, thus aiding transmission of the bacteria between hosts. The ecological benefits for an animal host harboring bioluminescent symbionts are obvious (*Section 10.5*) and selection pressure over many millennia must also have been a major evolutionary force in the development of such a complex process. Could other factors also be at work?

Recent studies suggest that bioluminescence may promote DNA repair. UV light causes the formation of pyrimidine dimers that prevent replication of the DNA. The enzyme DNA photolyase binds to the dimers and excises them in the presence of blue light (photoreactivation) and other enzymes restore the damaged segment of DNA. Induction of the DNA repair process is initiated by a complex regulatory system called SOS. Czyz *et al.* (2000) found that random mutagenesis of *Vibrio harveyi* led to many UV-sensitive mutants that were also unable to emit light. When UV-irradiated *V. harveyi* cells were incubated in the light, more cells survived than when they were grown in the dark. Czyz *et al.* found that when *luxA* or *luxB* mutants were cultivated in the dark after UV-irradiation, their survival was greatly reduced. Transfer of the *lux* operon from *V. harveyi* to *Escherichia coli* by cloning gave some protection to *E. coli* from the lethal effects of UV when cells were subsequently incubated in the dark. When wild-type (bioluminescent) bacteria and *lux*$^-$ mutants were grown in mixed cultures, the *lux*$^-$ mutants dominated the culture after a few days. However, when cultures were subject to low UV-irradiation, the wild-type bacteria predominated. In the absence of UV, cells possessing the *lux* genes reproduced a little more slowly (perhaps because of the additional genetic burden). However, under the selection pressure of UV-irradiation, possession of the *lux* genes is advantageous. Czyz *et al.* suggest that the cell's own light emission stimulates DNA repair.

Other research suggests that bioluminescence may protect bacteria against the toxic effects of O_2. As well as bacteria, bioluminescence is present in some fungi and diverse groups of animals. It appears to have multiple evolutionary origins because there is a very wide diversity of enzymes and substrates for the bioluminescent reaction mechanism. Apart from light emission, the only common feature is the requirement for oxygen. Rees *et al.* (1998) argue that bioluminescent reactions evolved primarily as a mechanism for detoxification of the highly toxic derivatives of molecular O_2 (singlet oxygen, superoxide anion, hydrogen peroxide and hydroxyl radical, see *Section 4.6.4*), with the initial evolutionary drive being the nature of the substrates rather than the luciferase enzymes. Czyz and Wegrzyn (2001) found that *luxA* and *luxB* mutants (which do not make functional luciferase) were more sensitive to hydrogen peroxide than the wild-type, whereas *luxD* mutants (which make luciferase but not the acyltransferase enzyme needed for fatty acid substrate synthesis) were not. This suggests that the luciferase enzyme may play a role in detoxification of hydrogen peroxide. In contrast to this hypothesis, Ruby and McFall-Ngai (1999) propose that bacterial luciferase may have evolved from a simpler reaction, which does not produce light but generates superoxide. These authors argue that the principal function of luciferases is the generation of superoxide leading to damage of host tissues and nutrient release, and that this is a common feature of many symbiotic and pathogenic bacteria. This idea is considered further in *Chapter 10.*

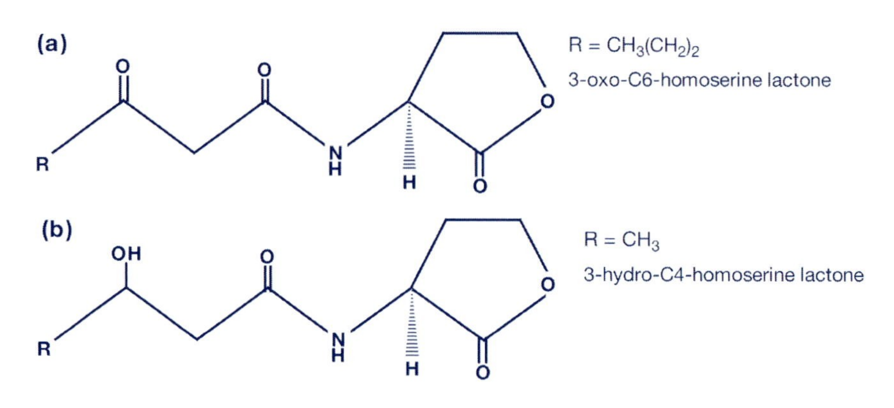

Figure 5.8

Homoserine lactone autoinducers from (a) *Vibrio fischeri* and (b) *V. harveyi*. A wide range of variations in structure of acyl homoserine lactones have been described in many different Gram-negative bacteria.

5.12.3 Regulation of bioluminescence

When grown in laboratory broth cultures, vibrios emit no light until the bacteria enter the late logarithmic or stationary phase and the population reaches a certain critical density (typically about 10^7 cells ml^{-1}). This is because a freely diffusible autoinducer molecule is synthesized by the bacteria and released into the medium. Low-density cultures can be induced to show bioluminescence by the addition of supernatants from high-density cultures. This mechanism of intercellular signaling between bacteria is called quorum sensing (QS). It was first discovered in *V. fischeri,* where the vibrio autoinducer molecule (VAI), encoded by the gene *luxI*, is an N-acyl homoserine lactone (AHL, *Figure 5.8*) synthesized from S-adenosylmethionine and an acyl–acyl carrier protein. When a certain threshold concentration is reached, the bioluminescence genes are expressed by activation of transcription, as shown in *Figure 5.9*. Positive regulation of the *lux* operon involves the product of another gene, *luxR*. LuxR is a polypeptide of about 250 amino acids comprising two domains, one of which (N-terminal) binds the autoinducer and the other (C-terminal) which activates transcription after binding to a palindromic sequence (*lux* box) upstream of the *lux* operon promoter.

This description of the QS mechanism in *V. fischeri* is somewhat simplified, as we now know that other regulatory factors are involved. Bioluminescence is also subject to catabolite repression (probably of *lux*R), because mutants defective in the cyclic AMP receptor protein produce very little light. Elucidation of the regulation of bioluminescence in another marine vibrio, *V. harveyi,* has revealed two separate systems. Like *V. fischeri, V. harveyi* synthesizes and responds to an AHL molecule (in this case termed AI-1, synthesized by the *luxL* and *luxM* genes). However, it also possesses a separate autoinducer (AI-2, synthesized by the *luxS* gene) which is not an AHL. The two autoinducers are recognized by sensor kinase proteins LuxN and LuxQ which have a

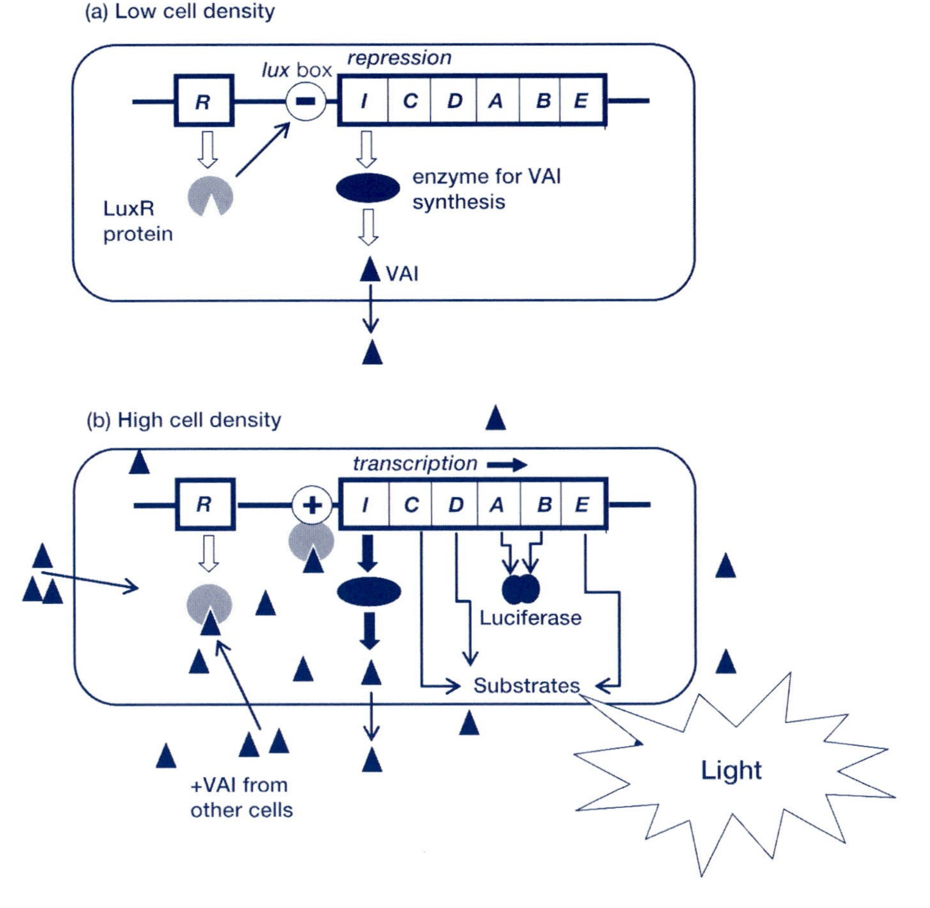

Figure 5.9

Quorum sensing in *Vibrio fischeri*. Bioluminescence is controlled by two regulatory genes *luxI* and *luxR* in two different operons. The *luxI* gene encodes the enzyme responsible for synthesis of the vibrio autoinducer (VAI). The upper diagram (a) represents a cell at low cell density, in which there is a low constitutive transcription of *luxI* and *luxR* (open arrows). As cell density increases, VAI accumulates in the area surrounding the cells. LuxR is a repressor of the *lux* operon promoter (−). When LuxR binds the autoinducer VAI, it binds strongly to the *lux* box upstream of the promoter and transcription of the right operon is enhanced (+). Production of VAI (solid arrows) increases exponentially, giving an autocatalytic feedback loop as well as initiation of bioluminescence. Genes *luxA* and *luxB* encode the α and β subunits of luciferase. Genes *luxC*, *luxD* and *luxE* encode enzymes for the synthesis of the aldehyde substrates. The LuxR–VAI complex also binds at the *luxR* promoter, but in this case it represses transcription, resulting in a compensatory negative feedback.

histidine kinase domain and a response regulator domain. The 'signal' is thus transduced by a phosphorylation mechanism to protein LuxU and on to LuxO, which is a negative regulator of the *luxCDABEGH* operon. Phosphorylation–dephosphorylation of LuxO determines the expression of the structural genes for bioluminescence. Therefore, these two integrated systems detect high cell density leading to the synthesis of luciferase and other enzymes needed for light production, as summarized in *Figure 5.10*.

Following its discovery and detailed investigation in marine vibrios, QS has emerged as one of the most important mechanisms of gene regulation in bacteria, and application of this knowledge extends across the whole field of microbiology (see *Box 16.1*).

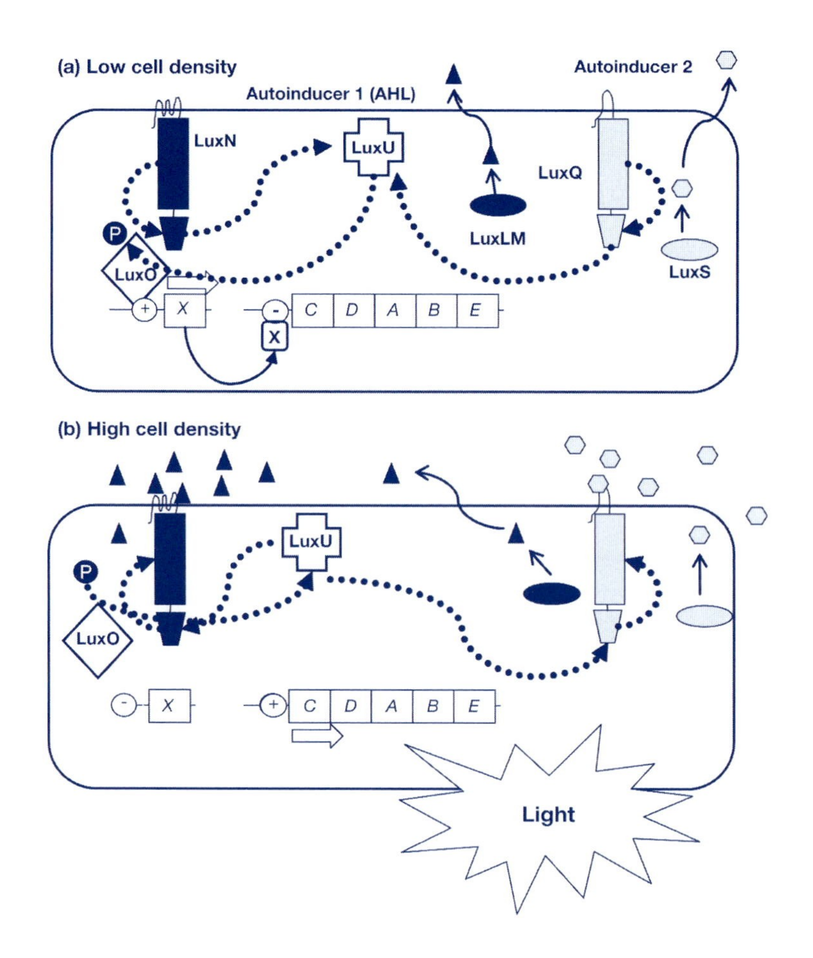

Figure 5.10

Quorum sensing in *Vibrio harveyi*. There are two separate QS systems linked by a phosphorelay system (dotted arrows). Signalling occurs via the two-component proteins LuxN and LuxQ which contain a sensor kinase domain (rectangle) and response regulator domain (trapezoid). Signals from both sensors are passed to the shared integrator LuxU protein (cross) to the LuxO protein (diamond). The upper diagram (a) represents a cell in a low density population. When phosphorylated (denoted by P), protein LuxO activates an unidentified repressor (shown as ([X]) of the *luxCDABE* operon so that the luciferase enzyme subunits and reaction substrates are not expressed. The lower diagram (b) illustrates a cell in a high-density population and shows the accumulation of the autoinducers (AI-1, AI-2) synthesized by proteins LuxLM and LuxS. These bind to the sensor proteins (in the case of AI-2 via a periplasmic binding protein) and switch the activities of the sensor kinases into phosphatases. LuxO is dephosphorylated so that repression of the *lucCDABE* operon is relieved.

5.13 Rickettsias

The rickettsias are small Gram-negative obligate intracellular parasites of animal cells. Rickettsias actively penetrate host cells and multiply within the cytoplasm before causing the host cell to lyse. *Piscirickettsia salmonis* is a major pathogen of salmon (*Section 14.3.5*) and several rickettsias have been isolated from diseased prawns in aquaculture (*Section 15.3.3*). However, this group has been poorly studied in marine environments, and is probably much more widespread as a pathogen of marine animals than currently realized.

Table 5.3 Properties of *Oceanospirillum* and related genera

Genus	Representative species	Characteristic properties[1]	Habitat/important properties
Oceanospirillum	*O. beijerincka* *O. maris* *O. linum* *O. multiglobuliferum*	Helical; bipolar flagellar tufts; OGT 25–32°C; ONC 0.5–8%; %GC 45–50	Found in gut of shellfish, coastal seawater, seaweed
Marinospirillum	*M. minutulum* *M. megaterium*	Helical; polar or bipolar flagellar tufts; OGT 15–25°C; ONC 2–3%; %GC 42–45	Found in nutrient-rich environments such as fish and shellfish guts
Alcanivorax	*A. borkumenis* *A. jadensis*	Rods; nonmotile; OGT 20–30°C; ONC 3–10%; %GC 53–64	Common in seawater or sediments contaminated with oil. Degrade n-alkanes
Marinobacter	*M. hydrocarbono-clasticus* *M. aquaeoli*	Rods; single polar flagellum (or nonmotile); OGT 30–32°C; ONC 3–6%; %GC 56–58	Isolated from oil-contaminated environments, hydrothermal vents, sulfide-rich sediments. Degrade n-alkanes
Neptunomonas	*N. napthovorans*	Rods; %GC 46	Isolated from oil-contaminated sediments. Degrade PAHs.
Marinomonas	*M. communis* *M. mediterranea*	Rods; single or bipolar flagella; OGT 20–25°C; ONC 0.7–3.5%; %GC 46–49	Isolated in open ocean and coastal seawater. Produces polyphenol oxidase (melanin biosynthesis)
Marinobacterium	*M. georgiense* *M. stanieri* *M. jannaschii*	Rods; single flagellum; OGT 37°C; ONC 0.6–2.9%; %GC = 55	Isolated from nutrient-rich coastal seawater. Degrades DMSP and other sulfur compounds

[1] Abbreviations: OGT = optimum growth temperature; ONC = optimum NaCl concentration for growth; %GC = DNA mol % Guanine + Cytosine ratio; PAHs = polyaromatic hydrocarbons; DMSP = dimethylsulfoniopropionate.

5.14 Spirilla

5.14.1 *Oceanospirillum* and related genera

As the name suggests, spirilla characteristically have spiral-shaped cells, but apart from this feature, they are very diverse in their physiology and ecology. There are many marine species of the genus *Oceanospirillum* and related genera. This group appears to consist of uniquely marine types, as reflected in the names given to the various genera shown in *Table 5.3,* with many of the species named in honor of famous microbial ecologists. Spirilla were originally distinguished from other cultivable marine bacteria by their characteristic shape, but electron microscopy of concentrated seawater reveals that a spiral shape is common among marine bacteria, and 16S rRNA analysis shows that types currently located in this genus probably belong to separate lineages of the β-*Proteobacteria*. *Oceanospirillum* spp. are aerobic and motile and undoubtedly play a major role in the heterotrophic cycling of nutrients in seawater. Some other members in this group are important in the sulfur cycle, especially through degradation of DMSP. Some are active in the biodegradation of hydrocarbons and may have applications in bioremediation (*Section 16.3.1*). There is wide variation in physiological properties such as optimum growth temperature, halophilicity and utilization of substrates. The taxonomy of these genera is in a state of

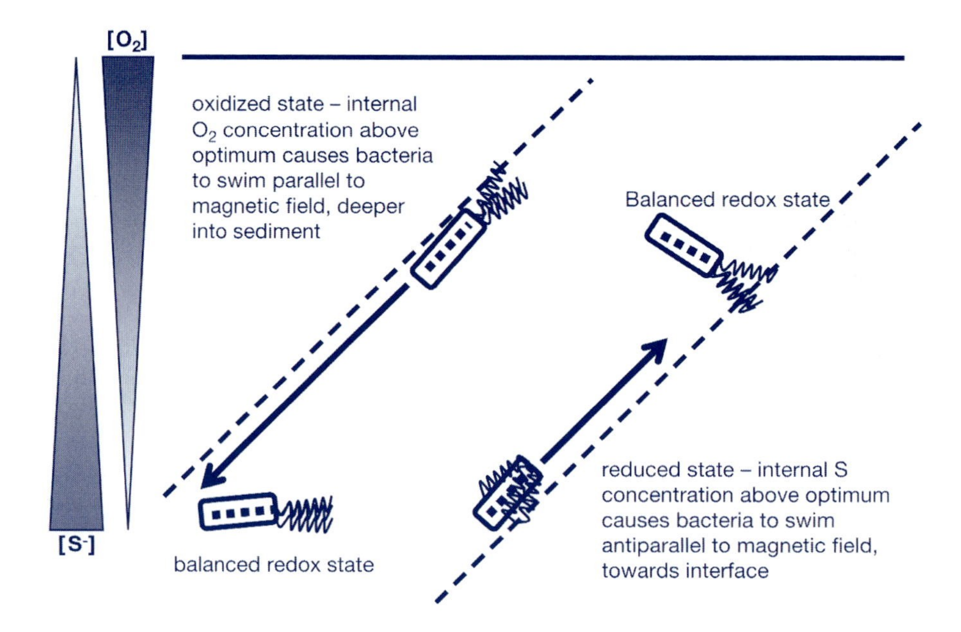

Figure 5.11

Movement in sediments by magnetotactic bacteria. Magnetotactic bacteria position themselves in conditions of optimum sulfide and O_2 concentration in sediments. In this model, cells avoid the waste of energy by constant aerotactic movement along gradients, but instead can attach to particles in preferred microniches until they reach an unfavorable internal redox state that triggers a magnetotactic response either parallel or antiparallel to the Earth's magnetic field lines. Adapted from Spring & Bazylinski (2000).

considerable flux, and 16S rRNA methods suggest that, although related, they represent a number of deeper phylogenetic groups. As shown in *Table 5.3*, even the 'spirilla' designation is not a reliable distinguishing feature, as several genera are rod-shaped rather than helical.

5.14.2 Magnetotactic bacteria

Cells of magnetotactic bacteria contain chains of magnetic particles comprised of Fe_3O_4 and Fe_3S_4 (magnetosomes) which permit the bacteria to orient themselves in the Earth's magnetic field and swim using polar flagella. The bacteria use this behavior in conjunction with aerotactic responses, to locate a favorable zone in sediments of optimal O_2 and sulfide concentrations, which they require for growth (*Figure 5.11*). Although the best-studied examples are found in freshwater mud, recent studies show that magnetotactic bacteria are widespread in salt marshes and other marine sediments and molecular studies indicate that magnetotaxis is not restricted to a small, specialized lineage as previously thought. Magnetotactic bacteria can be isolated quite easily by applying a magnetic field to mixed environmental samples, although they are difficult to cultivate. Microfossils of magnetite crystals found in deep-sea marine sediments are thought to originate from magnetotactic bacteria at least 50 million years ago.

5.14.3 *Bdellovibrio*

Bdellovibrio is a small spiral member of the δ-*Proteobacteria,* which has the unusual property of preying on other Gram-negative bacteria. *Bdellovibrio* attaches to its prey, burrows through the

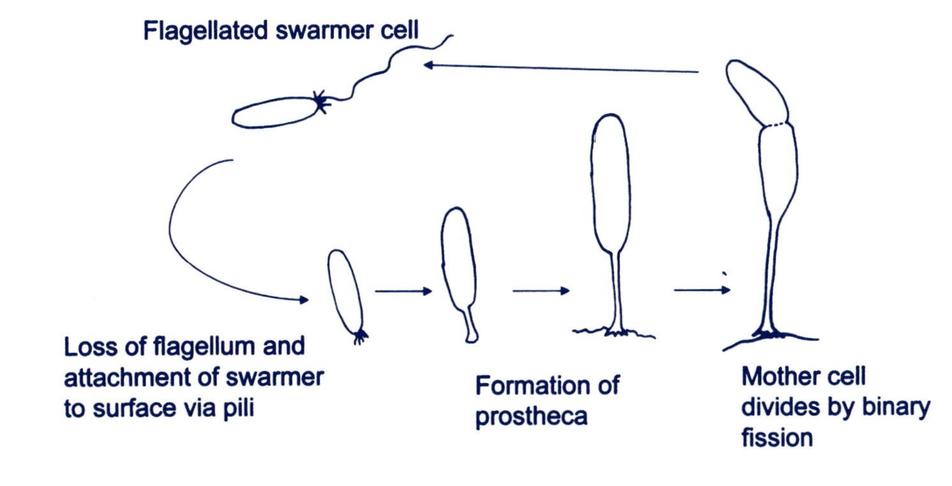

Figure 5.12

Life cycle of *Caulobacter crescentus.*

cell wall and replicates in the periplasmic space, causing the eventual lysis of the host cell and the release of up to 30 progeny *Bdellovibrio.* This organism is widespread in the marine environment, and is probably important in controlling populations of other bacteria, although its full ecological role is unknown.

5.15 Budding and stalked *Proteobacteria*

This group of bacteria is distinguished by the presence of extrusions or appendages of the cytoplasm, called prosthecae. Unlike almost all other bacteria, members of this group show a life cycle with unequal cell division, in which the 'mother cell' retains its shape and morphological features whilst budding off a smaller 'daughter' cell. This occurs as a result of polarization of the cell, that is, new cell wall material is grown from a single point rather than by intercalation as occurs in other bacteria. Some, such as *Hyphomicrobium* and *Rhodomicrobium*, bud off progeny cells from hypha-like extensions whilst others, such as *Caulobacter*, have distinctive stalks. This is a particular advantage in aquatic environments, because it means that these bacteria often attach firmly to algae, stones or other surfaces. The progeny cells are motile and swim away to colonize fresh surfaces (*Figure 5.12*). The increased SA:V ratio of stalked bacteria probably enables them to thrive in nutrient-poor waters and the prosthecae also enable these aerobic bacteria to remain in well-oxygenated environments by avoiding sinking into sediments. Phylogenetically, most representatives belong to the α-*Proteobacteria. Caulobacter* and *Hyphomicrobium* are chemoorganotrophic, whilst *Rhodomicrobium* and *Rhodopseudomonas* are phototrophic. These bacteria are often the first to colonize bare surfaces and they are especially important in the formation of biofilms, with significant consequences for larval settlement and biofouling (*Section 16.2*).

5.16 Planctomycetes – stalked bacteria

The planctomycetes are a distinct lineage of the *Bacteria,* but they differ from the stalked *Proteobacteria* just described because the stalk is a separate protein appendage rather than an extension of the cell. Unusually among the *Bacteria,* they also lack peptidoglycan with the cell

Box 5.3 RESEARCH FOCUS

Bacterial batteries at the bottom of the sea

Sulfur-reducing bacteria harvest energy from marine sediments

Marine sediments contain large amounts of organic material – could these provide a source of energy? If a graphite rod electrode (anode) is placed in anoxic marine sediments, and connected to a cathode in the O_2-containing seawater above, a significant electrical current is generated (Reimers *et al.* 2001). By comparing the communities that developed in the presence or absence of the electrical connection, Bond *et al.* (2002) have shown that the sediment becomes enriched with SRB. They used analysis of 16S rRNA to show that the most prevalent organism enriched during the electrical process is *Desulfuromonas acetoxidans,* which grows anaerobically by coupling the oxidation of acetate to reduction of sulfur. Using a quantitative PCR technique, Bond's group found that gene probes against *Desulfuromonas* are enriched by a factor of 100 on the anodes of current-generating batteries compared with unconnected control batteries. They also studied the generation of electricity with pure cultures and showed that the bacteria grow as a result of acetate oxidation coupled to reduction of the electrode. Marine sediments are unlikely to make a significant contribution to the world's energy problems, but the electricity produced in the initial studies has been sufficient to power small fuel cells on the seabed (Tender *et al.* 2002). An obvious and immediate application is for powering oceanographic monitoring equipment. The mechanism could almost certainly be harnessed in conjunction with bioremediation of sediments contaminated with organic compounds.

wall being an S-layer (*Section 3.7*) composed of cysteine- and proline-rich protein. Planctomycetes have a life cycle in which motile, flagellated swarmer cells attach to a surface and bud off a new cell from the opposite pole. An increasing number of marine planctomycetes are recognized from 16S rRNA analysis, including the genera *Planctomyces* and *Pirulella,* as well as many types which have not been cultured. They grow very slowly and are probably underrepresented in culture-based investigations. Because of their attachment mechanism, planctomycetes are especially associated with marine snow, where they probably play a major role in heterotrophic carbon cycling. Despite occupying a central phylogenetic position in the *Bacteria,* planctomycetes show some features that are reminiscent of those in eukaryotes; namely, they have membrane-bound compartments in the cell, in which metabolic and genetic components are separated. In some, the nucleus is bounded by a unit membrane confusing our traditional definition of prokaryotic cell structure.

5.17 Sulfur- and sulfate-reducing *Proteobacteria*

Most members of the sulfur- and sulfate-reducing bacteria (SRB) belong to the δ-*Proteobacteria* and their activities are highly significant in the sulfur cycle in anoxic marine environments. SRB acquire energy for metabolism and growth by utilizing organic compounds or hydrogen as electron donors and SO_4^{2-} or sulfur as electron acceptors (see *Box 5.3*). This is known as dissimilatory sulphate reduction, which is distinct from assimilatory reduction (found in many organisms) whereby reduced sulfur is incorporated into cell components such as the amino acid cysteine. SRB produce H_2S, resulting in the characteristic stench and blackening (due to deposits of FeS) of decomposition in anoxic mud, sediments and decaying seaweed. H_2S is highly toxic and adversely affects many forms of marine life, but it can be used by a wide range of chemotrophic and phototrophic bacteria, as described in earlier sections, completing the cycling of sulfur. A very large number of SRB has been isolated from marine sediments and

Table 5.4 Selected genera of marine sulfur- and sulfate-reducing bacteria

Genus	Morphology	Optimum temperature (°C)	DV[1]	DNA (mol% G+C)
Sulfate reducers; do not utilize acetate				
Desulfovibrio	Curved rods, motile	30–38[2]	+	46–61
Desulfomicrobium	Motile rods	28–37	–	52–57
Desulfobacula	Oval to coccoid cells	28	ND	42
Sulfate reducers: oxidize acetate				
Desulfobacter	Oval or curved rods, may be motile	28–32	–	45–46
Desulfobacterium	Oval, may have gas vesicles	20–35	–	41–59
Dissimilatory sulfur reducers; do not reduce sulfate				
Desulfuromonas	Rods, motile	30	–	50–63
Desulfurella	Short rods	52–57	–	31

[1]DV = desulfovoridin, a pigment used as a chemotaxonomic marker.
[2]One species is thermophilic.

these are classified on the basis of morphological and physiological properties as well as 16S rRNA typing. Over 25 species have been described (they also occur in soils, animal intestines and freshwater habitats), and representative examples and their properties are shown in *Table 5.4*, illustrating the diversity in physiological and morphological properties. The main division of the SRB is between: (a) those that couple the oxidation of substrates such as acetate or ethanol to reduction of elemental sulfur (but not SO_4^{2-}) to H_2S; and (b) those that can reduce SO_4^{2-}, using a range of organic compounds or H_2. It should be noted that some Gram-positive bacteria and *Archaea* (*Section 6.2.3*) also carry out sulfate reduction.

Recently, SRB have been shown to form syntrophic consortia with sulfide-oxidizing bacteria in an animal symbiosis (*Box 10.2*) and with *Archaea* in the anaerobic oxidation of methane (*Box 6.1*). Not all SRB are obligate anaerobes and many can coexist with oxygenic *Cyanobacteria* in microbial mats.

5.18 Gram-positive *Bacteria*

5.18.1 Endospore-formers – *Bacillus* and *Clostridium*

There are two major branches of Gram-positive *Bacteria* known as the *Firmicutes* and *Actinobacteria*. The *Firmicutes* are unicellular and have a low G+C ratio, whereas the *Actinobacteria* tend towards mycelial morphology and have a high G+C ratio. In the *Firmicutes*, the genera *Bacillus* and *Clostridium* contain a large number of diverse species and are best known as soil saprophytes, but they are also a major component of marine sediments. Although relatively little information exists on their abundance and distribution, some species have been named because of their initial isolation from marine sediments (e.g. *Bacillus marinus*) and it is likely that great diversity in marine representatives remains to be discovered. Their most distinctive feature is the production of extremely resistant endospores, which allow them to resist high temperatures, irradiation and desiccation and endospores may persist for thousands of years. *Bacillus* spp. are usually aerobic whilst *Clostridium* spp. are strict anaerobes. *Clostridium* has a wide range of fermentation pathways leading to the formation of organic acids, alcohols and hydrogen. Some types are also efficient N_2-fixers. Clostridia play a major role in decomposition and nitrogen

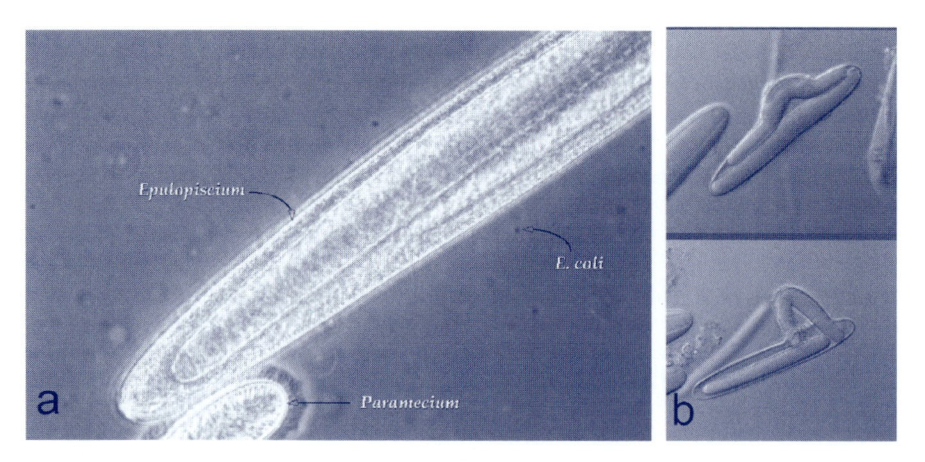

Figure 5.13

Epulopiscium – the largest prokaryote known. (a) Light micrograph showing comparative sizes of *Epulopiscium*, *Paramecium* (a protist) and *E. coli* (a 'typical' bacterium). (b) 'Viviparous' release of daughter cell during reproduction. Image courtesy of Esther R. Angert, Cornell University, USA.

cycling in anoxic marine sediments. One species, *C. botulinum,* is important as a source of toxins associated with fish products (*Section 11.2.4*).

5.18.2 Other *Firmicutes*

Staphylococcus, *Lactobacillus* and *Listeria* are aerobic, catalase-positive cocci and rods with typical respiratory metabolism. They are occasionally isolated in marine samples, but are probably very minor members of the marine bacterial community. They can be important as agents of fish spoilage and foodborne intoxication following processing (*Section 16.5*). *Streptococcus iniae* and some other species are important pathogens in warm-water fish and *Renibacterium salmoninarum* is an obligate pathogen of salmonid fish (*Section 14.3.6*).

5.18.3 *Epulopiscium fishelsoni*

This is one of the largest bacteria known. When originally discovered as a symbiont in the guts of herbivorous surgeonfish on the Great Barrier Reef and Red Sea, it was thought to be a protist because of its large size (*Figure 5.13*). In terms of cell volume, *Epulopiscium* is millions of times bigger than most marine bacteria. The giant cells can reach $600 \times 80\,\mu m$ and have a unique intracellular structure. Large amounts of DNA form a mesh of nuclear bodies around the periphery of the cell. The cytoplasm contains tubules, vacuoles and capsules which are like those seen in eukaryotic protists. These structures are believed to be involved in intracellular transport of nutrients and excretion of waste products. The mechanism of reproduction is also unique. *E. fishelsoni* is 'viviparous', meaning that new cells form inside the parent cell, which undergoes a localized cell lysis to release the active progeny. This process is reminiscent of the process of endospore formation, which involves the partition of the spore from the parent cell. Although it has not been cultured, 16S rRNA sequencing indicates that one of the closest known relatives of *Epulopiscium* spp. is *Metabacterium polyspora*, which typically produces multiple endospores as a means of propagation in its animal host, the guinea pig. Based on the phylogenetic affiliation and morphological observations, it has been hypothesized that daughter cell formation in *Epulopiscium* evolved from the process of endospore formation.

5.18.4 *Actinobacteria* – mycobacteria and actinomycetes

Mycobacteria are slow-growing, aerobic rod-shaped organisms distinguished by a high content of G+C in their DNA and unusual cell wall components which render them acid-fast in a staining procedure. They are widely distributed as saprophytes on surfaces such as sediments, corals, fish and algae. Some species (e.g. *Mycobacterium marinum*) have been identified as pathogens of fish and marine mammals which can be transmitted to humans (*Section 13.4*).

The actinomycetes and related genera are a large and diverse group of bacteria, again with a high G+C ratio. They have various cell morphologies, ranging from coryneform (club-shaped, e.g. *Corynebacterium, Arthrobacter*) to branching filaments with reproductive conidiospores (e.g. *Streptomyces* and *Micromonospora*). Actinomycetes are widely distributed in marine sediments. Because of their high abundance in soil, it is possible that actinomycetes in coastal sediments are derived from terrestrial run-off, but they are also found in deep-sea samples. Only one exclusively marine species, *Rhodococcus marinonacens,* has been cultured and studied in detail, but 16S rRNA studies suggest that there are many diverse marine actinomycetes. Their main ecological importance is in decomposition and heterotrophic nutrient cycling, due to their production of extracellular enzymes which break down polysaccharides, proteins and fats. The actinomycetes are also an exceptionally rich source of secondary metabolites. Many of the most widely used antibiotics come from actinomycetes. Several pharmaceutical companies are surveying the diversity of marine actinomycetes and identifying unique compounds with enormous biotechnological potential (*Section 16.8*).

5.19 The *Cytophaga-Flavobacterium-Bacteroides* (CFB) group

This collection of morphologically diverse, aerobic or facultatively anaerobic chemoheterotrophs is now recognized as one of the major branches of the *Bacteria* (Phylum *Bacteroidetes*). Many of the key genera (*Cytophaga, Flavobacterium, Bacteroides, Flexibacter* and *Cellulophaga*) are polyphyletic and their taxonomy is very confused. Based on analysis of the *gyrB* gene, the new genus *Tenacibaculum* has recently been created to accommodate marine types formerly known as *Flexibacter.* Many marine isolates of *Cytophaga* and *Flavobacterium* have unusual flexirubin and carotenoid pigments and can be easily isolated as colored colonies on agar medium inoculated from sediments, marine snow and the surfaces of animals and plants and incubated aerobically at ambient temperatures. Their most distinctive properties are gliding motility and the production of various extracellular enzymes, which are responsible for degradation of polymers such as agar, cellulose and chitin. Agar is normally resistant to bacterial degradation, which is why it is an ideal gelling agent for culture plates, but softening or the formation of craters on agar plates is often observed with marine isolates belonging to the CFB group. The production of hydrolytic enzymes is of major ecological significance in the degradation of complex organic materials such as the cell walls of phytoplankton and exoskeletons of crustacea. Some species are pathogenic for fish and invertebrates. Many are psychrophilic, being commonly isolated from cold-water marine habitats and sea ice. The normal habitat of the genus *Bacteroides* is the gut of mammals and they may be present in sewage-polluted waters and persist for quite long periods in the sea.

5.20 *Verrucomicrobia*

This phylum of *Bacteria* is poorly characterized, but 16S rRNA sequence analysis has shown that members have diverse habitats. However, only a very few strains from soil have been isolated in culture. Despite the occurrence of sequences from marine sediments and aggregates, nothing is known of the physiological properties and importance of this group in marine habitats.

5.21 *Spirochaetes*

The *Spirochaetes* are Gram-negative, tightly coiled flexuous bacteria distinguished by very active motility. This occurs due to the possession of internal flagella, located in the periplasmic space between the CM and cell wall. The internal flagella rotate as a rigid helix, like other bacterial flagella, causing the protoplasmic cylinder to rotate in the opposite direction leading to flexing and jerky movement. Genera such as *Cristispira* and *Spirochaeta* are widespread in marine habitats, but little is known about their ecological role. Many are strictly anaerobic and found in sediments and it is likely that they are also important components of the gut microbiota in marine animals. *Cristispira* occurs in the digestive tract of certain molluscs, but has not been cultured.

5.22 'Deeply branching' hyperthermophiles

5.22.1 *Aquifex*

As can be seen in *Figure 5.1, Aquifex* and its relatives form a deeply branching root of the phylogenetic tree of *Bacteria.* Analysis of sequences of 16S rRNA and several other genes confirms that this group is closest to the hypothetical ancestor of all *Bacteria.* Because *Aquifex* spp., such as *A. pyrophilus* and *A. aeolicus,* are extremely thermophilic (maximum growth temperature can be as high as 95°C) and chemolithotrophic, they have a major role in primary production in marine hydrothermal vents. They grow using H_2, thiosulfate or sulfur as the electron donor and O_2 or nitrate as the electron acceptor. They fix carbon via an unusual process known as the reductive citric acid cycle. The extremely thermophilic properties of these bacteria are obviously of great interest from a biotechnological perspective, and the full genome sequence of *A. aeolicus* has recently been published. It has a very small genome, only about one third that of *E. coli.* A related species, *Hydrogenothermus marinus* has also been isolated from deep-sea and coastal hydrothermal systems.

5.22.2 *Thermotoga*

Thermotoga also forms a deeply branching phylogenetically distant group of the *Bacteria.* As well as evidence from gene sequences, the function of the ribosome is very different from other *Bacteria,* and is not affected by rifampicin and other antibiotics that affect protein synthesis. The name of the genus derives from a unique outer membrane ('toga') which balloons out from the rod-shaped cells. The cells are Gram-negative, but the amino acid composition of the peptidoglycan is unlike that of other *Bacteria* and there are unusual long-chain fatty acids in the lipids. *Thermotoga* is widespread in geothermal areas and occurs in shallow and deep-sea hydrothermal vents. Different species vary in their temperature optima, with a range from 55°C up to 80–95°C for the hyperthermophilic species *T. maritima* and *T. neapolitana.* These are fermentative, anaerobic chemoorganotrophs and utilize a wide range of carbohydrates. They also fix N_2 and reduce sulfur to H_2S. Like *Aquifex,* these organisms have considerable biotechnological potential and the *T. maritima* genome has been sequenced. A large number of the genes are involved in the transport and utilization of nutrients, in keeping with its ability to utilize a wide range of substances for growth.

References and further reading

Adams, D.G. (2001) How do cyanobacteria glide? *Microbiol Today* **28**: 131–133.

Amann, R.I., Ludwig, W., and Schleifer, K.H. (1995) Phylogenetic identification and in situ detection of individual microbial cells without cultivation. *Microbiol Rev* **59**: 143–169.

Angert, E.R., Brooks, A.E., and Pace, N.R. (1996) Phylogenetic analysis of *Metabacterium polyspora*: clues to the evolutionary origin of daughter cell production in *Epulopiscium* species, the largest bacteria. *J Bacteriol* **178**: 1451–1456.

Berman-Frank, I., Lundgren, P., Chen, Y.B., Kupper, H., Kolber, Z., Bergman, B., and Falkowski, P. (2001) Segregation of nitrogen fixation and oxygenic photosynthesis in the marine cyanobacterium *Trichodesmium*. *Science* **294**: 1534–1537.

Bond, D.R., Holmes, D.E., Tender, L.M., and Lovley, D.R. (2002) Electrode-reducing microorganisms that harvest energy from marine sediments. *Science* **295**: 483–485.

Cottrell, M.T., and Kirchman, D.L. (2000) Community composition of marine bacterioplankton determined by 16S rRNA gene clone libraries and fluorescence in situ hybridization. *Appl Environ Microbiol* **66**: 5116–5122.

Curtis, T.P., Sloan, W.T., and Scannell, J.W. (2002) Estimating prokaryotic diversity and its limits. *Proc Natl Acad Sci USA* **99**: 1494–1499.

Czyz, A., and Wegrzyn, G. (2002) On the function and evolution of bacterial luminescence. In: Case, J.F., Herring, P.J., Robinson, B.H., Haddock, S.D.H., Kricka, L.J., and Stanley, P.E. (eds) *Bioluminescence and Chemiluminescence*, pp. 31–34. World Science Publishing Company, Singapore.

Czyz, A., Wrobel, B., and Wegrzyn, G. (2000) *Vibrio harveyi* bioluminescence plays a role in stimulation of DNA repair. *Microbiology* **146**: 283–288.

Dworkin, M. (ed.) 2001 *The Prokaryotes: an Evolving Electronic Resource for the Microbiological Community*, 3rd edition, Springer-Verlag, New York. http://www. prokaryotes.com (accessed March 31 2003).

Fuhrman, J.A., and Campbell, L. (1998) Marine ecology – microbial microdiversity. *Nature* **393**: 410–411.

Garrity, G.M., Boone, D., and Castenholz, R. (eds) (2001) *Bergey's Manual of Systematic Bacteriology*, 2nd edition, Vol. 1. Springer-Verlag, New York.

Garrity, G.M., Johnson, K.L., Bell, J.A., and Searles, D.A. (2002) Taxonomic outline of the prokaryotes. In: Garrity, G.M., Boone, D., and Castenholz, R. (eds) *Bergey's Manual of Systematic Bacteriology*. 2nd edition, release 3.0. Springer-Verlag, New York. http://www.springer-ny.com/bergeysoutline/outline/ (accessed March 31 2003).

Giovannoni, S., and Rappé, M. (2000) Evolution, diversity and molecular ecology of marine prokaryotes. In: Kirchman, D.L. (ed.) *Microbial Ecology of the Oceans*, pp. 47–84. Wiley-Liss Inc., New York.

Hagström, A., Pommier, T., Rowher, F., Simu, K., Stolte, W., Svensson, D., and Zweifel, U.L. (2002) Use of 16S ribosomal DNA for delineation of marine bacterioplankton species. *Appl Environ Microbiol* **68**: 3628–3633.

Hugenholtz, P., Goebel, B.M., and Pace, N.R. (1998) Impact of culture-independent studies on the emerging phylogenetic view of bacterial diversity. *J Bacteriol* **180**: 4765–4774.

Jørgensen, B.B., and Gallardo, V.A. (1999) *Thioplaca* spp.: filamentous sulfur bacteria with nitrate vacuoles. *FEMS Microbiol Ecol* **28**: 301–313.

Madigan, M.T., Martinko, J.M., and Parker, J. (2003) *Brock Biology of Microorganisms*. Prentice-Hall, New Jersey.

Miller, M.B., and Bassler, B.L. (2001) Quorum sensing in bacteria. *Ann Rev Microbiol* **55**: 165–199.

Moore, L.R., Rocap, G., and Chisholm, S.W. (2001) Physiology and molecular phylogeny of coexisting *Prochlorococcus* ecotypes. *Nature* **393**: 464–467.

Morris, G.M., Rappe, M.S., Connon, S.A., Vergin, K.L., Siebold, W.A., Carlson, C.A., and Giovannoni, S.J. (2002) SAR11 clade dominates ocean surface bacterioplankton communities. *Nature* **420**: 806–810.

Overmann, J. (2001) Diversity and ecology of phototrophic sulphur bacteria. *Microbiol Today* **28**: 116–118.

Partensky, F., Hess, W.R., and Vaulot, D.D. (1999) *Prochlorococcus*, a marine photosynthetic prokaryote of global significance. *Microbiol Molec Biology Rev* **63**: 106–127.

Rees, J.F., De Wergifosse, B., Noiset, O., Dubuisson, M., Janssens, B., and Thompson, E.M. (1998) The origins of marine bioluminescence: turning oxygen defence mechanisms into deep-sea communication tools. *J Exp Biol* **201**: 1211–1221.

Reid, R.P., Visscher, P.T., Decho, A.W., *et al.* (2000) The role of microbes in accretion, lamination and early lithification of modern marine stromatolites. *Nature* **406**: 989–992.

Reimers, C.E., Tender, L.M., Fertig, S., and Wang, W. (2001) Harvesting energy from the marine sediment–water interface. *Environ Sci Technol* **35**: 192–195.

Ribosome Database Project. Centre for Microbial Ecology, Michigan State University. http://rdp. cme.msu.edu (accessed March 31 2003).

Ruby, E.G., and McFall-Ngai, M.J. (1999) Oxygen-utilizing reactions and symbiotic colonization of the squid light organ by *Vibrio fischeri*. *Trends Microbiol* **7**: 414–420.

Scanlan, D. (2001) *Cyanobacteria*: ecology, niche adaptation and genomics. *Microbiol Today* **28**: 128–130.
Schulz, H.N. (2002) *Thiomargarita namibiensis*: giant microbe holding its breath. *ASM News* **68**: 122.
Schulz, H.N., and Jørgensen, B.B. (2001) Big bacteria. *Ann Rev Microbiol* **55**: 105–137.
Spring, S., and Bazylinski, D.A. (2000) Magnetotactic bacteria. In: Dworkin, M. (ed) The Prokaryotes: an evolving electronic resource for the microbiological community. 3rd edition, Springer-Verlag, New York. http://www.springer-ny.com (accessed March 31 2003).
Tender, L.M., Reimers, C.E., Stecher, H.A., *et al.* (2002) Harnessing microbially generated power on the seafloor. *Nature Biotechnol* **20**: 821–825.
Torsvik, V., Ovreas, L., and Thingstad, T.F. (2002) Prokaryotic diversity – magnitude, dynamics and controlling factors. *Science* **296**: 1064–1066.

Marine *Archaea*

6

6.1 Phylogenetic groups in the domain *Archaea*

As discussed in *Chapter 1*, the application of 16S rRNA methods led to the recognition of the *Archaea* as a completely separate domain of prokaryotic life. Other information supports the concept that the *Archaea* are a monophyletic group, especially the nature of the ribosomes and mechanisms of transcription and translation (*Section 3.8*). We now know that the previous view of 'archaeobacteria' as a rather uncommon, specialized subset of bacteria is completely wrong. In the marine environment, members of the *Archaea* are very abundant and very diverse. Currently, we recognize three major branches (phyla) of the phylogenetic tree, namely the *Euryarchaeota*, *Crenarchaeota* and *Korarchaeota*, as shown in *Figure 6.1*. Many of the marine *Archaea* are known only by gene sequences isolated from environmental samples and have not yet been cultured. Although many of the cultured *Archaea* are hyperthermophiles found in areas of marine hydrothermal activity, crenarchaeotes are widely distributed in cold waters. Recently, discovery of a very small *Archaeal* symbiont of the archaeon *Igniococcus* suggests that there may be an additional phylum of the *Archaea*, provisionally called *Nanoarchaeota* (*Box 6.1*).

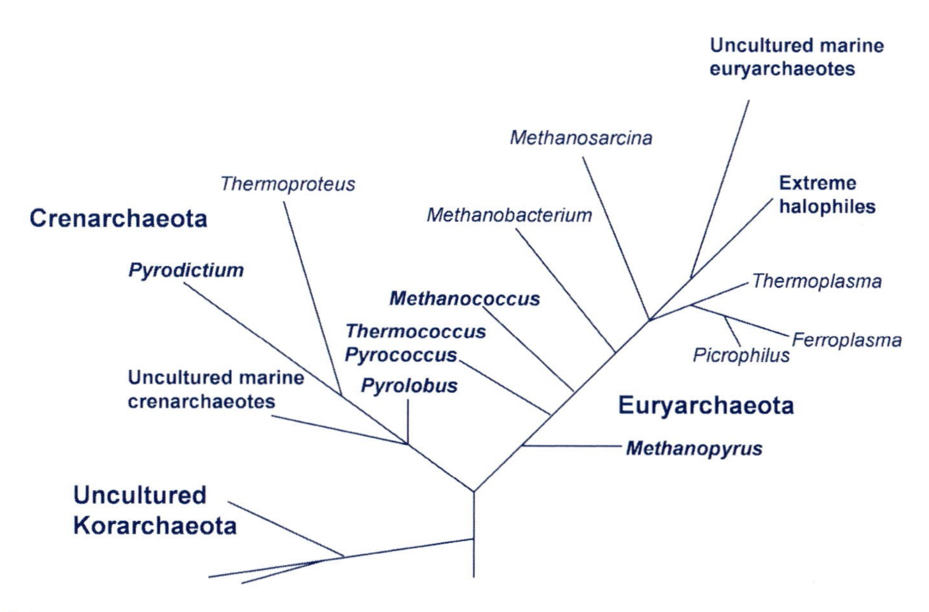

Figure 6.1

Simplified phylogenetic tree of the *Archaea*. Representative types found in marine habitats are shown in bold type. Redrawn from Madigan *et al.* (2003).

Box 6.1 RESEARCH FOCUS

More surprises from the *Archaea*

Discovery of novel metabolic consortia and symbioses

Recent research has produced surprising insights into members of the *Archaea* and their importance in marine processes. This box highlights two particularly exciting discoveries of syntrophic and symbiotic interactions.

For many years, the fate of the large amounts of methane produced in marine sediments has been something of a mystery. Huge reservoirs of methane, largely in the form of crystalline, frozen methane hydrate have been discovered in offshore marine sediments, and their exploitation will be a major feature of our search for new energy resources to replace dwindling stocks of oil. As well as its importance as a fuel source, understanding the fate of methane is important, because it is a powerful 'greenhouse gas' and periodic release from the ocean floor may be important in climatic variation in the Earth's history. We know that production of methane is an anaerobic process carried out by particular types of *Archaea* (*Section 6.2.1*) and, until recently, all microorganisms capable of using methane (methanotrophs) were thought to be aerobic. Until the discovery described below, no organism that can utilize methane in anaerobic environments had ever been found. However, for many years, geochemists suspected that methane oxidation does occur in deep marine sediments and that it somehow involved electron transfer to sulfate. Hinrichs *et al.* (1999) used a stable isotope method combined with 16S rDNA studies to provide evidence that unknown *Archaea* are involved in this process. Lipids characteristic of the *Archaea* were shown to be synthesized using methane as a carbon source. Could a consortium of microorganisms operating together (syntrophy) be capable of carrying out the metabolic transformations needed for the anaerobic oxidation of methane? To answer this question, Antje Boetius and colleagues at the Max Planck Institute for Marine Microbiology used the FISH technique in conjunction with CLSM to

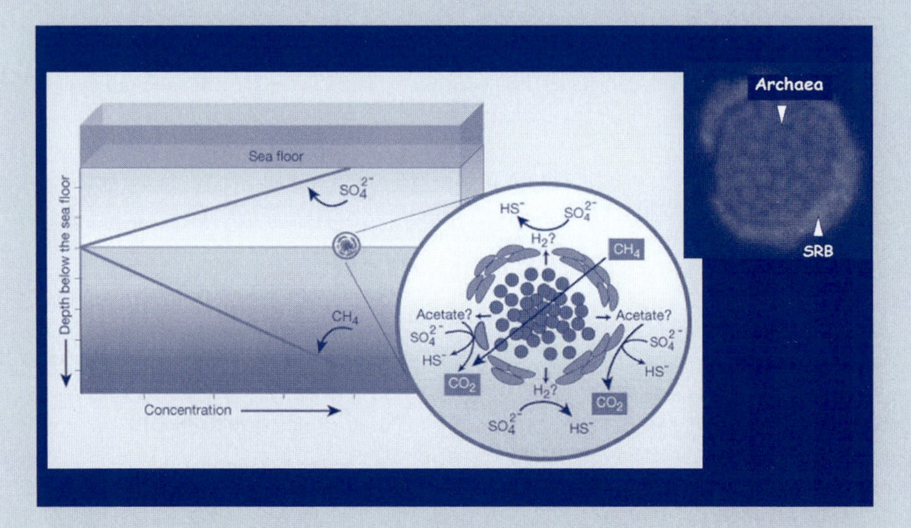

Model showing anaerobic oxidation of methane in marine sediments. From De Long (2000), reproduced with permission from Nature-Macmillan Ltd. Inset shows confocal micrograph of metabolic consortium of about 100 methane-oxidizing archaea surrounded by sulfate-reducing bacteria, visualized by FISH. Image courtesy of K. Knittel, A. Gieseke and A. Boetius, Max Planck Institute for Microbiology, Germany. (See Boetius *et al.* (2001) for color image.)

investigate marine sediments. With probes directed against known groups of *Archaea* and sulfate-producing *Bacteria* (SRB), Boetius *et al.* (2001) produced amazing images of small groups of archaeal cells tightly surrounded by a shell of SRB, as shown in the figure below. This is the first observation of a closely integrated syntrophic consortium (symbiosis?) between *Archaea* and *Bacteria*. Together, the partners carry out the reaction

$$CH_4 + SO_4^{2-} \longrightarrow HCO^{3-} + HS^- + H_2O$$

The biochemistry and thermodynamics of this methano*trophic* process have not been fully resolved, but it is believed to function as a reversal of methano*genesis*. The products of the

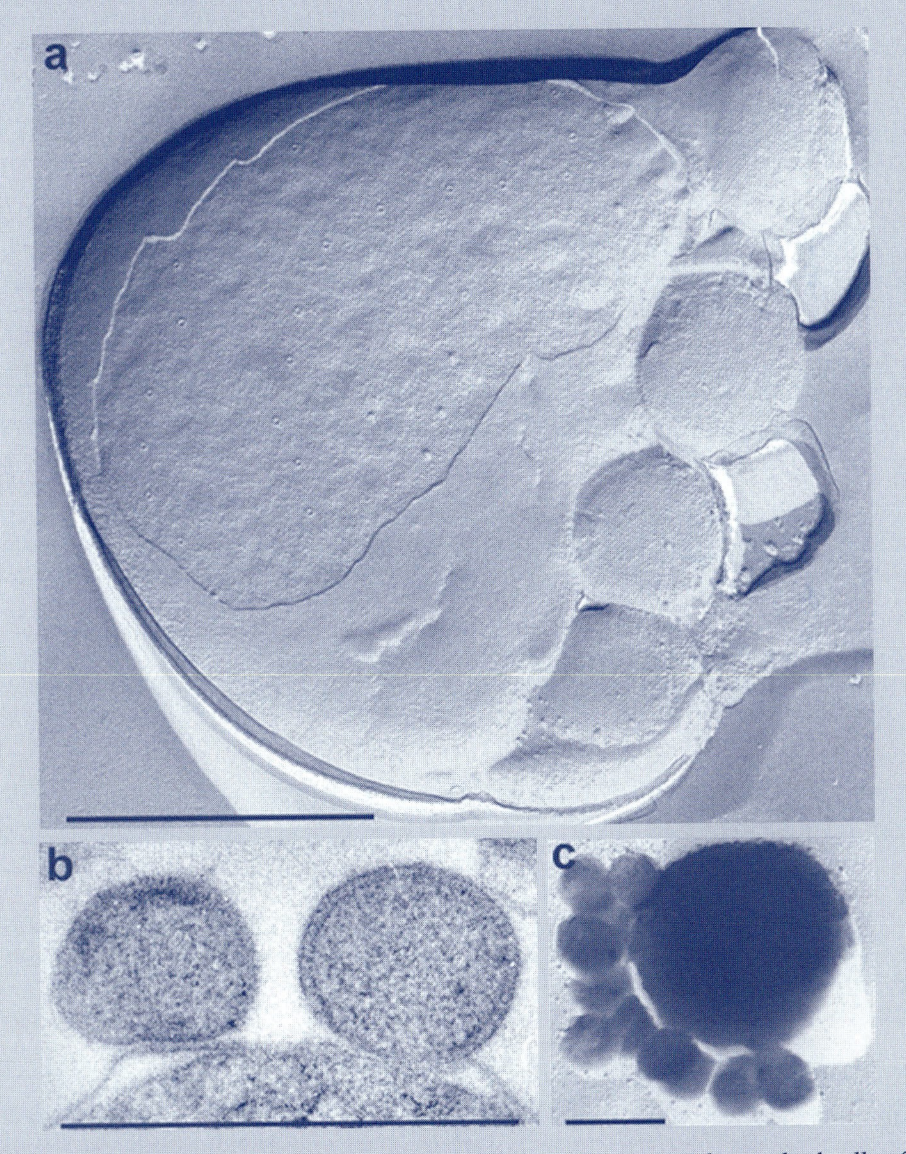

Electron micrographs (a–c) showing larger cells of *Igniococcus* with attached cells of *Nanoarchaeum*. Image courtesy of H. Huber, R. Rachel, M.J. Hohn, C. Wimmer and K.O. Stetter, University of Regensburg, Germany (from Huber *et al.* (2002), reproduced with permission from Nature-Macmillan Ltd.).

reaction are very important. For example, Michaelis *et al.* (2002) have shown the presence of these consortia in massive microbial reef-like mats covering 4-metre-high reefs of precipitated carbonate at methane seeps in the Black Sea. The sulfide can be used in chemosynthetic symbioses involving tubeworms, clams or giant sulfide-oxidizing bacteria, which form thick carpets on the seabed above the methanotrophic consortia. This type of methanotrophic metabolism may have evolved very early on Earth, possibly predating the development of O_2 in the atmosphere due to oxygenic photosynthesis (about 2.2 billion years ago).

Another exciting recent discovery has been that of a symbiosis between two different archaeal types. Harald Huber and colleagues at the University of Regensburg were investigating hyperthermophilic microbial communities near a submarine hot vent in Icelandic waters. After anaerobic incubation at 90°C (using specialized high-temperature fermentors) in the presence of sulfide, H_2 and CO_2, they observed large spherical cells identified as *Ignicoccus* (Huber *et al.* 2002). Groups of tiny cocci were attached to the surface of the *Ignicoccus* cells and these were shown to contain DNA using fluorescence microscopy. Electron micrographs (see figure on previous page) showed these to be only about 400 nm in diameter. All attempts to grow the small cells in pure culture failed, even using extracts of *Ignicoccus*. However, the growth rate of *Ignicoccus* was not affected by presence of the small cells, and it could grow in their absence. The real surprise came when Huber's group tried to identify the cells using 16S rDNA methods. Although the RNA genes from the *Ignicoccus* could be successfully amplified, this could not be achieved with the smaller cell, even when primers considered 'universal' were used. However, two kinds of SSU rRNA could be detected in the cell mixture using Southern blot hybridization and the rRNA from the small cells was subsequently isolated and sequenced. This showed that the sequence is close to those of known *Archaea*, but very different from the existing groups *Crenarchaeota*, *Euryarchaeota* and the '*Korarchaeota*' (the latter known only from DNA sequences in environmental samples, *Figure 6.1*). It therefore seems that the small cells (which the authors call *Nanoarchaeum equitans*) belong to a new archaeal phylum, called the *Nanoarchaeota* in recognition of their small size. The tiny cells and small size of the genome (0.5 Mb, about 1/10 the size of the *E. coli* genome) seem to be near the lower limit predicted for cellular life. At present, we do not know if *Nanoarchaeum* is a 'primitive' form of life (i.e. closely resembling those believed to have evolved in the early history of the Earth) or whether it is a 'highly evolved' symbiont, which has undergone reduction in cell size and genetic complement, as is observed in other obligate symbionts and parasites. Similar sequences were subsequently discovered in an abyssal vent and terrestrial hot springs, suggesting that these organisms are widely distributed in thermal systems (Hohn *et al.* 2002). At the time of writing, the genome sequence of *N. equitans* has been completed by the Celera Genomics and Diversa companies, but not yet published. Such an unusual organism could clearly have important properties for biotechnology and may help to answer fundamental questions about the nature of life. Craig Venter and Hamilton Smith of Celera have announced plans to build a living organism 'from scratch' by removing all genetic material from an existing organism and replacing it with a synthetic chromosome containing the minimum number of genes necessary for life (Mullen 2003). Until the discovery of *N. equitans*, their plans were to use the bacterium with the smallest known genome, *Mycoplasma genitalium*. If the *N. equitans* genome turns out to be even smaller, it may well be a preferred choice for this project.

6.2 The *Euryarchaeota*

6.2.1 Methanogens

A large number of members of the *Euryarchaeota* produce methane as the final step in the anaerobic biodegradation of organic material. Methanogens are mesophilic or thermophilic and can be isolated from a wide range of habitats including the gut of animals, anoxic sediments and decomposing material. They also occur as endosymbionts of anaerobic protozoa found especially in the hindgut of termites. It is likely that shipworms and other marine invertebrates which digest wood and cellulose also harbor ciliates with archaeal endosymbionts. Thermophilic methanogens are also important members of the microbial community of hydrothermal vents. Methanogenic *Archaea* in anoxic marine sediments are responsible for the production of huge amounts of methane, much of which has been sequestered over millennia as methane hydrate. This is of global significance as a future energy source. The fate of methane is also important because of its role as a 'greenhouse gas' affecting climate and its metabolism by other *Archaea* in syntrophic association with sulfate-reducing *Bacteria* is discussed in *Box 6.1*.

Methanogens show high morphological and physiological diversity, and are grouped into a number of genera using these criteria. Most genera utilize H_2, with CO_2 serving as both the oxidant for energy generation and for incorporation into cellular material. The energy-yielding reaction used to generate ATP (via generation of a proton motive force) is

$$CO_2 + 4H_2 \longrightarrow CH_4 + 2H_2O$$

Some methanogens also use methyl compounds (such as methanol, methylamine and dimethyl sulfide), formate, acetate, pyruvate or carbon monoxide. Molecules such as sugars and fatty acids are not directly used for methane generation, but because methanogens exist in syntrophic communities with bacteria, virtually any organic compound can eventually be converted to methane. Bacterial fermentation produces H_2, CO_2 and acetate as end products, which are then utilized by methanogens. The process of methane formation depends on a group of coenzymes which act as carriers for the C_1 unit from the substrate (CO_2) to the product (CH_4). Other coenzymes transfer the electrons from hydrogen or other donors, if used. These compounds are unique to this group of *Archaea*.

Although they are strict anaerobes, methanogens can also be found in surface microbial mats and ocean waters, which can contain high levels of dissolved methane. Presumably, they exist in the anoxic interior of particles in which oxygen has been depleted by respiratory activity of other organisms. Methanogenesis is also significant in deeper waters with upwelling of nutrients. Here, intense heterotrophic oxidation of sinking organic matter leads to O_2 depletion.

The extremely thermophilic archaeon *Methanococcus jannaschii* has been recognized as one of the most important members of hydrothermal vent communities and has been studied extensively. It is a primary consumer of H_2 and CO_2 produced by geochemical activity at the vents. Analysis of the complete genome sequence reveals a 1.66 Mb circular genome containing about 1700 genes. Sequence analysis of the genes encoding central metabolic pathways and cellular processes shows that these are similar to those found in *Bacteria,* whereas those encoding protein synthesis and DNA replication show more similarity to eukaryotic genes. This supports the hypothesis that all three domains of life evolved from a common universal ancestor, as discussed in *Section 1.2.4*. However, a large number of genes in *M. jannaschii* have little homology to bacterial or eukaryotic genes, suggesting that many novel cellular processes remain to be discovered.

Methanopyrus is one of the most thermophilic organisms known, with rapid growth (generation time of 1 h) at a maximum temperature of 110°C. It is found in the walls of black smoker chimneys in hydrothermal vents. Although it is a methanogen (utilizing H_2 + CO_2 only), it is phylogenetically distant from the rest of this group and has very unusual membrane lipids and thermostabilizing compounds in the cytoplasm.

6.2.2 Hyperthermophilic chemoorganotrophs – *Thermococcus* and *Pyrococcus*

These genera of *Euryarchaeota* are hyperthermophiles found in hydrothermal vents, with optimum growth temperatures in excess of 80°C. Their phylogenetic position as a deeply branching group resembles that of the hyperthermophilic genera *Aquifex* and *Thermotoga* in the tree of *Bacteria* (*Figure 5.1*). It is interesting that extreme thermophiles in both domains branch very near the root of the tree and this is taken as evidence that these organisms most closely resemble those which evolved first in the hotter conditions of the early Earth. Indeed, some scientists suggest that life may have evolved in submarine hydrothermal vents or subsurface rocks. The adaptations of hyperthermophiles to life at high temperatures is discussed in *Section 4.6.2*, and *Table 4.2* includes the growth temperatures of some hyperthermophilic *Archaea*.

Thermococcus celer forms highly motile spherical cells about 0.8 μm in diameter and is an obligately anaerobic chemoorganotroph, utilizing complex substrates such as proteins and carbohydrates, with sulfur as the electron acceptor. It has an optimum growth temperature of 80°C. *Pyrococcus furiosus* has similar properties to *Thermococcus*, but has an optimum growth temperature of 100°C and a maximum of 106°C. Both organisms have been investigated extensively because of their biotechnological potential.

6.2.3 Hyperthermophilic sulfate-reducers and iron-oxidizers – *Archaeoglobus* and *Ferroglobus*

Archaeoglobus spp. are also extreme thermophiles (optimum temperature 83°C) found in sediments of shallow hydrothermal vents and around undersea volcanoes. They are strictly anaerobic organisms which couple the reduction of sulfate to the oxidation of H_2 and certain organic compounds, resulting in the production of H_2S. Although *Archaeoglobus* forms a distinct phylogenetic group in the *Euryarchaeota*, it makes some of the key coenzymes used in methanogenesis and analysis of its genome sequence shows that it shares some genes with the methanogens. However, it lacks genes for one of the key enzymes, methyl-CoM reductase, so the origin of the small amount of methane produced is unknown. *Archaeoglobus* also occurs in oil reservoirs and has caused problems with sulfide 'souring' of crude oil extracted from the North Sea and Arctic oilfields.

Members of the related genus *Ferroglobus*, also found in hydrothermal vents, do not reduce sulfate but are iron-oxidizing/nitrate-reducing chemolithoautotrophs.

6.2.4 Extreme halophiles

Extreme halophiles grow in concentrations of NaCl greater than 9% and many can grow in saturated NaCl solutions (35%). They are found in salt lakes such as the Great Salt Lake, Utah and the Dead Sea in the Middle East. Extreme halophiles also occur in coastal regions in solar salterns, which are lagoons in which seawater is allowed to evaporate to collect sea salt. They operate as semicontinuous systems and maintain a fairly constant range of salinity throughout the year. Both conventional microbiological and 16S rRNA methods show that the overall diversity of microorganisms decreases as salinity increases. Up to about 11% NaCl, the range of bacteria is similar to that found in coastal seawater (most marine *Bacteria* are moderately halophilic) whilst *Archaea* are scarce. However, above 15% salinity, culturable members of the *Archaea* such as *Halorubrum, Halobacterium, Halococcus* and *Halogeometricum* become dominant, as well as *Archaea* with previously unidentified gene sequences. The halophilic *Archaea* are Gram-negative rods or cocci, which may contain very large plasmids constituting up to 30% of the genome. They are chemoorganotrophs that usually use amino acids or organic acids as an energy source. The requirement for very high Na^+ concentrations is achieved by pumping large amounts of K^+ across the membrane, so that the internal osmotic pressure remains high and protects the cell from dehydration. Cells lyse in the absence of sufficient concentrations of Na^+ because Na^+ stabilizes the high levels of negatively charged acidic amino acids in the cell wall.

Halobacterium salinarum and some other species use a membrane protein called bacteriorhodopsin to synthesize ATP using light energy. This compound is so called because it is structurally similar to the rhodopsin pigment in the eye of animals. Bacteriorhodopsin is complexed with a carotenoid pigment, retinal, which absorbs light and generates a proton motive force for the production of ATP. Light is not used for photosynthesis, but it provides energy for the proton pump required to pump Na^+ out and K^+ into the cell and can give enough energy for a low metabolic activity when organic nutrients are scarce. *H. salinarum* also contains other types of rhodopsin. Halorhodopsin captures light energy used for pumping Cl^- into the cell to counterbalance K^+ transport. Two other rhodopsin molecules act as light sensors that affect flagellar rotation, enabling chemotaxis towards light. As discussed in *Box 2.2*, an analog of bacteriorhodopsin has recently been discovered in oceanic *Proteobacteria*.

6.3 The *Crenarchaeota*

6.3.1 The diversity of *Crenarchaeota*

The phylum *Crenarchaeota* is phylogenetically distinct from the *Euryarchaeota* described above, although many have similar physiological properties including extreme hyperthermophily. Most cultured representatives are known from extensive study of terrestrial hot springs, but several other species occur in submarine hydrothermal vents. They use a wide range of electron donors and acceptors in metabolism and can be either chemoorganotrophic or chemoheterotrophic. Most are obligate anaerobes.

One of the greatest surprises of recent years has been the discovery that, far from being restricted to high-temperature habitats, members of the *Crenarchaeota* are ubiquitous in the marine environment. The discovery of archaeal gene sequences in Antarctic waters and sea ice and the subsequent realization that the *Crenarchaeota* are among the most numerous life forms in deep ocean waters has led to radical reappraisal of the diversity and ecology of marine prokaryotes.

6.3.2 Hyperthermophiles – the *Desulfurococcales*

Several species isolated from shallow and deep-sea hydrothermal areas belong to the order *Desulfurococcales*. The type genus, *Desulfurococcus*, is an obligate anaerobe with coccoid cells which uses the following reaction for energy generation.

$$H_2 + S^0 \longrightarrow H_2S$$

Cells of *Pyrodictium* spp. are disk-shaped and are connected into a mycelium-like layer attached to crystals of sulfur by very fine hollow tubules. Most *Pyrodictium* spp. are chemolithoautotrophs which gain energy by sulfide reduction. However, *P. abyssi* is a heterotroph that ferments peptides to CO_2, H_2 and fatty acids.

Pyrolobus fumarii holds the distinction of being the organism with the highest growth temperature known (113°C) and is found in the walls of black smoker chimneys in hydrothermal vents. In this very extreme environment, it is probably a significant source of primary productivity. Cells are lobed cocci and the cell wall is composed of protein. It is a facultative aerobic obligate chemolithotroph, using the reactions

$$4H_2 + NO_3^- + H^+ \longrightarrow NH_4 + 2H_2O + OH^- \quad \text{(a)}$$

$$5H_2 + S_2O_3^{2-} \longrightarrow 2H_2S + 3H_2O \quad \text{or} \quad \text{(b)}$$

$$H_2 + \tfrac{1}{2}O_2 \longrightarrow H_2O \quad \text{(c)}$$

Igniococcus is a sulfide-reducing chemolithotroph with a very different structure to other *Archaea*. An outer membrane, reminiscent of that seen in Gram-negative *Bacteria*, surrounds the cell as a loose sac enclosing a very large periplasmic space containing vesicles, which may function in transport. Recently, a new species of *Igniococcus* was discovered which is covered with tiny symbiotic cocci of another archaeal cell (*Box 6.1*).

Species of the genus *Pyrobaculum* show various modes of nutrition. Some species link sulfur reduction to anaerobic respiration of organic compounds, whilst others are chemolithotrophic autotrophs using the reactions

$$H_2 + NO_3 \longrightarrow NO_2^- + H_2O \quad \text{or} \qquad \text{(a)}$$

$$H_2 + 2Fe^{3+} \longrightarrow 2Fe^{2+} + H^+ \qquad \text{(b)}$$

Staphylothermus marinus forms aggregates of cocci and is a chemoorganotroph which, like *Pyrodictium abyssi,* ferments peptides to CO_2, H_2 and fatty acids. It is widely distributed in shallow- and deep-sea hydrothermal systems and is a major decomposer of organic material. It is the largest known member of the *Archaea*. Although normally about 1 μm diameter, it can form very large cells up to 15 μm diameter in high nutrient concentrations.

6.3.3 The uncultured psychrophilic marine *Crenarchaeota*

Analysis of marine waters using primers for archaeal 16S rRNA led to some very unexpected results. Until recently, crenarchaeotes were known only as extreme thermophiles, but sequences corresponding to this group were found to be widespread in oligotrophic ocean waters, including the Antarctic and very deep waters with temperatures as low as −2°C. As shown in *Figure 6.2*, crenarchaeotes have been shown to comprise a large fraction of the picoplankton. The total number of prokaryotes in ocean waters decreases with depth, from 10^5–10^6 ml^{-1} near the surface to 10^3–10^5 cells ml^{-1} below 1000 m. *Bacteria* are most prevalent in the upper 150 m of the ocean, but below this depth the fraction of *Archaea* equals or exceeds that of *Bacteria*. The pattern is consistent throughout the seasons of the year. Combining the figures

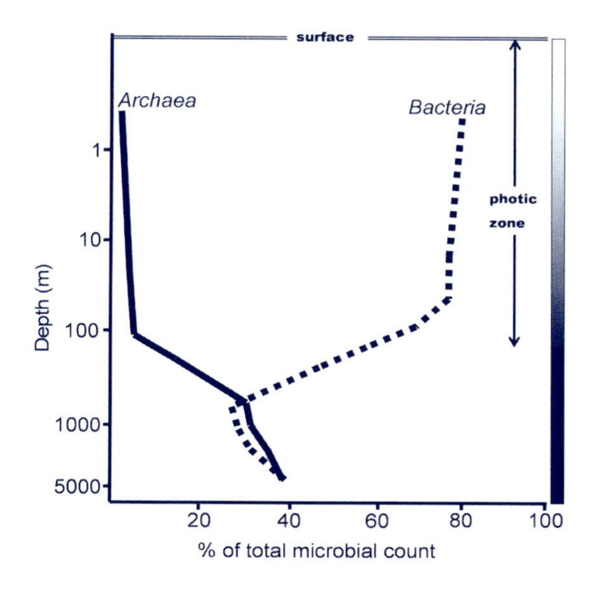

Figure 6.2

Distribution of *Bacteria* and *Archaea* in the ocean, showing mean annual depth profiles of the North Pacific subtropical ocean gyre. Total cell abundance was measured using the DAPI nucleic stain and *Bacteria* and *Archaea* measured using whole-cell rRNA targeted FISH with fluorescein-labeled polynucleotide probes. The counts of *Archaea* were almost entirely due to the crenarchaeote clade; counts of euryarchaeotes were a few percent of the total throughout the water column. Redrawn from data in Karner *et al.* (2001), with permission from Nature-Macmillan Ltd.

for cell density with oceanographic data for the volume of water at different depths gives the result that the world's oceans contain approximately 1.3×10^{28} cells of *Archaea* and 3.1×10^{28} cells of *Bacteria* (also see *Box 1.1*). About 1.0×10^{28} cells, i.e. 20% of all the picophytoplankton, is dominated by the single clade of the pelagic *Crenarchaeota*, suggesting that a common adaptive strategy has allowed them to radiate through the entire water column of the oceans.

The physiology of these uncultured psychrophilic crenarchaeotes is of course completely unknown because they are phylogenetically different to the cultured thermophilic crenarchaeotes. We might guess that they share the common features of hydrogen- and sulfur-based chemolithotrophy, but they could equally well be chemoorganotrophs like some of the cultured types. Genomic analysis using 'chromosome walking' (*Section 2.6.9*) and searching for genes encoding specific metabolic enzymes could provide the answer. It may also be possible to collect sufficient biomass from concentrated water samples to perform biochemical analyses, but efforts to bring these organisms into culture using new methods (*Box 2.1*) will probably pay the greatest dividends.

One species of crenarcheote, *Crenarchaeum symbiosium*, forms a symbiotic association with the cold-water sponge *Axinella mexicana* and accounts for 65% of the microorganisms associated with the tissues of the sponge. It grows well at 10°C. A number of other sponge species have now been found to harbor *Archaea* as symbionts and it is likely that this is a very widespread phenomenon. This may have significant biotechnological potential, since sponges are rich sources of natural products. Evidence is accumulating that many of these compounds are actually synthesized by microorganisms colonizing the tissue, and isolation and characterization of these organisms could yield rich rewards.

References and further reading

Boetius, A., Ravenschlag, K., Schubert, C.J., *et al.* (2000) A marine microbial consortium apparently mediating anaerobic oxidation of methane. *Nature* **407**: 623–626.

Danson, M.J., and Hough, D.W. (1998) Structure, function and stability of enzymes from the Archaea. *Trends Microbiol* **6**: 307–314.

DeLong, E.F. (1997) Marine microbial diversity: the tip of the iceberg. *Trends Biotechnol* **15**: 203–207.

DeLong, E.F. (2000) Resolving a methane mystery. *Nature* **407**: 577–579.

DeLong, E.F. (2001) Archaeal means and extremes. *Science* **280**: 542–543.

Dworkin, M. (ed.) 2001 *The Prokaryotes: an Evolving Electronic Resource for the Microbiological Community*, 3rd edition. Springer-Verlag, New York. http://www. prokaryotes.com (accessed March 31 2003).

Fuhrman, J.A., and Davis, A.A. (1997) Widespread archaea and novel bacteria from the deep sea as shown by 16S rRNA sequences. *Mar Ecol Prog Ser* **150**: 275–285.

Garrity, G.M., Boone, D., and Castenholz, R. (eds) (2001) *Bergey's Manual of Systematic Bacteriology*, 2nd edition, Vol. 1. Springer-Verlag, New York.

Hinrichs, K., Hayes, J.M., Sylva, S.P., Brewer, P., and DeLong, E.F. (1999) Methane-consuming archaebacteria in marine sediments. *Nature* **398**: 802–805.

Hohn, M.J., Hedlund, B.P., and Huber, H. (2002) Detection of 16S rDNA sequences representing the novel phylum 'Nanoarchaeota': Indication for a wide distribution in high temperature biotopes. *Syst Appl Microbiol* **25**: 551–554.

Huber, H., and Stetter, K.O. (1998) Hyperthermophiles and their possible potential in biotechnology. *J Biotechnol* **64**: 39–52.

Huber, H., Hohn, M.J., Rachel, R., Fuchs, T., Wimmer, V.C., and Stetler, K.O. (2002) A new phylum of *Archaea* represented by a nanosized hyperthermophilic symbiont. *Nature* **417**: 63–67.

Karner, M.B., DeLong, E.F., and Karl, D.M. (2001) Archaeal dominance in the mesopelagic zone of the Pacific Ocean. *Nature* **409**: 507–510.

Madigan, M.T., Martinko, J.M., and Parker, J. (2003) *Brock Biology of Microorganisms*. Prentice-Hall, NJ.

Michaelis, W., Seifert, R., Nauhaus, K., *et al.* (2002) Microbial reefs in the Black Sea fueled by anaerobic oxidation of methane. *Science* **297**: 1013–1015.

Mullen, L. (2003) Life from scratch. *Astrobiology Magazine*. http://www.astrobio.net/news/article319.html (accessed April 14 2003).

Marine eukaryotic microbes

7

7.1 Introduction to the protists and fungi

The term protist was introduced to accommodate those eukaryotic organisms that did not fit into the plant, animal or fungal kingdoms. Two major divisions of the protists are recognized, called the protozoa and algae, with the commonly held belief that these represent primitive forms of animals or plants respectively. Thus, protozoa and algae have traditionally been the province of zoologists or botanists respectively, and separate classification schemes for each group have been developed. However, both ultrastructural and molecular studies (largely based on SSU rRNA sequencing) show that neither the algae nor the protozoa are monophyletic groups. Different algal lineages appear to have evolved independently on several different occasions through the independent acquisition of chloroplasts by endosymbiosis events. Furthermore, many protists have nutritional features typical of both plants and animals, because they can carry out photosynthesis or engulf particulate food (phagotrophy). Therefore, the terms protozoa and algae should not be used to designate a formal taxonomic status, but they remain useful to encompass groups of organisms that share some biological and biochemical characteristics. Whilst molecular methods are beginning to shed new light on classification schemes (*Figure 7.1*), the picture remains very confused. Traditional classification schemes used by many marine scientists still recognize distinct groups of the protozoa and algae. In the sections that follow, no attempt is made to present a formal classification scheme and only a brief description of the properties of microbial marine protists that have particular importance in marine processes is given. Many protozoa are important parasites of marine animals, but space does not permit their inclusion here. The larger multicellular algae (seaweeds) are also not covered.

The fungi are believed to constitute a monophyletic group (true fungi, *Eumycota*) and may exist as unicellular types (yeasts) and filamentous or mycelial forms (molds), some of which may be multicellular or coenocytic, with visible fruiting bodies. All fungi are saprotrophic, absorbing nutrients from the environment, and photosynthesis and phagotrophy do not occur. The study of marine fungi (mycology) is considerably less developed than that of the protists and it is commonly assumed that they do not play a significant role in marine ecosytems. This is probably a reasonable conclusion in pelagic environments, but is almost certainly an oversimplification in coastal habitats and sediments.

7.2 Overview of eukaryotic cell structure and function

It is assumed that most readers will be familiar with the basic biology of the eukaryotic cell, so only a brief overview is given here. Protists and fungi possess a defined nucleus bounded by a double nuclear membrane, which is continuous with a membranous system of channels and vesicles called the endoplasmic reticulum. This is the site of fatty acid synthesis and metabolism and is also lined with ribosomes responsible for protein synthesis. The Golgi apparatus (a series of flattened membrane vesicles) processes proteins for extracellular transport and is also responsible for the formation of lysosomes, membrane-bound vesicles containing digestive enzymes that fuse with vacuoles in the cell, either for the digestion of food in phagocytic vacuoles or for the recycling of damaged cell material.

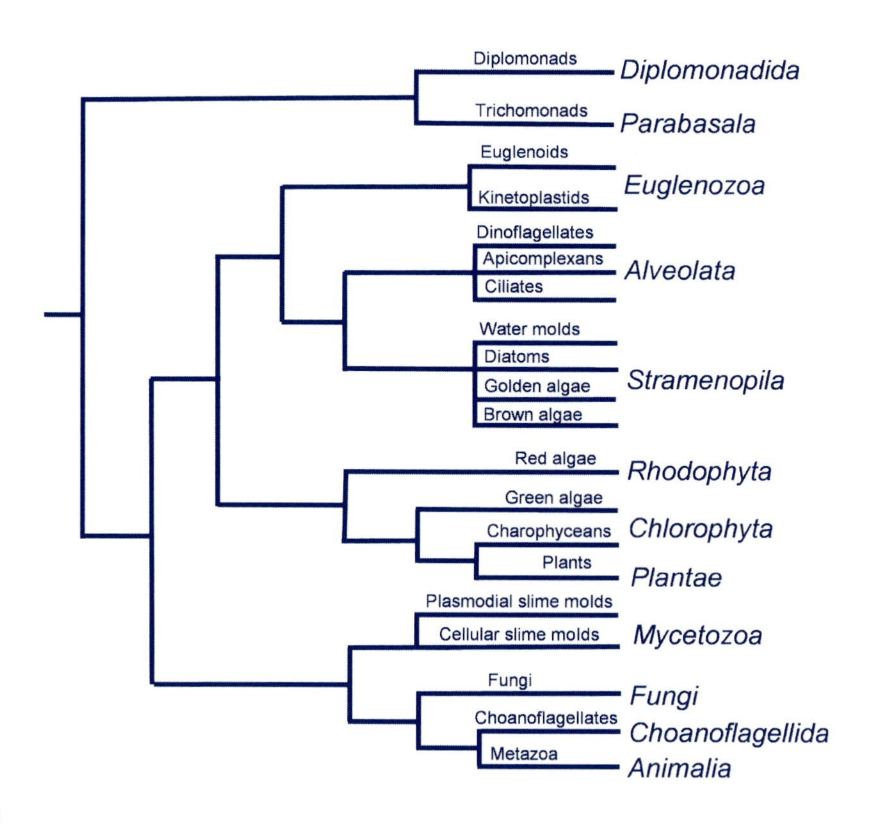

Figure 7.1

Possible phylogenetic relationships of major eukaryotic groups. Note that the arrangement of groups into kingdoms, phyla and lower taxonomic units is in a state of flux. The phylogeny of the prymnesiophytes, radiolarians and foraminiferans is uncertain and these are omitted.

Most protists and fungi demonstrate some form of sexual recombination as a regular component of their life cycle in which the genomes of two cells and their nuclei is followed by meiosis. Unlike multicellular organisms, which are usually diploid (i.e. they contain two copies of each chromosome and reproduce by forming haploid gametes), protists and fungi vary greatly in whether the diploid or haploid state is the predominant phase in the life cycle. Some are polyploid, with multiple copies of genes.

The cytoskeleton is a system of hollow microtubules about 24 nm diameter (composed of the protein tubulin) and microfilaments about 7 nm diameter (composed of two actin monofilaments wound around each other). Protists often display a variety of other filaments. Microtubules often extend under the surface and provide the basic shape of the cell, or they may be grouped into bundles to support extensions of the cell such as pseudopodia. Microfilaments provide a strengthening framework for the cell. The cytoskeleton provides a mechanism for movement of intracellular structures within the cell, for example in the separation of chromosomes during nuclear division.

Flagella occur in almost half of the major lineages of the protists. The major marine 'flagellates' occur in the euglenid, dinoflagellate and choanoflagellate groups. Eukaryotic flagella are composed of nine pairs of peripheral microtubules and two pairs of central microtubules. One pair of the tubules in each pair contains the protein dynein, which moves along an adjacent microtubule; rapid and repeated bending causes the flagellum to beat. Cilia have the same basic structure as flagella, but flagella tend to be longer and have various patterns of movement, whereas cilia often cover the cell surface or are arranged in groups and beat synchronously to provide movement or to create water currents to channel particulate food into the cell. The cilia or flagella are anchored in the cell via a basal body (kinetosome), which varies greatly in its ultrastructure and can be used in classification of different groups. Flagella or cilia do not occur in fungi.

Another function of the cytoskeleton, which is of particular importance in the marine radiolarians and foraminifera, is amoeboid movement due to changes in the crosslinking state of the protein actin. Rearrangement of the membrane also enables the process of phagocytosis, which brings food particles or prey organisms into the cell, enclosed within a food vacuole, which subsequently fuses with lysosomes.

All fungi and most protists contain mitochondria, organelles with an outer membrane and an extensively folded inner membrane. The nature of the infoldings, termed cristae, is distinctive of different eukaryotic groups. There are three types of cristae: (a) lamellar (fungi, plants and animals); (b) tubular (alveolates and stramenopiles); and (c) tubular (euglenids and kinetoplastids). Some protists do not contain mitochondria. Most of these are animal parasites, but free-living amitochondrial protists may be important in marine anaerobic sediments, microbial mats and particles. Molecular studies have supported the view that mitochondria are the descendants of endosymbiotic bacteria, almost certainly *Proteobacteria.*

Photosynthetic protists contain chloroplasts. The arrangement of the internal membranes and the nature of photosynthetic pigments in chloroplasts is one of the main methods used in conventional schemes for the classification of algae. The evolutionary origin of chloroplasts is thought to be the ancestors of the modern *Cyanobacteria.* SSU rRNA analysis shows that chloroplasts have arisen on several occasions during evolution, even within groups regarded as monophyletic, such as the dinoflagellates.

7.3 Nanoplanktonic flagellates

Traditionally, the heterotrophic flagellates have been classified as 'zooflagellates' in the protozoan subphylum *Mastigophora.* However, many of these organisms are chloroplast-containing 'phytoflagellates' and these have been classified in various algal divisions. The group known as the euglenids provides the biggest challenge to traditional 'zoo-' and 'phyto-' based classification, since they may be phototrophic, phagotrophic or saprotrophic and in the past have been classified as both *Euglenozoa* and *Euglenophyta.* This section will give a brief review of only the heterotrophic and mixotrophic flagellates in the nanoplankton (2–20 μm) size range. Since the introduction of ELM and FCM, these types have been shown to be abundant in the plankton and to play a major role in microbial food webs (*Section 9.5.3*). The nanoplanktonic flagellates are the most efficient of all grazers of bacteria, which they encounter as they swim through the water. In turn, they are preyed upon by larger protists (e.g. ciliates and dinoflagellates) and provide a link to metazoan plankton in the food web. Many of these organisms can be isolated and cultured in the laboratory. Some representative species are illustrated in *Figure 7.2.*

Some of the commonest heterotrophic and mixotrophic flagellates in seawater are the chrysomonads, such as *Paraphysomonas* and *Ochromonas.* These have two flagella, one short and smooth and one long with hairs. The long flagellum is used for cell motility, whilst the shorter one is probably used to entrap bacteria and other particles. The bicosoecids have similar arrangement of flagella, but these types are often attached to particles or contained within a cup-shaped structure known as a lorica (e.g. *Bicosoeca* and *Pseudobodo*).

The kinetoplastids are unicellular flagellates currently classified in the *Euglenozoa,* which possess a unique structure termed the kinetoplast near the base of the flagellum, easily visible in the light microscope after applying DNA stains. This is a large single mitochondrion containing a few interlocked large circular chromosomes (maxicircles), which encode mitochondrial enzymes, plus thousands of smaller minicircles of DNA. The minicircles do not appear to encode any complete genes, but are involved in posttranscriptional editing of mRNA. The bodonids (e.g. *Bodo*) are kinetoplastids with oval cells about 4–10 μm long and are very common in coastal waters, often moving over surfaces using a trailing flagellum. A shorter flagellum propels water currents towards a cytopharynx lined with microtubules.

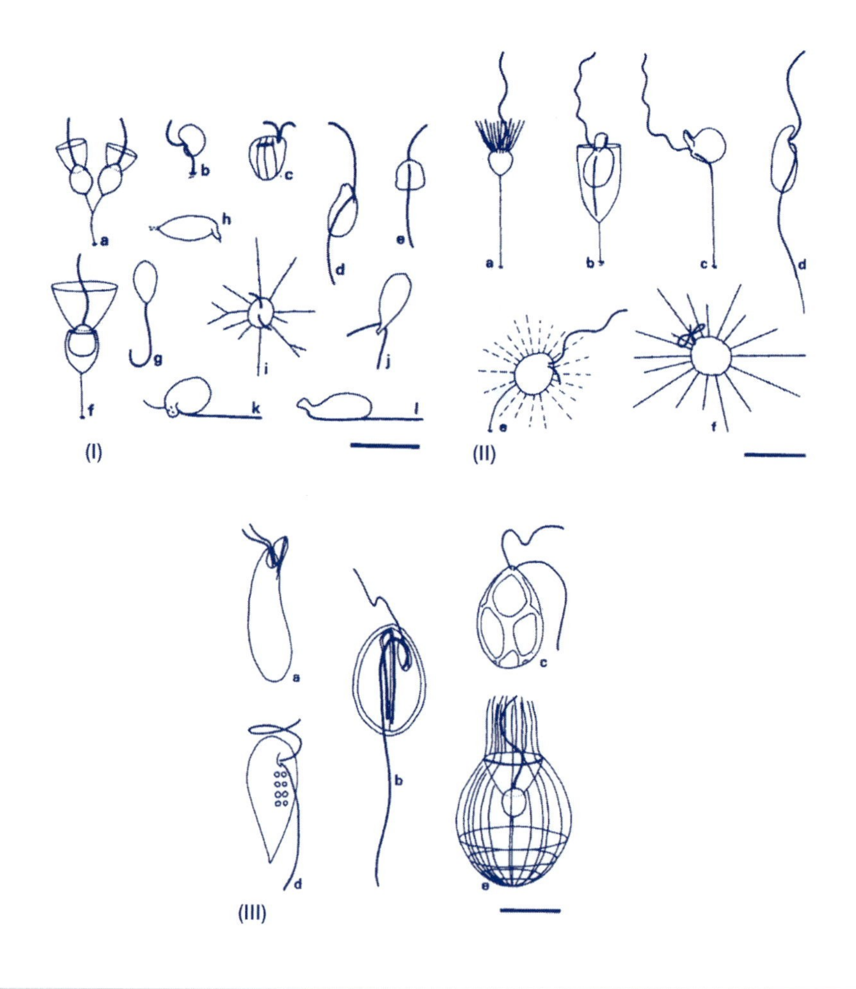

Figure 7.2

Examples of marine heterotrophic flagellated protists in the nanoplankton size range (bar = 10 μm). Reproduced from Sherr & Sherr, 2000; with permission from Wiley-Liss Inc. Key to species: **I** – (a) *Codosiga botrytis;* (b) *Cafeteria roenbergensis;* (c) *Goniomonas pacifica;* (d) *Bordnamonas tropicana;* (e) *Caecitellus parvulus;* (f) *Salpingoeca infusorium;* (g) *Metromonas simplex;* (h) *Amastigomonas debrunei;* (i) *Massisteria marina;* (j) *Telonema subtile;* (k) *Ancyromonas sigmoides.* **II** – (a) *Pteridomonas danica;* (b) *Bicosoeca conica;* (c) *Pseudobodo tremulans;* (d) *Bodo designis;* (e) *Paraphysomonas imperforate;* (f) *Ciliophrys infusionum.* **III** – (a) *Diplonema ambulator;* (b) *Ploeotia costata;* (c) *Ebria tripartite;* (d) *Leucocryptos marina;* (e) *Diaphanoeca grandis.*

The choanoflagellates are a distinctive group, which have a single flagellum that draws a flow of water through a ring of tentacle-like filaments around the top part of the cell. Bacteria are trapped and taken into food vacuoles in the cell. Some choanoflagellates produce a delicate basket-like shell around the cell. Attached choanoflagellates can form dense colonies on surfaces. In rRNA-based phylogenetic trees, the choanoflagellates appear to be the closest link with the animals and resemble the feeding cells found in sponges.

7.4 Dinoflagellates

At least 2000 species of dinoflagellates are known, of which many are marine. The most distinctive feature of the dinoflagellates is the presence of two flagella, one of which is a transverse flagellum

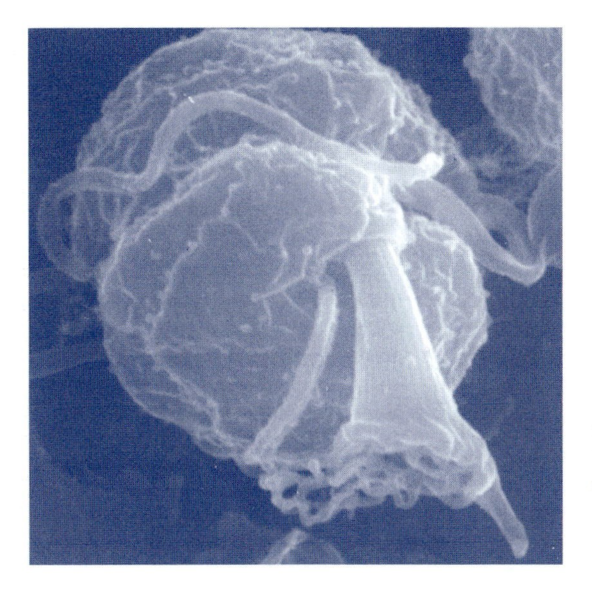

Figure 7.3

Zoospore of the dinoflagellate *Pfiesteria piscicida,* with its feeding peduncle extended. Image courtesy of Howard Glasgow, Associate Director of North Carolina State University Center for Applied Aquatic Ecology.

that encircles the body in a groove, and a longitudinal flagellum that extends to the posterior of the cell. This gives rise to a distinctive spinning motion during swimming, from which the name dinoflagellate is derived (from the Greek *dinein* = 'to whirl'). Another distinctive feature of their cell structure is the presence of a layer of vesicles (alveoli) underlying the cell membrane. In many species of dinoflagellates, these alveoli contain cellulose, which forms a protective system of plates that fits together like a suit of armor. There are numerous forms of such armored dinoflagellates, mainly in the genera *Dinophysis* (200 marine species) and *Protopteridinium* (280 marine species), often with unusual morphology with spines, spikes or horn-like projections. The geometry of the arrangement of plates is an important factor in identification. Many other naked or unarmored dinoflagellates also occur, including those that resemble species in the genera *Gymnodinium, Gyrodinium* and *Katodinium.* Dinoflagellates are traditionally regarded as photosynthetic and therefore classified as algae in the phylum *Pyrrophyta*; they make a significant contribution to CO_2 fixation and primary productivity in the oceans. However, heterotrophy occurs in about one half of known dinoflagellates and such species possess a variety of feeding mechanisms, including simple absorption of organic material, engulfment of other microbes by phagotrophy or the presence of a tube that is used to suck out the contents of larger prey (*Figure 7.3*). Such dinoflagellates can prey on phytoplankton and bacteria of all sizes, as well as zooplankton eggs, larvae, or even fish; they therefore have major effects on food web structure. Many types are mixotrophic and possess a number of very different forms in their life cycle. The most dramatic example of this occurs with the predatory dinoflagellate *Pfiesteria piscicida,* which has been described as having 24 distinct morphological stages in the life cycle, ranging from photosynthetic to an 'ambush predator' of fish (*Figure 12.4*), although this interpretation is controversial (*Box 14.1*). Most dinoflagellates are in the 20–200 μm size range, although several mixotrophic genera have been identified that appear to be below 20 μm. *Noctiluca* can reach a size of 2 mm.

Photosynthetic dinoflagellates contain chlorophylls *a* and *c,* plus carotenoids and xanthophylls. The latter pigment confers the golden-brown color typical of many photosynthetic types and gives rise to the name zooxanthellae, used to describe the dinoflagellates that form important symbioses in invertebrates such as corals, anemones and clams (*Section 10.2.2*). Endosymbioses also occur within other protists, such as ciliates, foraminiferans, and colonial radiolarians. Dinoflagellates are responsible for the formation of deleterious blooms, some of which may lead to the accumulation of toxins in shellfish and fish, causing illness and death in humans and marine animals (*Chapter 12, Sections 13.2, 14.6*).

7.4.1 Bioluminescence and biological clocks

About 2% of dinoflagellate species found in coastal waters are bioluminescent, with the best-known genera being *Noctiluca* and *Gonyaulax*. They occur worldwide, but exceptionally high densities occur in certain tropical coastal waters and produce spectacular displays of 'phosphorescence' when the surface of water is broken at night. Bioluminescence in dinoflagellates usually consists of brief flashes of blue-green light (wavelength about 475 nm), containing 10^8 photons and lasting about 0.01 second. *Gonyaulax* may also emit red light at 630–690 nm. The stimulus for light emission is deformation of the membrane due to shear forces, such as agitation of water by fish, breaking waves or the wake of a boat. In the laboratory, bioluminescence can be stimulated by lowering the temperature or pH of the medium. The bioluminescent flash is preceded by an action potential, during which the inside of the membrane becomes negatively charged. This leads to acidification of vesicles in the vacuolar membrane containing the enzyme luciferase and the substrate luciferin. There are about 400 such vesicles (scintillons) in each cell and they occur as spherical evaginations of cytoplasm into the cell vacuole, containing luciferin complexed with a special binding protein. A transient pH change results from the opening of membrane proton channels in the scintillons; this activates the reaction by release of the luciferin from its complexed state so that it can be oxidized in an ATP-mediated reaction similar to that in bacteria (*Figure 5.7*). Bioluminescence is regulated by a circadian rhythm and is much more intense during the night than the day. Dinoflagellates such as *Gonyaulax* migrate vertically in the water column more than 30 m in each direction between the surface and lower levels each day (a remarkable two million times the cell diameter). The onsets of both ascent and descent are regulated by a biological clock so that the cells anticipate sunrise and are in the optimum position to start photosynthesizing at the surface as soon as light is available. As light declines, they migrate to lower levels in the water column to take advantage of higher nutrient levels for the dark phase. During their daily migration, the cells encounter gradients of light (amount and spectral quality), temperature and nutrients and all of these can act as input signals for regulating the circadian rhythm. The circadian expression of bioluminescence involves the daily synthesis and destruction of the scintillons and component proteins; this is regulated at the translational level and may involve a clock-controlled repressor molecule that binds to mRNA. There are two main theories for the ecological function of luminescence in dinoflagellates. The most likely function is that the brief, bright flash startles potential predators and leads to their disorientation. An alternative possibility, called 'the burglar alarm hypothesis', is that light produced by luminescent prey attracts grazing predators, which in turn sends a signal to larger predators, so that the grazers themselves become prey. Grazing on bioluminescent dinoflagellates will decrease if the risk of consuming them results in reduction of the net benefit that consumers receive. Thus, bioluminescence could reduce grazing pressure by reduction of feeding efficiency and the species as a whole could benefit, even though some mortality of individuals occurs.

7.5 Ciliates

Classification and identification of groups within the ciliates is based largely on morphology, particularly the cell form and arrangement of cilia, found in at least one stage in the life cycle. At least 8000 species are known, many of which are marine. Marine ciliates are generally in the size range 15–80 μm, with some up to 200 μm. One group of ciliates, the tintinnids, produce a 'house' called a lorica, which is constructed from protein, polysaccharides and accumulated particulate debris collected from the water. As a result, tintinnid loricae are large enough to be collected in fine-mesh plankton nets and were therefore one of the first groups of marine ciliates to be studied. There has been intense interest in the ciliates in recent years because of growing recognition of the essential role that they play in the microbial loop of ocean food webs (*Section 9.5.3*) by ingesting other small protists and bacteria and being preyed upon in turn by larger protists and zooplankton. Selective grazing on particular prey types is an important factor in structuring the composition of microbial communities in food webs. As a consequence, there have been numerous studies of the abundance and activity of marine ciliates, with a wide range of results. Typically, there are about 1–150 ciliates ml^{-1} in seawater, with the highest numbers in coastal waters. Abundance varies greatly with water depth, temperature and nutrient concentration. Water stratification during the different seasons is a major factor affecting ciliate numbers.

The most common marine ciliates are spherical, oval or conical cells with a ring of cilia surrounding the cytostome ('mouth'), which they use to filter bacteria and small flagellates from the surrounding seawater. Upon entering the cytostome, ingested food particles are engulfed by phagosomes, which then fuse with lysosomes in the cytoplasm. The phagosomes become acidified and enzymes contained in the lysosome lead to digestion. Nutrients pass into the cytoplasm and undigested waste material is egested. In traditional classification schemes, the ciliates are classified as the phylum *Ciliophora* in the protozoa. Although most genera are phagotrophic, some are strict photosynthetic autotrophs. The most important of these is *Mesodinium rubrum,* which can occur in large blooms in coastal waters and may make a sizable contribution to primary productivity. This ciliate contains functional photosynthetic cryptomonads as endosymbionts. *Laboea spiralis* is a mixotroph that retains chloroplasts from ingested algal cells, a process known as kleptoplastidy (*Section 10.6*).

A distinctive feature of the ciliates is the possession of two types of nuclei. The larger macronucleus consists of multiple short pieces of DNA and is concerned largely with transcription of mRNA and growth processes. The smaller micronucleus is diploid and is responsible for an unusual type of sexual reproduction (conjugation), in which two cells fuse for a short period and exchange haploid nuclei derived from the micronuclei by meiosis. The macronuclei disappear during this process. The result of conjugation is that each partner ends up with one of its own haploid nuclei and one from its partner; these then fuse and the cells separate. In each cell, the diploid nucleus thus formed divides and differentiates into micro- and macronuclei.

7.6 Diatoms

Diatoms are usually regarded as members of the algal group *Chrysophyta,* whilst in molecular-based classification schemes they form part of the stramenopiles, which also includes the multicellular brown algae and the nonphotosynthetic oomycetes (water 'molds'). The stramenopiles are also known as heterokont algae, because of the presence of two different types of flagella. The major pigments in this group are usually chlorophylls *a* and *c* and the carotenoid fucoxanthin. It is thought that the chloroplast in stramenopiles arose by a secondary endosymbiosis event involving a red alga ancestor. Carbohydrates are stored as a β-1,3-linked glucose polysaccharide. The diatoms are one of the largest groups of the protists and over 10 000 species have been described in both fresh and marine waters. For many years, diatoms have been recognized as the dominant member of the marine phytoplankton and therefore attributed with the major role in primary productivity. In some habitats (especially coastal waters containing high levels of nutrients) this presumption is still valid, but we now realize that *Cyanobacteria* are dominant in many parts of the world's oceans. The complex ecological factors that determine community composition of the phytoplankton are discussed in *Chapter 9.* The most distinctive feature of diatoms is enclosure of the cell within a frustule composed of two overlapping plates (thecae) of silicon dioxide (silica). These glass structures form a variety of shapes of great architectural beauty and are highly distinctive for identification (*Figure 7.4*). The nature of the frustule enclosing the diatom cell results in an interesting reproductive cycle. Reproduction is usually asexual, and each daughter cell constructs a new theca within the old one, resulting in a progressive reduction in cell size with each division cycle. At a critical point, usually about one third of the original size, sexual reproduction occurs. The diploid vegetative cell forms gametes via meiosis and these fuse to form a zygote, which increases in size and synthesizes new full-size thecae. Hence, sexual reproduction accomplishes genetic variability as well as allowing the cell line to regain maximum size

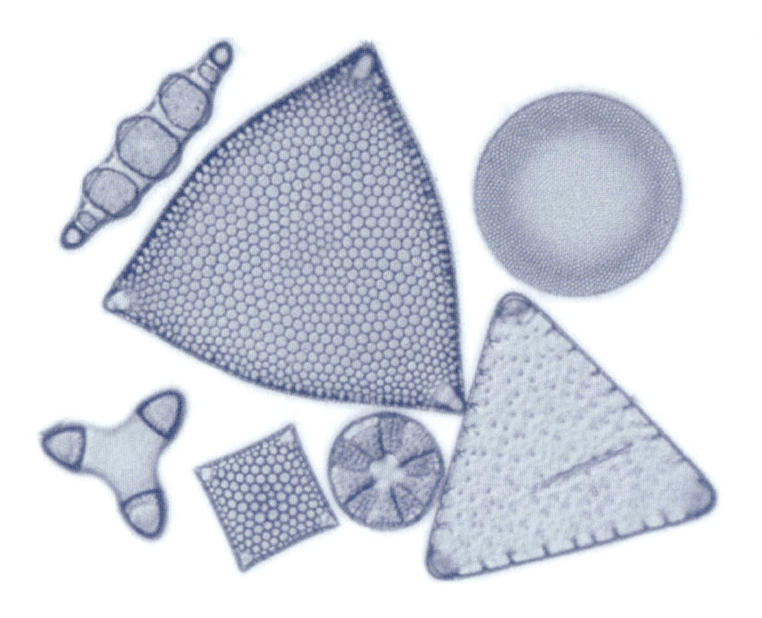

Figure 7.4

Representative marine diatoms, acid cleaned to show their silica cell walls. Image courtesy of the National Institute of Water & Atmospheric Research, New Zealand (Healey Collection) @ http://www.niwa.co.nz/.

before the next round of vegetative cell division. Diatoms possess either radial or bilateral symmetry; these forms are termed centric or pennate, respectively. Most diatoms are free-living in the plankton, but many attach to surfaces such as marine plants, molluscs, crustaceans and larger animals. The skin of some whales has been shown to have dense colonies of diatoms. Locomotion of diatoms occurs in contact with surfaces, probably via the action of minute filaments protruding through slits or pores in the frustule. Some diatoms have pairs of thin spines projecting from the ends of the cells, which link with those of other cells to form long chains, thereby increasing buoyancy and forming large mats.

Diatoms are largely responsible for the spring bloom along the continental shelf in temperate waters and for seasonal blooms in regions of nutrient upwelling. A key factor determining the size of diatom blooms is the availability of nutrients (*Section 9.4*). For diatoms, the concentration of silica is a limiting factor and most diatoms will not grow at silica concentrations below about 2 μM. Some species make frustules with very high silica contents and in some waters (e.g. the Antarctic circumpolar current) the shells are very resilient to dissolution as they settle through the water. Blooms are generally followed by the exhaustion of nutrients and aggregation and sinking of diatoms. The dynamics of diatom production and settlement are still poorly understood, but are important in understanding ocean biogeochemistry. *Box 7.1* describes recent evidence linking bacterial enzyme activity in the dissolution of diatom frustules and *Box 9.1* discusses open ocean experiments that have shown the importance of iron in stimulating diatom blooms.

Diatoms began to accumulate on the seabed about 100 million years ago, and reached a peak of abundance in the middle part of the Cenozoic period. This led to thick deposits and formation of sedimentary rocks, which are mined as diatomaceous earth. This has many industrial uses such as filtration compounds, abrasives, insulating agents, pharmaceutical products and insecticides.

Few diatoms are toxic, but an important exception is the genus *Pseudonitzschia*, which produces domoic acid. This is responsible for human illness associated with shellfish consumption and for mortalities in marine mammals and seabirds (*Section 12.2.4, Box 13.1*).

Box 7.1 RESEARCH FOCUS

Breaking glass

Bacteria aid in the dissolution of diatom shells and the cycling of silica

The availability of silicon (in the form of silicic acid) is a critical factor in the growth of diatoms, which incorporate it into their shells (*Section 7.6*). When diatoms die, they aggregate and sink through the water column to the seabed. A fraction of the silica dissolves to form silicic acid, which is then available through upwelling to support more diatom growth. The input of silicic acid from terrestrial run-off is balanced by deposition in sediments, most of which occurs in the iron-limited areas of the ocean. Recent work by Kay Bidle and Farooq Azam at the Scripps Institute of Oceanography, California, shows that bacteria play a crucial role in the recycling of silicon. Bidle and Azam (1999) argue that the rate of supply of silicic acid by upwelling is the limiting factor for diatom growth in the oceans, rather than iron or nitrate. The ratio of silica dissolution to production is very variable, but in some studies has been shown to be higher than can be explained by purely chemical and physical factors. Bidle and Azam suspended debris from diatoms, lysed by freezing and thawing, in fresh seawater or in seawater passed through various pore-size membrane filters. In 0.2-μm-filtered seawater, the dissolution of silica was very low at about 0.3% per day, but increased to between 8 and 17% per day when suspended in seawater that was passed through 0.7-μm filters, or larger. To test the conclusion that bacteria are responsible for this effect, the experiments were repeated using a variety of antibacterial compounds, including sodium azide and several antibiotics. Inhibition of bacterial activity closely paralleled reduction in dissolution of the diatom shells. Diatom debris was rapidly colonized by bacteria, to levels over 10 million times higher than in the bulk seawater. Bidle and Azam used two different diatom species in their study, and the time-course of dissolution was very different, reflecting differences in the cell walls of the two species. Diatom species differ in both the thickness of their silica shells and the degree to which they are protected by glycoprotein coats, and these factors determine the extent to which the shells of different species remain intact during grazing by zooplankton and during their descent in the water column. For example, about half of the diatom ooze under the Antarctic circumpolar current is composed of shells of just one species, *Fragilariopsis kerguelensis* (Smetaceck 1999a). Bacteria denature the protective coat through the production of extracellular proteases. A wide variety of bacterial types are involved in interactions with diatom debris, but *α-Proteobacteria* and members of the CFB group seem to be most commonly associated with enzymic activity against the shells (Bidle & Azam, 2001). In other experiments, Bidle *et al.* (2002) showed that temperature strongly affects the process of bacterial dissolution of silica, affecting the amount of organic material that remains attached to diatom shell debris. In colder waters, such as the polar regions, slow bacterial activity means that more carbon will be sequestered in sediments with sinking diatom shells. Understanding this process is important for modeling the role of the oceans in the removal of CO_2 from the atmosphere, and the effects of global warming on geochemical cycles.

7.7 Coccolithophorids

Coccolithophorids are members of the algal group *Haptophyta,* also commonly referred to as the prymnesiophytes. They are major components of the phytoplankton, especially in the open ocean. There are about 500 living species of prymnesiophytes in 50 genera, with many additional species identified from fossils. Most species are marine and occur worldwide, with the greatest diversity and abundance in tropical waters. They are flagellated unicells, which usually contain a

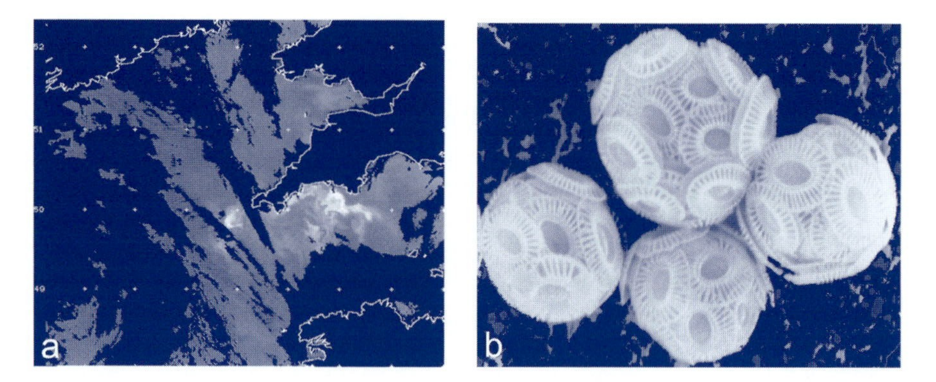

Figure 7.5

The coccolithophorid alga *Emiliana huxleyi*. (a) Satellite image of a high reflectance *E. huxleyi* bloom in the English Channel, south of Plymouth, UK. Image courtesy of SeaWiFS, processed by the Plymouth Marine Laboratory Remote Sensing Group using data from the Natural Environmental Research Council of the UK (NERC), Dundee Satellite Receiving Station, July 30 1999. (b) Scanning electron microscope image of *E. huxleyi*. Each cell is approximately 5 μm in diameter. Image courtesy of Toby Colins and Willie Wilson, Marine Biological Association of the UK, Plymouth.

haptonema; this is a thin structure reminiscent of a flagellum, but with a different structure and unknown function. The haptonema is composed of six or seven microtubules in a ring or crescent, with a fold of endoplasmic reticulum extending out within the flagellum. A particular characteristic is the complex architecture and variety of shapes shown by the external plates that typically cover the cell surface. In the coccolithophorids, of which the best-known example is *Emiliana huxleyi,* the scales are calcified (*Figure 7.5b*). Accumulation of coccolithophorid plates on the seabed contributes to the formation of ocean sediments and rocks such as the Mesozoic limestones and chalks. Fossils of this group first appeared in the Jurassic period and reached their greatest abundance in the Late Cretaceous period, at the end of which a mass extinction of many genera occurred. *E. huxleyi* has a global distribution and is intensively studied because of its role in the global carbon cycle, being the largest global producer of calcium carbonate and, hence, a major sink for CO_2. In addition, its role in the production of DMS is highly significant in global climatic processes (*Box 8.1* and *Section 9.7*). Blooms of *E. huxleyi* occur frequently in summer in the nutrient-rich coastal waters of northern Europe and Scandinavia and are highly visible as white areas in satellite images due to the coccolith plates on the cell surface (*Figure 7.5a*). Although prymnesiophyte blooms are nontoxic, they can sometimes be harmful. For example, the large amount of mucilage surrounding the cells of *Phaeocystis* is responsible for the foam that commonly affects seashores in Europe during summer; this may cause fish mortalities through clogging of the gills. Excessive production of DMS is also suspected of affecting the migration behavior of fish, which seek to avoid it.

7.8 Radiolarians and foraminifera

These two protozoan groups comprise mostly marine species. They are characterized by an amoeboid body form, using pseudopodia for locomotion and feeding. Many have diameters less than 100 μm, but some species are among the largest unicellular protists known, with diameters up to several cm.

Radiolarians are characterized by stiff needle-like pseudopodia and silicaceous internal skeletons, which can take a variety of forms used in species identification. Larger species are often associated with surfaces and may contain algal symbionts that provide some nutrients to the cell, whilst the smaller types occur throughout the water column and in deep-sea sediments. Densities vary greatly, ranging from 10 000 m^{-3} in some parts of the subtropical Pacific Ocean to less than 10 m^{-3} in the Sargasso Sea. The silicaceous skeletons of radiolarians deposit as microfossils and are second only to diatoms as a source of silica in sediments. The cell body consists of a central mass of cytoplasm surrounded by a capsular wall. This contains pores through which the cytoplasm extrudes into an extracapsular cytoplasm and forms the stiffened pseudopodia. The cytoplasm moves by streaming and captures other protists and small zooplankton, which are then surrounded and digested in food vacuoles. Hydrated, polymerized silicon dioxide is deposited within a framework of the cytoplasm. Reproduction occurs asexually by binary fission or sexually via the production of haploid gametes.

Foraminifers secrete shells (known as tests) composed of calcite (calcium carbonate) and live mainly in the surface waters of open oceans and deep marginal seas, but are largely absent from shelf areas. Foraminifers make up only a very small component of the plankton, but their discarded skeletons are the major source of calcareous deposits, accumulating as the 'globigerian ooze' over vast areas of the ocean floor. Massive deposits accumulated during the Tertiary period, about 230 million years ago, and these have become uplifted over time and exposed as limestone beds in Europe, Asia and Africa. The famous White Cliffs of Dover in southern England are composed almost entirely of foraminiferan shells. Global production of calcite by foraminifers today is about two gigatons per year, but only 1–2% of this reaches the ocean floor. The preservation of shells depends on the biogeochemistry of the water column and on the rate of sinking (*Section 1.7.2*). The single-celled organism forms a multichambered shell within a cytoplasmic envelope produced by pseudopodia; the shapes of these are highly distinctive for species identification. The chambers are connected by openings and have sealed pores that face the external environment, through which cytoplasmic spines stretch for long distances to form a net for the capture of prey, which can include bacteria, phytoplankton and small metazoan animals.

7.9 Fungi

Fungi are generally regarded as being primarily terrestrial organisms and, as noted above, have been relatively little studied in marine habitats. Of nearly 100 000 described species of fungi, only about 500–1500 are marine. There appear to be some obligately marine organisms; for example the *Halosphaeriales,* an order of the fungal division *Ascomycetes,* is composed almost entirely of marine types (43 genera and 133 species). Many other fungi associated with freshwater and terrestrial habitats can grow in seawater, for example, when decomposing detritus finds its way into the sea.

Fungi are heterotrophic and the majority are saprophytes playing a major role in decomposition of complex organic materials in the environment. The best-characterized marine fungi are those associated with decomposition of plant material (e.g. decaying wood, leaves and intertidal grasses). Recent studies have shown that salt marsh plants and tropical mangroves are particularly rich sources of marine fungi and it is likely that further investigation will reveal many more novel species. These could have biotechnological potential as sources of enzymes and pharmaceuticals. Other fungal habitats include estuarine muds, the surface of algae, corals, sand and the intestinal contents of animals. The role of fungi in the decomposition of wood is particularly important, especially in low-oxygen environments such as estuarine muds and mangroves, which inhibit wood-boring invertebrates. Few bacteria are active in degradation of lignocellulose. As well as their importance in decomposition of plant detritus, mangrove roots or tree branches washed downstream from rivers, fungi cause damage through attack of wooden structures such as piers and pilings. The enzymes involved in degradation of cellulose, hemicellulose and lignin have been characterized from a few types of marine fungi.

There has been considerable recent progress in the isolation and identification of marine yeasts (unicellular fungi). The isolation frequency of yeasts falls with increasing depth, and yeasts in the order *Ascomycetes* (e.g. *Candida, Debaryomyces, Kluyveromyces, Pichia,* and *Saccharomyces*) are commoner in shallower water, whilst yeasts belonging to the *Basidiomycetes* are commonest in deep waters (e.g. *Rhodotorula* has been isolated at a depth of 11 000 m). Marine yeasts have been investigated for their potential for use in aquaculture feeds (as a replacement for live larval food and as probiotic supplements) and for the production of fuel ethanol or single-cell protein from the breakdown of chitin in fisheries waste.

Marine lichens are a familiar sight on the rock surfaces in intertidal zones. Lichens consist of an intimate mutualistic symbiotic association between a fungus and an alga or cyano-bacterium. The fungus produces the mycelial structure that allows the lichen to attach firmly to rocks and it also produces compatible solutes that help to retain water and inorganic nutrients from the atmosphere, receiving organic material from the photosynthesis of the algal partner. Marine fungi are also involved in symbiotic associations with salt marsh plants, through the formation of mycorrhizae.

Some fungi are pathogens of marine animals (such as crustaceans, corals, mollusks and fish) or plants (such as seaweeds, intertidal grasses and mangrove roots), but have generally been poorly investigated.

References and further reading

Anderson, O.R. (2001) Protozoa, radiolarians. In: Steele, J., Thorpe, S., and Turekian, K. (eds) *Encyclopedia of Ocean Sciences,* pp. 2315–2319. Academic Press, New York.

Bidle, K.D., and Azam, F. (1999) Accelerated dissolution of diatom silica by marine bacterial assemblages. *Nature* **39**: 508–512.

Bidle, K.D., and Azam, F. (2001) Bacterial control of silicon regeneration from diatom debris: significance of bacterial ectohydrolases and species identity. *Limnol Oceanog* **46**: 1606–1623.

Bidle, K.D., Manganelli, M., and Azam, F. (2002) Regulation of oceanic silicon and carbon preservation by temperature control on bacteria. *Science* **298**: 1980–1984.

California Academy of Sciences. *Diatom Collection.* http://www.calacademy.org/ research/diatoms/ (accessed March 31 2003).

Capriulo, G.M. (1997) *Ecology of Marine Protozoa.* Oxford University Press, Oxford.

Emiliana huxleyi website. University of Southampton. http://www.soes.soton.ac.uk/ staff/tt/ (accessed March 31 2003).

Hyde, K.D., Jones, E.B.G., Leano, E., Pointing, S.B., Poonyth, A.D., and Vrijmoed, L.L.P. (1998) Role of fungi in marine ecosystems. *Biodiv Conserv* **7**: 1147–1161.

Moss, S.T. (ed.) (1986) *The Biology of Marine Fungi.* Cambridge University Press, Cambridge.

Patterson, D.J., and Larsen, J. (eds) (1991) *The Biology of Free-Living Heterotrophic Flagellates.* Clarendon Press, Oxford.

Planktonic Ciliate Project. University of Liverpool. http://www.liv.ac.uk/ciliate/ index.htm (accessed March 31 2003).

Pointing, S.B., and Hyde, K.D. (2000) Lignocellulose-degrading marine fungi. *Biofouling* **15**: 221–229.

Protist Image Database. University of Montreal. http://megasun.bch.umontreal.ca/protists/ protists.html (accessed March 31 2003).

Reid, P.C., Turley, C.M., and Burkill, P.H. (1991) *Protozoa and Their Role in Marine Processes. Proceedings of Advanced Study Institute, Plymouth (UK).* Springer-Verlag, Berlin.

Round, F.E., Crawford, R.M., and Mann, D.G. (1990) *The Diatoms.* Cambridge University Press, Cambridge.

Schiebel, R., and Hemieben, C. (2001) Protozoa, plankton foraminifera. In: Steele, J., Thorpe, S., and Turekian, K. (eds) *Encyclopedia of Ocean Sciences,* pp. 2308–2314. Academic Press, New York.

Sherr, E., and Sherr, B. (2000) Marine microbes: an overview. In: Kirchman, D.L. (ed.) *Microbial Ecology of the Oceans,* pp.13–46. Wiley-Liss Inc., New York.

Smetaceck, V. (1999a) Bacteria and silica cycling. *Nature* **397**: 475–476.

Smetaceck, V. (1999b) Diatoms and the ocean carbon cycle. *Protist* **150**: 25–32.

Sze, P. (1997) *A Biology of the Algae,* 3rd edn. William C Brown, Dubuque, IA.

Wilson, T. and Hastings, J.W. (1998) Bioluminescence. *Ann Rev Cell Devel Biol* **14**: 197–230.

Marine viruses

8.1 The nature of marine viruses

Viruses are the smallest and most abundant members of marine ecosystems. Viruses are small particles, typically between 20 and 200 nm in size, which consist of nucleic acid (either DNA or RNA) surrounded by a protein coat (capsid). They are incapable of independent metabolism and growth and replicate as obligate intracellular parasites by taking over the biosynthetic machinery of a host cell. All forms of prokaryotic and eukaryotic organisms are susceptible to virus infection. The true importance and abundance of viruses in marine ecosystems and global processes has only been discovered quite recently and during the 1990s the field has developed to become one of the most exciting branches of marine microbiology. *Table 8.1* shows a list of representative marine virus families and their hosts. In this chapter, the focus will be on the properties and activities of viruses indigenous to the marine environment with respect to their interactions with other members of the plankton, especially prokaryotes and protists. Viruses

Table 8.1 Examples of viruses infecting marine organisms

Virus family	Nucleic acid[1]	Shape	Size (nm)	Host
Myoviridae	dsDNA	Polygonal with contractile tail	80–200	Bacteria
Podoviridae, Siphoviridae	dsDNA	Icosahedral with noncontractile tail	60	Bacteria
Microviridae	ssDNA	Icosahedral with spikes	23–30	Bacteria
Leviviridae	ssRNA	Icosahedral	24	Bacteria
Corticoviridae, Tectiviridae	dsDNA	Icosahedral with spikes	60–75	Bacteria
Cystoviridae	dsRNA	Icosahedral with lipid coat	60–75	Bacteria
Lipothrixviridae	dsDNA	Thick rod with lipid coat	400	Archaea
SSV1 group	dsDNA	Lemon shaped with spikes	60–100	Archaea
Parvoviridae	ssDNA	Icosahedral	20	Crustacea
Caliciviridae	ssRNA	Spherical	35–40	Fish, marine mammals
Totiviridae	dsRNA	Icosahedral	30–45	Protozoa
Reoviridae	dsRNA	Icosahedral with spikes	50–80	Crustacea, fish
Birnaviridae	dsRNA	Icosahedral	60	Molluscs, fish
Adenoviridae	dsDNA	Icosahedral with spikes	60–90	Fungi
Orthomyxoviridae	ssRNA	Various, mainly filamentous	20–120	Marine mammals
Baculoviridae	dsDNA	Rods, some with tails	100–400	Crustacea
Phycodnaviridae	dsDNA	Icosahedral	130–200	Algae
Iridoviridae	dsDNA	Icosahedral	125–300	Fish
Rhabdoviridae	ssRNA	Bullet-shaped with projections	100–430	Fish

[1] ds = double-stranded, ss = single-stranded.

which infect animal hosts such as invertebrates, fish or marine mammals will be considered in other chapters, as will human viruses introduced via sewage pollution.

8.2 Viruses infecting prokaryotes

Even though studies of marine viruses that infect bacteria (bacteriophages) have been made since the 1950s, the significance of early findings was not appreciated and it was not until the late 1980s that serious attention was paid to this field. It should be noted that use of the term bacteriophage does not distinguish viruses which infect *Bacteria* and those which infect *Archaea* since it was introduced before the recognition that these are distinct domains of prokaryotes. An increasing number of viruses which infect *Archaea* are now being discovered, but no specific term to denote them has yet been coined. The abbreviated term 'phage is therefore used in this chapter, except where reference is made to viruses infecting a specific group.

The relatively slow development of this area of research may be linked to the long-held fallacy that microbial populations in the oceans are insignificant and that, by association, viruses must also be unimportant. As discussed in *Chapter* 5, early marine microbiologists seriously underestimated the abundance and diversity of bacteria because of reliance on inappropriate culture methods. The classical method of enumerating 'phages relies on the formation of plaques of lysis in lawns of susceptible hosts grown on agar plate (as shown in *Figure 8.1*). This method can be used with 'phages which infect *Vibrio* spp. and other easily cultivated marine bacteria. Plaque assays can also be used to enumerate some 'phages which infect *Cyanobacteria* such as *Synechococcus* (these are called cyanophages). However, these methods are limited by the fact that less than 1% of known marine prokaryotes can be cultured. The development of methods for direct observation and molecular analysis led to breakthroughs in the study of both marine prokaryotes and their 'phages and a subsequent realization that, far from being insignificant players in marine systems, 'phages play a critical and central role in ocean food webs. This ecological importance is considered further in *Section 9.5.4*.

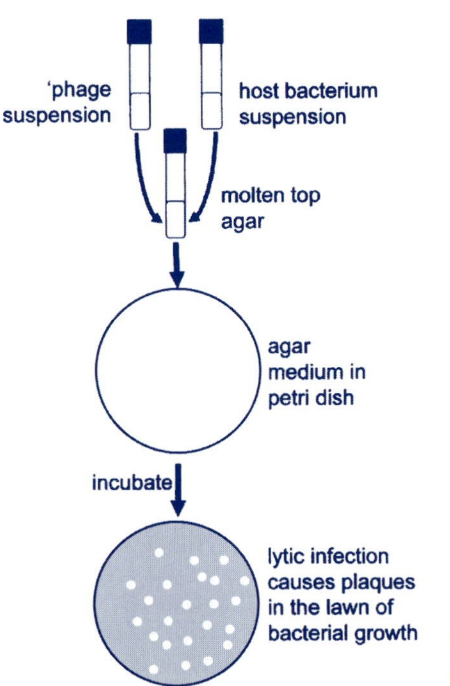

Figure 8.1

Plaque assay for lytic bacteriophages.

8.3 Enumerating viruses and virus-like particles

The most productive method of studying free marine viruses is by direct microscopic observation of seawater samples, which avoids the need for infection of a cultured host. However, strictly we should refer to structures observed in this way as 'virus-like particles' (VLP) until it has been demonstrated that they can infect and cause the death of a suitable host. In order to visualize viruses, it is necessary to concentrate seawater samples by high-speed centrifugation or ultrafiltration through hollow-fiber filters. Although technically straightforward, observing and enumerating viruses in seawater samples requires great care and control of variables such as the density of suspended particles. Typically, seawater volumes of 20 liters or more are concentrated for microscopic examination. Concentrated samples may be placed onto grids and negatively stained with uranyl acetate (or similar heavy metal stain) for examination by TEM. TEM is of great value because it shows the morphology of viruses, although preparation of samples is time-consuming and electron microscopes cannot be used on board ships at sea. A more commonly used approach for counting is an adaptation of ELM, in which the sample is treated with a fluorochrome which binds to nucleic acid (see *Section 2.2.4*). Stains include DAPI and the recently introduced fluorochromes Yo-Pro-1 and SYBR Green. VLP appear as very small bright dots that can be distinguished from the larger stained cells of bacteria and other members of microscopic plankton (*Figure 8.2*). Using modern digital imaging systems linked to the light microscope, good agreement between TLM and ELM counts can be obtained, although TEM counts are usually somewhat lower than those obtained by ELM. The reasons for this are not entirely clear, but some small filamentous viruses may not be visible in TEM because they are obscured by sediment particles. Since viruses are below the limits of resolution of the light microscope, ELM provides no information about the morphology of viruses (the reason that they can be seen with this method is because of the bright speck of light emitted by the fluorochrome).

FCM (*Section 2.3*) can also be used to enumerate viral populations and study community dynamics. Recent improvements include the use of immunolabeling or virus-specific probes for detection of infected cells.

Figure 8.2

Epifluorescence light micrograph of filtered seawater stained with SYBR Green. The numerous smallest bright specks are virus-like particles, the larger units are bacteria. Image courtesy of Jed Fuhrman, University of Southern California.

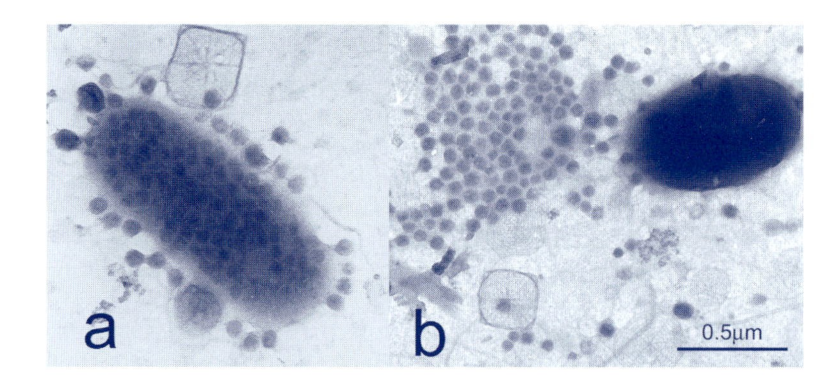

Figure 8.3

Virus-like particles infecting bacteria in natural seaweater. Transmission electron micrographs, stained with uranyl acetate. (a) shows mature phages inside the host cell before lysis; (b) shows free VLP. Image courtesy of Mikal Heldal and Gunnar Bratbak, University of Bergen.

PFGE (*Section 2.6.8*) can be used to separate large DNA fragments and has been used to study viral genomes in water samples. With careful control of the concentration steps, it can provide quantitative estimates of the abundance of virus genomes. If specific oligonucleotide probes are available for a virus, these can be applied to Southern blots of the PFGE gels to detect particular viruses.

PCR (*Section 2.6.3*) can also be used to amplify and detect specific gene sequences if these have been obtained from the target virus. Obviously, this method depends on the availability of sequence data and it is necessary to choose a well-conserved and unique sequence. Quantitative PCR has been applied successfully to the enumeration of cyanophages.

8.4 Morphology of marine viruses

Using TEM, the majority of free VLP observed in seawater have capsids with pentagonal or hexagonal icosahedral three-dimensional symmetry (*Figure 8.3*). They vary in size, with diameters usually in the range of 30–100 nm. Some very large virus-like structures (up to 750 nm) have also been observed. Both tailed and nontailed forms occur and appendages such as capsid antennae or tail fibers can sometimes be seen. The method of sample preparation has a great influence on the observed morphology. Generally, details of morphology are much easier to obtain in viruses from laboratory-cultured hosts than in free VLP observed directly in seawater samples.

8.5 Estimates of virus abundance

Using these methods, many studies have now been carried out to determine virus densities in seawater from various geographic sites and different depths. Counts obtained in these studies vary from about 10^4–$10^8\,ml^{-1}$. As a general approximation, a consensus value of about 10^6–$10^7\,ml^{-1}$ may be taken as representing the typical density of VLP in most seawater samples, although it is hard to generalize to all depths and locations. In the open ocean, virus density declines rapidly below a depth of 250 m to a relatively constant value of about $10^6\,ml^{-1}$. A rough calculation, using the volume of water in the oceans, produces the staggering result that there may be as many as 10^{30} marine viruses. Counts in coastal waters are usually higher than in the open ocean and are not so dependent on depth. The distribution of viruses in the water column generally mirrors the productivity and density of host populations of bacterioplankton and, to a lesser extent, phytoplankton. Typically, there are about 5–15 times as many viruses as bacteria in most seawater samples. Changes in virus density are very dynamic and large fluctuations can occur over short timescales due to synchronized lysis of host cells and degradation of released virus particles. This emphasizes the important role of viruses as an active component of marine microbial communities. This will be discussed further in *Chapter 9.*

8.6 Observing phage-infected cells

It is possible to use high-speed centrifugation to obtain a pellet of plankton which can be embedded in resin and sectioned for TEM. Examination of such thin sections of bacterioplankton from various locations reveals that, typically, 1–4% of cells contain mature, fully assembled virus particles. Viruses can only be seen within infected host cells in the final stages of the lytic cycle of infection. Since this stage usually represents about a quarter of the total lytic cycle, it is possible to estimate the total proportion of bacterioplankton which are infected at any one time. This value varies from about 10 to 40%.

The 'burst size' of virus infection is a term used to describe the number of progeny virus particles released from the host cell at the time of lysis. Knowing this value is important for modeling the dynamics of viral infection of host populations. The larger the burst size, the smaller the number of host cells that need to be lysed to support viral production. Burst size can be measured directly by observing cells using TEM, or it can be calculated using models of virus:host ratios and theoretical rates of contact and infection required to maintain virus production. Lysis can also be induced artificially by the addition of the antibiotic streptomycin. There is a wide variation in results, depending on the location of the study and the methods used, but values in the range 10–50 progeny viruses per cell are typical for *in situ* studies. For those phages which can be propagated in laboratory cultures of bacteria, the burst sizes are larger (average about 180) because *in-vitro*-grown bacteria are bigger and support greater virus densities. In *Synechococcus*, a large burst size of about 100–300 occurs. Virus production can also be estimated directly by measuring the rate of incorporation of radioactively labeled ^{3}H-thymidine or ^{32}P-phosphate into viral nucleic acid.

8.7 Virus inactivation

Loss of infectivity of viruses arises due to irreparable damage to the nucleic acid or protein capsid. Usually, a virus will lose its infectivity before showing obvious signs of degradation. However, since most marine viruses are studied by direct TEM or ELM observation, the term 'virus decay' reflects the observation of a decline in numbers of VLP over time, in the absence of new viral production. Many of the studies of virus inactivation in water have been carried out in connection with the health hazards associated with sewage-associated viruses (such as enteroviruses or coliphages, *Section 11.3*) in waters for swimming or cultivation of shellfish. Subsequently, the results of these studies have been applied to the population dynamics of indigenous marine viruses. A wide range of physical, chemical and biological factors can influence virus infectivity. Different studies have produced various estimates of decay rates, but a value of about 1% per hour is typical in natural seawater kept in the dark. Visible light and UV irradiation are by far the most important factors influencing survival, and in full-strength sunlight the decay rate may increase to 3–10% per hour, and can be as high as 80%. Light will have its greatest effect in the upper part of the water column, but is probably still effective down to about 200 m in clear ocean water. Even in very turbid coastal waters, virucidal effects can be observed down to several meters. Such high rates of inactivation would lead to the conclusion that there are no, or very few, infective viruses in the top layer of water. However, this conclusion is incorrect, probably because repair of UV-induced damage can occur by mechanisms encoded either by the host or the virus (one likely mechanism is photoreactivation, see *Box 5.2*). Another important factor in decay is the presence of particulate matter and enzymes such as proteases and nucleases produced by bacteria and other members of the plankton. Here, there are complex interactions, because adsorption of viruses to particles can also afford some protection. Virus inactivation does not proceed at a constant rate, and it appears that there is variation in the resistance of viral particles to damaging effects, presumably because of minor imperfections in the capsid. Thus, over time, inactivation will lead to a slowly decaying low level of infective particles.

8.8 Host specificity

For those marine phages which have so far been propagated in cultivated bacteria, all show specificity for particular bacterial species and sometimes for particular strains. These findings are similar to those seen in other well-known viruses and are usually thought to be due to the molecular specificity of virus receptors on the host surface, the presence of restriction enzymes or compatibility of the replication processes. However, there are indications that these highly specific interactions may be something of an artefact introduced by the assay system *in vitro* and some marine viruses (especially cyanophages) may infect related, but not identical, hosts. If correct, this has significant implications for the possibility of genetic exchange between different organisms.

8.9 Lysogeny

Virulent phages take over the host cell, replicate their nucleic acid and cause lysis of the host cell after assembly of the virus particles. However, another outcome is seen when 'phages known as temperate viruses infect the cell (*Figure 8.4*). Bacteria can enter into a state in which the virus genome replicates along with the host DNA, but is not expressed. Often, the silent viral genome is stably integrated into the bacterial genome and this latent state is known as a prophage. Bacteria infected with these phages are known as lysogenic, because under certain conditions the bacteria spontaneously lyse and release infective virus particles. Experimental conditions such as exposure to UV light, temperature shifts, treatment with antibiotics or certain other chemicals will often induce the lytic cycle. The genetic control of mechanisms which determine whether (a) the phage enters the lytic or lysogenic cycle and (b) the process by which expression of the prophage genes controlling the lytic pathway switch is repressed have been well-studied in λ-bacteriophage of *Escherichia coli* and a few other examples. However, the molecular events in lysogeny in marine bacteria are largely unexplored.

Integration and excision of the phage genome has very important evolutionary consequences, because it provides a natural mechanism (transduction) by which host genes can be transferred from one cell to another when part of the host DNA becomes incorporated into the mature virus particle. The presence of a prophage also confers resistance to infection of the host bacterium by viruses of the same type (phage immunity). This is because the virus repressor proteins which prevent replication of the prophage genome also prevent replication of incoming genomes of the same virus. Prophage genes can also affect the phenotypic characteristics of the host cell in other ways, including general reproductive fitness. An important example of phenotypic modification is the role of phages in the acquisition of virulence factors enabling colonization and toxicity in *Vibrio cholerae*, a bacterium found in coastal and estuarine waters and responsible for the human disease cholera (*Section 11.2.1*). Recent research shows that phage-mediated virulence may be very widespread in bacteria (*Box 15.1*).

Lysogenic immunity and phage conversion both carry a strong selective value for the host cell and could provide viruses with a strategy to survive periods of low host density or low metabolic activity. Since many marine environments contain low levels of slow-growing bacterioplankton

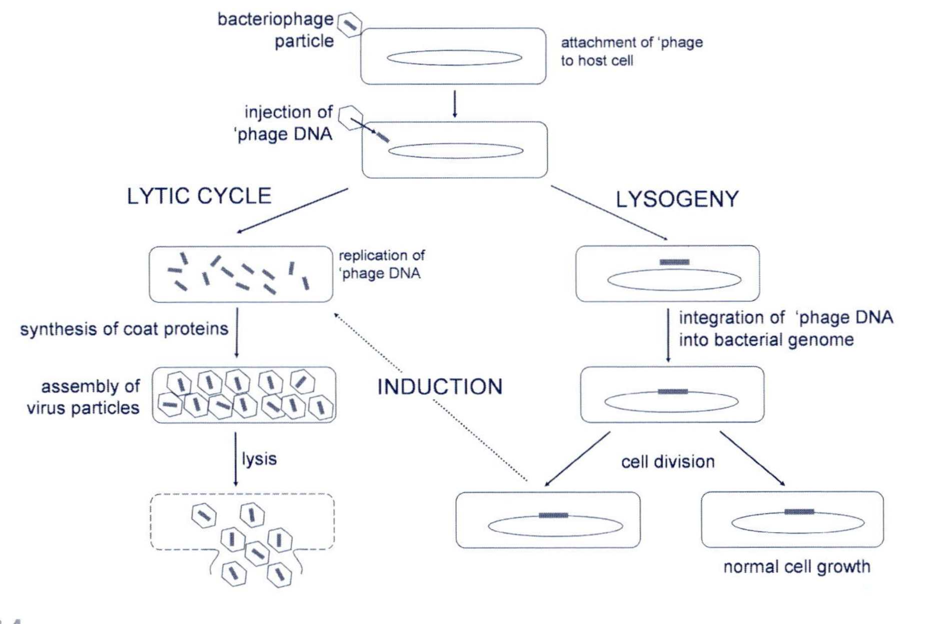

Figure 8.4

Possible outcomes of infection of a bacterial cell by a temperate DNA bacteriophage.

and because free viruses are inactivated quite quickly, virulent (lytic) phages could become rapidly depleted. For all these reasons, it would be reasonable to consider lysogeny to be the normal state of virus infections in nature. Indeed, some studies with cultivable bacteria have shown that lysogeny is more common in bacteria (up to 50%) in samples from offshore environments than nutrient-rich coastal environments. However, some other studies have produced conflicting results and we have no knowledge at all about the importance of lysogeny (or pseudolysogeny) in microorganisms which have not yet been cultured. Pseudolysogeny is a form of latent infection in which the viral nucleic acid remains in the host cell for an extended period, but it is thought that the lysis of the host cell is delayed. This may be due to nutrient deprivation and low metabolism of the host cell and might also be a common trend in oligotrophic marine waters. It is hoped that the application of new techniques will reveal the true importance of lysogeny and pseudolysogeny in marine virus–host interactions.

8.10 Effect of viruses on plankton mortality

Many studies on the effects of viruses on populations of heterotrophic marine bacteria have shown that 'phages are responsible for about 10–50% of the total mortality and may be more important in regulating bacterial communities than grazing by flagellates (see *Section 9.5.4*). One way of measuring the impact of viruses is to enrich seawater with artificially added viruses. This can be achieved by preparing concentrated viral suspensions by ultrafiltration with submicron filters. Using this method, the proportion of infected bacteria can be increased from about 10 to 35%. Photosynthetic rates in seawater can be reduced by as much as 50% following addition of viruses, but virus enrichment of oligotrophic water seems to stimulate overall bacterial growth and carbon fixation rates. This is probably due to bacterial lysis leading to increased availability of nutrients, which can be used by the noninfected cells. Interpreting data from such virus enrichment experiments may be complicated by the possibility that unknown inhibitory

Box 8.1 RESEARCH FOCUS

Do microbes fly with the clouds ... do viruses control the global climate?

The importance of DMS production in the oceans

One of the most exciting discoveries about the role of viruses in the oceans is the discovery of their role in the formation of the gas DMS from DMSP. DMSP is present in many types of phytoplankton and probably functions as an osmolyte, protecting microalgal cells from changes in salt concentration. In polar regions, it may also act as an antifreeze compound. DMSP is converted enzymatically by a group of enzymes called DMSP lyases, which are widespread in many marine microorganisms. DMS is highly volatile and escapes to the atmosphere where it leads to the formation of sulfates, which act as nuclei for water vapor causing the formation of clouds. As cloud increases, the amount of photosynthetic activity declines, less DMS is produced, until cloud formation diminishes again. Thus, there is a constant feedback mechanism. A diagrammatic representation of the role of DMS in controlling global climate is shown below. The idea that DMS production may control the global climate is a central tenet of James Lovelock's 'Gaia Hypothesis' (Charlson *et al.* 1987). Another fascinating hypothesis is that DMS production by microorganisms may have evolved as a dispersal mechanism (Hamilton & Lenton 1998). Many types of microorganisms are found in the atmosphere and it is likely that they are transported there during formation of surface convection wind and cloud formation. These authors suggest that DMS production by microbes may aid in this process.

Research into the proposed role of microbes in the ocean's carbon cycle is vital to our understanding of global warming and climate change. CO_2 levels are increasing due to the burning of fossil fuels and ozone depletion is also altering the amount of UV radiation hitting the Earth's surface, whilst emission of sulfur compounds from burning petroleum and coal adds to the naturally produced sulfates in the atmosphere. Many researchers are studying the production of DMSP and DMS, and factoring microbial processes into models of global climate change.

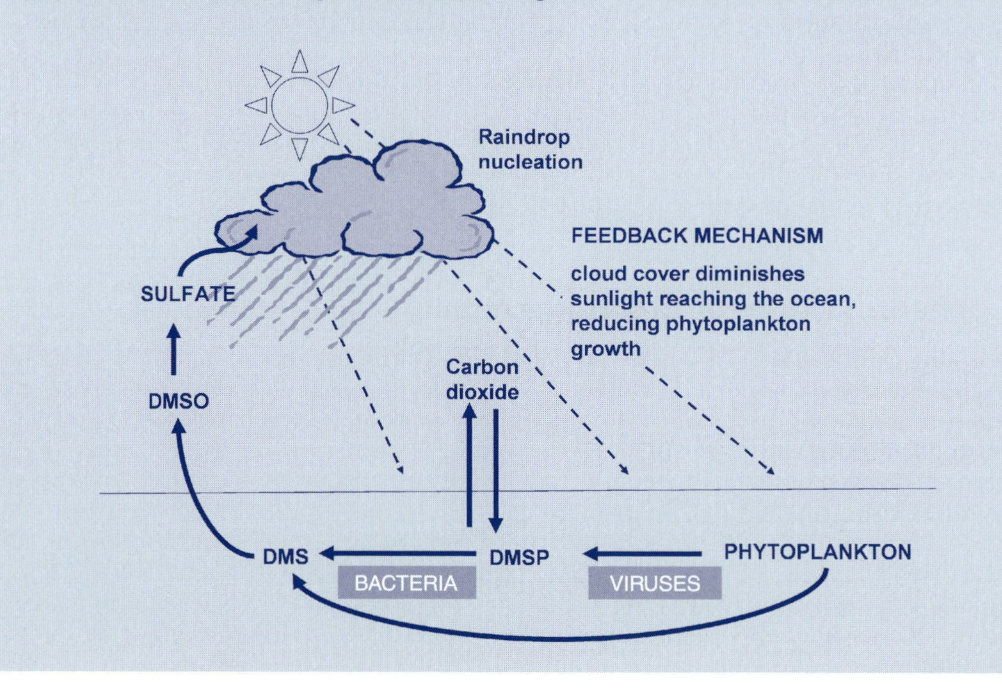

One of the most prolific producers of DMSP and DMS is the prymnesiophyte (coccolithophorid) alga *Emiliana huxleyi,* which is found worldwide and forms regular blooms in temperate marine waters. These frequently occur in summer in the nutrient-rich coastal waters of northern Europe and Scandinavia (see *Figure 7.5*). Specific viruses have been associated with natural blooms and observed during mesocosm experiments. Brussard *et al.* (1996) observed that blooms often show a spectacular 'crash' and up to half of the cells can be seen to be infected by VLP. Jacquet *et al.* (2002) describe a mesocosm experiment in which very large plastic bags (11 m^3, see *Figure 2.1*) were submerged in a fjord, filled with seawater and enriched daily with nitrate and phosphate. They monitored the development of *E. huxleyi* by FCM and observed a periodic pattern of virus production, linked to the daily light cycle and dependent on the concentration of added nutrients. The virus has now been identified by Schroeder *et al.* (2002). It is icosahedral, about 170–200 nm diameter and belongs to a new genus (named *Coccolithovirus*) in the family of algal viruses *Phycodnaviridae. E. huxleyi* possesses DMSP lyase activity, but strains vary in the amount of enzyme produced. *Coccolithovirus* seems to infect only the low lyase producers, and virus infection may not only lead to the release of DMSP from the algal cells, but may also trigger its breakdown by the algae to DMS. As well as the connection with DMS production, study of this alga and its lysis by viruses is particularly relevant to global warming studies because the calcite plates are deposited in sediments and CO_2 is released during this process. Elevated production of DMS has also been shown following viral infection of another alga, *Phaeocystis pouchetii* (Malin *et al.* 1996).

The recent discovery of such a critical role for viruses, with profound importance for the maintenance of the planet, is one of the most dramatic illustrations of the advances made in marine microbiology in the past few years.

substances are also present in the concentrated seawater. However, these experiments clearly show that viruses are capable of influencing microbial dynamics in seawater and sediments.

8.11 Viruses of eukaryotic plankton

Viruses are now being recognized increasingly in a range of algae, including coccolithophorids and dinoflagellates. There is particular interest in *Emiliana huxleyi*, where research indicates that the effects of viral infection on DMS production have major significance for global climatic processes (*Box 8.1*). Other microalgae responsible for nuisance or harmful algal blooms ('red tides' or 'brown tides', see *Chapter 12*) such as *Aureococcus anophagefferens, Phaeocystis globosa* and *Heterosigma akashiwo* have been the focus of particular attention, with research directed towards the possibility of biological control of blooms via viral infection. Viruses have also now been described from other marine protists such as flagellates and ciliates, and several research groups are studying the effects of viruses on community structure and population dynamics of these organisms.

References and further reading

Borsheim, K.Y. (1993) Native marine bacteriophages. *FEMS Microbiol Ecol* **102**: 141–159.

Brussard, C.P.D., Kempers, R.S., Kop, A.J., Riegman, R., and Heldal, M. (1996) Virus-like particles in a summer bloom of *Emiliana huxleyi* in the North Sea. *Mar Ecol Prog Ser* **128**: 133–142.

Charlson, R.J., Lovelock, J.E., Andreae, M.O., and Warren, S.G. (1987) Oceanic phytoplankton, atmospheric sulphur, cloud albedo and climate. *Nature* **326**: 655–661.

Fuhrman, J.A. (1999) Marine viruses and their biogeochemical and ecological effects. *Nature* **399**: 541–548.

Fuhrman, J.A. (2000) Impact of viruses on bacterial processes. In: Kirchman, D.L. (ed.) *Microbial Ecology of the Oceans,* pp. 327–350. Wiley-Liss Inc., New York.

Hamilton, W.D., and Lenton, T.M. (1998) Spora and Gaia: how microbes fly with their clouds. *Ethol Ecol Evol* **10**: 1–16.

Jacquet, S., Heldal, M., Iglesias-Rodriguez, D., Larsen, A., Wilson, W.H., and Bratbak, G. (2002) Flow cytometric analysis of an *Emiliana huxleyi* bloom terminated by viral infection. *Aquat Microb Ecol* **27**: 111–124.

Malin, G., Wilson, W.H., Bratbak, G., Liss, P., and Mann, N.H. (1998) Elevated production of dimethylsulfide resulting from viral infection of cultures of *Phaeocystis pouchetti. Limnol Oceanog* **43**: 1389–1393.

Proctor, L.M. (1997) Advances in the study of marine viruses. *Microscopy Res Tech* **137**: 136–161.

Schroeder, D.C., Oke, J., Malin, G., and Wilson, W.H. (2002) *Coccolithovirus* (*Phycodnaviridae*): characterisation of a new large dsDNA virus that infects *Emiliana huxleyi. Arch Virol* **147**: 1685–1698.

Thingstad, T.F. (2000) Elements of a theory for the mechanisms controlling abundance, diversity, and biogeochemical role of lytic bacterial viruses in aquatic systems. *Limnol Oceanog* **45**: 1320–1328.

van Etten, J.L., Lane, L.C., and Meints, R.H. (1991) Viruses and viruslike particles of eukaryotic algae. *Microbiol Rev* **55**: 586–620.

Wilson, W.H., and Mann, N.H. (1997) Lysogenic and lytic viral production in marine microbial communities. *Aquat Microb Ecol* **13**: 95–100.

Wommack, K.E., and Colwell, R.R. (2000) Virioplankton: viruses in aquatic ecosystems. *Microbiol Molec Biol Rev* **64**: 69–114.

The role of microbes in ocean processes

9.1 Changing paradigms

Preceding chapters have discussed some of the activities of individual types of marine microbes in major processes such as photosynthesis, chemolithotrophy, breakdown of organic material, nitrogen fixation, production/ utilization of methane and oxidation or reduction of sulfur. These processes occur in seawater and sediments, as well as in specialized habitats such as hydrothermal vents, cold seeps and epi- or endobiotic associations. In this chapter, the focus is largely on the role of planktonic microbes in processes in the upper layers of the pelagic zone and will emphasize the overall picture of nutrient cycling and its global biogeochemical significance. This exciting area of work, in which the activities of microbiologists, chemical and physical oceanographers come together has led to spectacular paradigm shifts in our view of the importance of oceanic microbes in planetary systems. Space does not permit a full historical treatment of the development of modern ideas, but it is instructive to outline some of the most important highlights from the past 30 years.

As indicated in *Chapter 2*, our altered perspective in marine microbiology has resulted almost entirely from the application of new techniques. Until the mid-1970s, marine bacteria were regarded as of little importance, other than as decomposers of detritus. The classic view of trophic interactions in the oceans was of a simple food chain in which primary production is due mainly to the algae large enough to be trapped in traditional plankton nets. These algae are consumed by copepods, which are in turn consumed by larger zooplankton, eventually reaching fish at the end of the food chain. Phytoplankton was thought to be consumed as rapidly as it is produced, with all primary production going through herbivorous macrozooplankton. Bacteria did not feature at all in this food chain. Estimates of bacterial abundance were orders of magnitude lower than is now known to be the case. This fitted with the view then prevalent that most of the oceans are biological 'deserts' with low nutrient fluxes, low biomass of phytoplankton and low productivity. Until the early 1980s, the perceived small populations of bacteria were thought to be largely dormant. As discussed in *Chapter 2,* the development of ultraclean analytical techniques, measurement of the incorporation of radiolabeled precursors and the measurement of ATP levels led to the realization that metabolic activities and productivity in seawater were much higher than previously envisaged. The development of controlled pore-size filters and ELM in combination with fluorescent DNA stains revealed that the oceans are, in fact, teeming with microorganisms (typically 10^6–10^7 per ml) in the picoplankton size class (<2 μm). ELM also led to the discovery of *Synechococcus,* a previously overlooked cyanobacterium now recognized as one of the major contributors to primary productivity. These new methods also revealed the importance of the consumption of bacteria by small phagotrophic protists, providing evidence of a direct link between bacterial production and higher components of the food web. A paper by Lawrence Pomeroy of the University of Georgia, entitled *The Ocean's Food Web: A Changing Paradigm* (Pomeroy 1974), is widely credited as being the most significant advance in our thinking about the role of microbes in marine systems. The main arguments in this paper were: (a) that the main primary producers in the oceans are 'nanoplankton' (small phototrophs less than 60 μm in size) rather than the 'net' phytoplankton previously recognized; (b) that microbes are responsible for most of the metabolic activity in seawater; and (c) that dissolved and particulate organic matter forms an important source of nutrients in marine food webs, which is consumed by heterotrophic microbes. Evidence about the role of the heterotrophic bacterioplankton steadily accumulated until a series of seminal papers was published in the early 1980s by Farooq Azam and co-workers at the Scripps Oceanographic Institute, California. They developed the concept of the 'microbial loop' to explain the flow and cycling of dissolved organic material (DOM) in the oceans. DOM is a term used to describe dissolved monomeric, oligomeric and polymeric compounds plus colloids and small cell fragments. Some measurements of DOM focus exclusively on the carbon content, leading to the term dissolved organic carbon, DOC, whilst some studies focus on particulate organic matter (POM). In fact, there is no clear cut-off between DOM and POM and *Section 1.7.1* introduced modern views about the formation and fate of marine snow particles and the gel-like nature of seawater due to polymeric substances. The development of the microbial loop concept represents a true paradigm shift, as thinking about the role of microbes in mineral cycling and food webs underwent a sudden and dramatic change. The original microbial loop model is now realized to be an oversimplification and it has been refined to encompass a broader range of trophic interactions, but the concept remains a major breakthrough in oceanography.

The next major technical advance came with the development of flow cytometers that could be used on research vessels, leading to the discovery of photosynthetic *Prochlorococcus* (*Sections 2.3, 5.4*). This organism is now known to be one of the major primary producers in the oceans, and pos-sibly the most abundant photosynthetic organism on Earth, yet it escaped detection until 1988. Since the late 1980s the use of molecular techniques has revealed unexpected diversity among prokaryotes and demonstrated the previously unimagined importance of *Archaea* as components of the picoplankton (*Chapters 5* and *6*). In the last few years, we have witnessed the discovery of previously unknown types of photosynthesis (*Box 1.1*) and other energy-generating mechanisms and appreciated the role of protists and viruses in controlling community structure in the plankton. Marine microbiology is sure to produce many more surprises in the next few years.

9.2 Carbon cycling in the oceans

The fate of carbon dominates consideration of the microbial ecology of the oceans. The unique chemical properties of carbon mean that it is present in many different inorganic and organic forms. The major reservoir of carbon is as carbonate minerals in sedimentary rocks in the Earth's crust and in organic compounds such as coal, oil and natural gas. The turnover of this carbon by natural geological processes is very slow, occurring over timescales of millennia. However, in the last few hundred years, human activities have accelerated the flux of carbon to the biosphere due to the use of fossil fuels and changes in land use.

The oceans (excluding sediments) are the largest reservoir of biologically active carbon, containing about 4×10^{13} tonnes of carbon (47 times more than the atmosphere and 23 times more than the terrestrial biosphere). Gaseous CO_2 is highly soluble in water and forms carbonic acid (H_2CO_3). At the natural pH of seawater (7.8–8.2), this mostly dissociates rapidly to form bicarbonate (*Section 1.6.1*). Absorption of CO_2 at the interface between ocean and atmosphere and its transfer to deeper layers is driven partly by physical processes. Circulation of water occurs due to turbulence created by surface winds and from temperature gradients. Towards the poles, cold water with a high density sinks vertically to a depth of 2000–4000 m and then distributes through the ocean basins. The sinking mass of water displaces deep ocean water to the surface again through upwelling. As the water warms, CO_2 escapes to the atmosphere again. This model constitutes the so-called solubility pump for CO_2 circulation in the oceans (*Figure 9.1a*).

In addition to purely physical factors, a biological pump is responsible for massive redistribution of carbon in the oceans (*Figure 9.1b*). Marine phytoplankton are responsible for about half of the global CO_2 fixation. Organic matter is released when these organisms die, or it may be released as dissolved substances. Heterotrophic organisms provide a *sink* for fixed carbon through assimilation of DOM and remineralization as CO_2, as well as a *link* to higher trophic

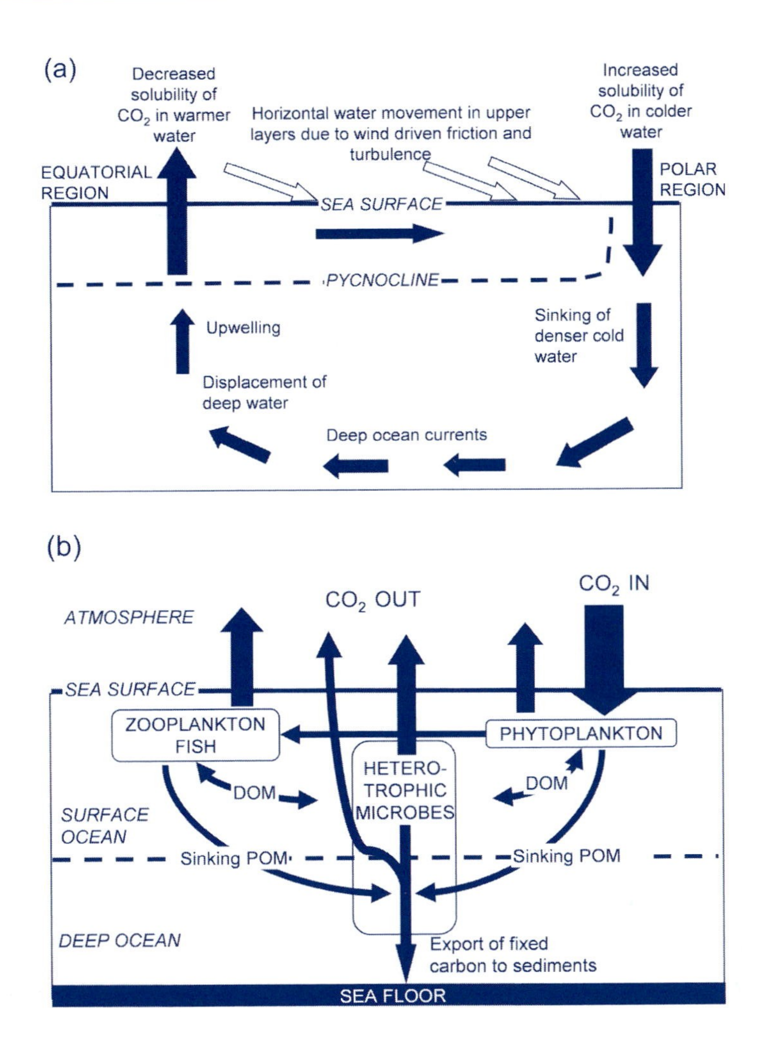

Figure 9.1

Carbon cycling processes in the oceans. (a) The solubility pump. Arrows show the circulation of water and CO_2. The dotted line represents the boundary between the upper ocean layer and deeper waters, caused by differences in temperature and density. (b) The biological pump. Arrows show the movement of CO_2 and fixed carbon compounds. DOM = dissolved organic matter, POM = particulate organic matter. The dotted line represents a depth of about 150–200 m.

levels. Although photosynthesis by the phytoplankton in the upper layers of the oceans is by far the most important source of marine primary production, the reader should bear in mind the significant local importance of primary production by photoautotrophs and chemolithotrophs in benthic microbial mats, symbiotic interactions, vents and seeps, as discussed in other chapters. Traditionally, studies of primary productivity have been concerned only with the 'net' phytoplankton (principally diatoms and dinoflagellates). Phototrophic prokaryotes and (to a lesser degree) mixotrophic protists are now known to contribute a large fraction of primary productivity. (Note: despite the inaccuracy of the prefix *phyto-*, because neither protists nor bacteria are plants, the term phytoplankton is so firmly entrenched in popular usage that it must be extended to include these groups and it is used in this broad context in this book.)

9.3 Photosynthesis and primary productivity

How much atmospheric CO_2 is fixed and incorporated into cellular material? This has been the central question in biological oceanography for over a century. Various methods are available for measuring primary productivity and these produce different results depending on the parameter measured and the timescale employed. The most common laboratory experiments measure $H^{14}CO_3^-$ incorporation into organic material, comparing the effects of incubation in light and dark conditions. Changes in O_2 concentration can be measured by high-precision titration, which indicates O_2 evolution by oxygenic photosynthesis and O_2 consumption by respiration. Alternatively, specialized equipment can be used to follow $^{18}O_2$ evolution from $H_2{}^{18}O$, indicating photosynthesis without interference from respiratory processes. All methods have certain drawbacks and it is necessary to make allowances for isotope effects and differences in the photosynthetic quotient (ratio of HCO_3^- assimilated to O_2 generated), which varies according to species and nutrient availability. It is important to recognize that growth of phytoplankton is not a simple process concerned only with these processes; growth also depends on assimilation of other nutrients and many metabolic transformations to produce macromolecules in the cell. Clearly, the discovery of widespread anoxygenic phototrophic bacteria (*Box 2.2*) means that earlier assumptions based on the use of these methods need to be reconsidered. Measuring the physiological rates of photosynthesis in the natural situation is difficult. Most experiments are conducted by enclosing seawater containing natural communities in bottles, which are then held at various depths in the field. Photosynthetic rates are measured over time under different environmental conditions, including light. Bottle experiments obviously have inherent limitations because nutrients are quickly depleted in the absence of diffusion and artifacts can be introduced during deployment and retrieval of samples, by contamination or by alteration of the spectral qualities of light entering the containers. Mesocosm experiments overcome some of these problems (*Section 2.1*).

Primary production can be defined in terms of the total amount of CO_2 fixed into cellular material (*gross* primary production) or the fraction that remains after losses due to respiration by the phytoplankton (*net* primary production). A further concept, net community production, has been introduced to encompass the effects of respiration by heterotrophic organisms. Oceanographers also use various other ways of measuring productivity, in terms of new, export and regenerated production. Both the biomass of phytoplankton and the rate of gross and net production are affected by a range of factors, which vary greatly over both long-term and short-term timescales, due to the interplay of hydrographic conditions (currents, upwelling and diffusion) with light and nutrient availability. Temperature increases both photosynthetic and respiratory rates, but this effect is complicated in the oceans by the fact that although most high-nutrient upwelling waters have a low temperature, the stimulation of photosynthetic rates by increased nutrient levels overcomes the inhibitory effects of cooling.

As discussed in *Section 1.6.2,* the penetration of water by light of different wavelengths limits photosynthesis to the upper 100–200 m of clear ocean waters, and considerably less in the presence of suspended material. Different phototrophic organisms are adapted to utilize light of different wavelengths by the use of different types of chlorophyll and ancillary pigments (*Box 2.2* and *Section 5.4.4*) and this affects their distribution in the water column. Indeed, the traditional definition of the limit of the photic zone as the depth at which light intensity is reduced to 1% of its surface level is no longer valid, since some types of *Prochlorococcus* are photosynthetic at light levels less than 0.1% of surface irradiance. Very high irradiances can inhibit photosynthesis, especially due to damage to the photosystem by UV light, and organisms possess a number of photoprotective mechanisms to overcome these effects (*Section 4.6.5*). The water depth at which incorporation of CO_2 by photosynthesis and its loss due to respiration are in balance (compensation depth) varies according to species composition of the phytoplankton, geographic region, season, light penetration during the day and nutrient availability. Clouds and atmospheric dust have a marked effect on light penetration and the density of phytoplankton itself is a major factor. It is a fallacy to think that maximum primary production occurs at the surface of the ocean; instead, it may occur some distance below the surface where light and nutrient levels are optimal.

Since the distribution of primary production is highly dynamic and influenced by many factors, reliable estimates of productivity have only been possible since the advent of remote sensing, using the CZCS and SeaWiFS spectral scanners on satellites (*Section 2.7.4*). By measuring the different absorbance properties of chlorophyll and other photosynthetic pigments, it is possible to map phytoplankton biomass (standing crop) on a daily basis (cloud cover permitting). Mathematical models linking chlorophyll measurements, photosynthesis rates and irradiance are used to extrapolate experimental values to different depths and to estimate the effects of diurnal variation. The total net annual primary production of the world's oceans is close to 5×10^{16} g carbon fixed, which is very similar to the total productivity of all terrestrial ecosytems. On an area basis the annual global marine productivity is about 50 g carbon fixed m^{-2}, which is only one third of that on land. This discrepancy is due to the lower utilization of solar radiation by ocean phytoplankton than terrestrial plants, largely because of nutrient limitation and the effect of suspended particles (including the phytoplankton cells themselves) in absorbing light.

9.4 Productivity and nutrients

9.4.1 Nutrient limitation

As well as light as an energy source, photosynthesis obviously depends on the availability of nutrients. As shown in *Table 9.1a* there is great variation in the distribution of primary productivity in the world's oceans. The most productive areas of open ocean are found in the major upwelling regions of the north-west Atlantic off Africa, the eastern Pacific off Peru and the US west coast, the western Pacific off Namibia and the Arabian Sea. The least productive areas are the central gyres of the ocean basins. Although, on a global basis, productivity is fairly constant across the year (*Table 9.1b*), productivity is highly seasonal in some parts of the world. High-latitude temperate regions (especially the North Atlantic) characteristically show a spring bloom. Strong mixing occurs during the cold, dark, windy winter; this brings nutrients to the

Table 9.1 Estimates of primary productivity of the oceans

	Primary production ($\times 10^{15}$ g carbon fixed)	Percent
(a) Annual production in different oceans		
Pacific	19.7	42.8
Atlantic	14.5	31.5
Indian	8.0	17.3
Antarctic	2.9	6.3
Arctic	0.4	0.9
Mediterranean	0.6	1.2
Total global annual production	**46.1**	**100.0**
(b) Seasonal estimates of global ocean productivity		
March–May	10.9	23.0
June–August	13.0	28.2
September–November	12.3	26.7
December–February	11.3	22.1
Total global annual production	**47.5**	**100.0**

Estimated from SeaWiFS remote sensing data using a vertically generalized production model. Note: there is a slight discrepancy in the total productivity estimates in (a) and (b). Adapted from Ocean Primary Productivity Study, 2000: see this website for maps showing distribution of primary production.

surface layers, which promotes a rapid increase in photosynthesis as light levels increase during spring. Increased stratification (*Section 1.6.2*) and nutrient depletion lead to reduction in productivity during summer, even though light levels are at their highest. During autumn, mixing occurs again and a small secondary peak of production results. In tropical seas, seasonal effects are much less pronounced, except where there is a seasonal upwelling of nutrients, such as occurs in the Arabian Sea. Around coastlines, productivity is generally high due to the input of nutrients from rivers and wind-blown dust.

Carbon is rarely a limiting factor in productivity, since HCO_3^- is abundant in seawater. However, nitrogen, phosphorus, silica, trace metals (especially iron) and some trace organic compounds such as B vitamins all have the potential to be a limiting nutrient, thus affecting productivity. The concept of a single limiting nutrient originates from Libeig's Law of the Minimum, a principle which holds that the single chemical factor in shortest supply will act to limit chemical reactions (and by inference, plankton growth). Study of the relationship between the elemental composition of plankton and seawater led to the concept of the Redfield ratio, which is the ratio of the elements C:N:P. This is normally 106:16:1 (on a molar basis). The ratio holds good in many waters, leading to the concept of a universal biochemical stoichiometry in the oceans, but there are many circumstances in which the N:P ratio varies considerably from 16:1, and microbial processes (especially nitrogen fixation and nitrification) are particularly important under these circumstances. Different species of phytoplankton may be limited by different nutrients or may be affected differently at different stages of their life cycle. Furthermore, the utilization of these and other nutrients is closely linked and in mixed natural communities, it is difficult to untangle these complex interactions. In the last few years, there has been a radical reappraisal of nutrient cycling in the oceans.

Until recently, it was widely held that nitrogen was the most important limiting nutrient in marine waters. This belief arises from application of the Law of the Minimum to studies of the spring bloom in temperate North Atlantic waters. Investigators observed that the increase in algal biomass and decrease in concentration of nitrate were inversely proportional. Phytoplankton growth in seawater increases with experimental addition of ammonia, but not of phosphate. In addition, if the concentrations of phosphate and nitrate in surface seawater are compared as they are utilized, some phosphate remains even when the nitrate levels reach zero. At the onset of the spring bloom in temperate waters, phytoplankton biomass (as indicated by chlorophyll assays) is uniform throughout the mixed layer, but as stratification develops the maximum chlorophyll levels occur in deeper waters where nitrate levels are higher. This led to the concept of new production, which is dependent on N from such deep-water nitrate pools or from terrestrial runoff, and regenerated production, which is dependent on the recycling of N (largely as ammonium ion) by bacterial decomposition within the photic zone. This rather simple view is challenged by detailed study of other ecosystems. In particular, remote sensing has confirmed the existence of high-nutrient, low-chlorophyll (HNLC) expanses of the oceans, especially in the eastern equatorial and subarctic regions of the Pacific and the Southern Ocean. These areas occupy about 20% of the surface area of the oceans and have high ($>2\ \mu M$) concentrations of nitrate, but support some of the lowest levels of phytoplankton biomass ($<0.5\ g\ l^{-1}$ chlorophyll). As discussed below, there is growing evidence that this anomaly is largely explained by low levels of iron. The oligotrophic gyres occupy 70% of the oceans' area and are characterized as low-nutrient, low-chlorophyll (LNLC) biomes. HNHC and LNHC regions are typical of coastal waters and occupy about 5% area each. In some areas, such as the eastern Mediterranean, phosphorus appears to be the key limiting nutrient, whilst silicon availability can be limiting for diatom growth in coastal waters and may account for imbalance in population dynamics, leading to harmful algal blooms (*Section 12.6*). Details of the emerging hypotheses about the complex interconnections between productivity and nutrient dynamics in the oceans and the experimental evidence to test them are beyond the scope of this book. However, two key aspects of microbial processes that are leading to new paradigms in ocean ecology, namely nitrogen cycling and the connected role of iron, will be considered in some detail.

9.4.2 Microbial aspects of nitrogen cycling

As noted above, dissolved inorganic N (DIN) was assumed to limit new and regenerated production through nitrate and ammonia respectively. Nitrate is formed from ammonia by the action of nitrifying bacteria (*Section 5.5*) and this process occurs mainly in deep waters, since it is inhibited by light. Ammonia is produced from the breakdown of organic material, derived directly or ultimately from phytoplankton, by microbial loops (heterotrophic bacteria and grazing protists) and zooplankton. Microbes take up DIN via membrane transport systems, with a preference for ammonium assimilation (*Section 4.4.5*), which is less energy-demanding as the nitrate/nitrite reductase system requires large amounts of NADPH and ATP. Indeed, some microbes cannot take up nitrate at all. By comparison with terrestrial plants, it has long been assumed that phytoplankton use nitrate as their main source of nitrogen. However, the recent use of molecular probes directed against genes for the key enzyme, nitrate reductase, shows that this assumption is not always valid. Various members of the phytoplankton community differ in their ability to assimilate different DIN compounds, especially via inducible high-affinity uptake systems. In particular, the high-light-adapted ecotype of *Prochlorococcus* (found in the upper layers of the photic zone, see *Section 5.4.4*) appears to lack genes for both nitrate reductase and nitrite reductase, meaning that it can grow only by using ammonium ion. Thus, it depends entirely on regenerated nitrogen. The low-light-adapted ecotype (found in deeper waters) lacks only nitrate reductase and can therefore assimilate nitrite and ammonium, but not nitrate. Since *Prochlorococcus* is now thought to contribute the majority of primary production in the vast tropical and sub-tropical regions of the oceans, this finding has great bearing on nutrient dynamics. In the microbial loop concept presented in *Figure 9.4,* heterotrophic microbes are shown as contributing to higher trophic levels in the food web through recycling of DOM. However, heterotrophs also assimilate DIN as well as organic nitrogen compounds. Variations in species composition of the plankton species will affect the balance between ammonia production, nitrification and DIN uptake. Bacterial nitrogen assimilation genes in both photoautotrophs and heterotrophs appear to be expressed only under conditions of nitrogen limitation and extensive research on the distribution and expression of these genes in various plankton species, under various conditions, is currently underway using immunological assay of expressed surface proteins (*Section 2.4*) and RT-PCR or microarray techniques (*Section 2.6*). For example, the ammonia-oxidizing nitrifier *Nitrosococcus* has been shown to be widespread and abundant in plankton using these methods (*Section 5.5*). Furthermore, it is now recognized that the microbial loop affects primary productivity in a more complex way than previously envisaged, because: (a) a substantial component of DOM is derived from bacterial cell walls; (b) DOM production is influenced by the activity of extracellular enzymes and photochemical reactions; and (c) both algal and cyanobacterial phytoplankton can assimilate DOM directly.

Another factor affecting nitrogen (and phosphorus) distribution is the vertical migration of plankton. Some diatoms such as *Rhizosolenia* and the cyanobacterium *Trichodesmium* form large mats that can migrate across physical mixing barriers between the surface and deep pools of nutrients (80–100 m) in order to assimilate nitrate, before returning to shallower water with optimum light levels for photosynthesis. This process depends on the regulation of cell buoyancy by gas vesicles (*Section 3.3*). The daily migration of zooplankton between the surface and deeper waters also results in a two-way flux of nutrients, the net effect of which will depend on the relative nitrogen:phosphorus balance at different depths. Passive upward flux of low-density lipid-rich (hence phosphorus-rich) material could also be important. These processes must now be considered alongside the purely physical processes of upwelling and turbulent mixing.

The atmosphere contains abundant supplies of nitrogen in the form of gaseous N_2; high concentrations of dissolved N_2 (800 μg N atoms l^{-1}) occur in the open ocean compared with the usually low concentrations of DIN in the form of nitrate or ammonia (<0.25 μg N atoms l^{-1}). Although N_2-fixing organisms are common in coastal waters, sediments, microbial mats and coral reefs, surprisingly few known N_2-fixing organisms have been documented in open-ocean habitats. As described in *Section 4.4.6,* N_2 fixation is energetically very demanding and is restricted to certain species of *Bacteria* and *Archaea*. Special strategies are needed by aerobic bacteria to protect the process from inhibition by O_2. Free-living *Cyanobacteria* containing heterocysts (the most usual strategy for protecting nitrogenase) are rare in the open sea. However, heterocyst-forming cyanobacterial endosymbionts (such as *Richelia*) of diatoms (such as *Rhizosolenia* or *Hemiaulus*) do occur. Studies of N_2 fixation in the Sargasso Sea led to the original identification of *Trichodesmium*, a large free-living filamentous cyanobacterium that is able to fix N_2 without heterocyts. It achieves this largely by temporal separation of cellular processes (*Section 5.4.3*). Although *Trichodesmium* can occur as extensive blooms and can account for over half of the N_2 fixation in tropical waters, it does not appear to be numerous enough to account for the apparent rates of fixation, nor to explain the imbalance in nitrogen budgets. Evidence that N_2 fixation is extensive in tropical and subtropical oceans comes from the finding that the isotope composition of DIN pools in surface waters is similar to that of N_2 gas (i.e. [^{14}N] rather than [^{15}N]), as well as a departure from the Redfield ratio of N:P = 16:1. This mystery is now being resolved using molecular biological techniques, which show that diverse N_2-fixing bacteria do occur in the oceans, as discussed in *Box 9.1*. A schematic representation of the complex interactions in nitrogen cycling is shown in *Figure 9.2*.

9.4.3 The importance of iron

Iron is an essential component of many enzymes and electron transport proteins, but is present in extremely small concentrations in seawater. Photosynthesis and N_2 fixation are especially dependent on iron because the key components in these processes (photosystem–cytochrome complex and nitrogenase–enzyme complex, respectively) rely on iron-containing proteins. Over 99% of 'dissolved' iron is tightly bound to organic compounds; it occurs mainly as colloidal $Fe(OH)_3$, which rapidly coagulates and adsorbs to organic particles. Microbes need special mechanisms to acquire this iron and prokaryotes achieve this by production of siderophores (*Section 4.4.7*), which bring iron into the cell via binding to surface receptors. Some bacteria utilize iron bound to siderophores produced by other species. Photochemical reactions with α-hydroxy acid-containing siderophores (such as aquachelins) may lead to an increased bioavailability of the siderophore-complexed iron. Eukaryotic phytoplankton do not appear to synthesize siderophores, but can take up iron via a cell-surface ferrireductase enzyme that liberates iron bound to organic compounds such as porphyrins, which are released from cells through zooplankton grazing and viral lysis (see below). Phagotrophic eukaryotes such as flagellates can acquire iron from ingested bacteria. Prokaryotic and eukaryotic microbes could therefore be competing for iron, depending on the chemical nature of available iron complexes, and the outcome will affect separation of ecological niches and community composition, with consequences for the fate of carbon in low-iron waters (*Figure 9.3*). Siderophore-bound iron may be the major source of the element in regions dominated by cyanobacterial photosynthesis and microbial loops (such as the tropical and subtropical oceans), whilst porphyrin-complexed iron may be the major source in coastal regions dominated by diatoms and zooplankton grazing. Another important factor may be the production of storage proteins to sequester scarce elements like iron within cells. We know that many bacteria produce ferritin-like compounds to store iron when it is present at higher concentrations than those needed for immediate use, but these compounds have not yet been detected in aquatic bacteria. Microalgae are known to synthesize phytochelatins, which can store a range of trace metals, but these do not appear to sequester iron. An interesting observation is that domoic acid, a toxin produced by the diatom *Pseudo-nitzschia* (*Section 12.2.4*) binds iron and is produced in greater amounts when cultures are grown with high iron concentrations. This offers a possible explanation for the production of this compound (*Section 12.5*).

Could the low iron concentrations account for the low productivity of HNLC regions, i.e. could iron be a limiting nutrient? The major source of iron is terrestrial and it follows that coastal regions receive regular input from rivers and runoff from weathering of rocks. (Note, however, that the extent of this input depends on the geology of the land and some upwelling coastal regions can be iron-deficient.) By contrast, the HNLC regions occur at great distances from the

Box 9.1 RESEARCH FOCUS

Nifty techniques

New molecular methods reveal the importance of nitrogen fixation in the oceans

As noted in *Section 9.4.2,* recent biogeochemical evidence suggests that N_2 fixation (diazotrophy) in the large portion of the biosphere occupied by the tropical and subtropical oceans is more important than previously realized. Classical microbiological and physiological techniques have been successful in identifying *Trichodesmium*, a large filamentous cyanobacterium, as a diazotroph but have been largely unsuccessful in solving the mystery as to why there appear to be so few ocean organisms that can exploit the abundant supplies of gaseous N_2 under the evolutionary pressure of severe shortages of dissolved inorganic nitrogen compounds. A major breakthrough in this area has come from the work of Jonathan Zehr and colleagues (now at the University of California, Santa Cruz) who first used molecular biological approaches to detect the presence of nitrogenase genes in water samples. Zehr *et al.* (1998) exploited the fact that the nitrogenase enzyme contains two proteins that are highly conserved in a wide range of prokaryote species. They designed PCR primers that recognize conserved sequences of the *nifH* gene (which encodes the iron protein component of nitrogenase) and used these to amplify the gene from samples of picoplankton obtained by membrane filtration of up to 20 liters of seawater from the Atlantic and Pacific Oceans. Amplified fragments were cloned and the DNA was sequenced. A variety of *nif* sequences were identified, and comparison with GenBank data showed that they represented several diverse bacterial lineages, notably *Cyanobacteria* and α- and β-*Proteobacteria.* Some differences in the distribution of different phylotypes were observed between Atlantic and Pacific sources. Zehr *et al.* (2001) showed that nitrogenase genes from the subtropical north Pacific Ocean gyre (Station ALOHA, Hawaii) were closely related to genes from unicellular *Cyanobacteria.* The most abundant cyanobacterial genera, *Synechococcus* and *Prochlorococcus,* do not appear to fix N_2 but there is evidence that larger unicellular *Cyanobacteria* may do so. Zehr and colleagues cultured large (3–10 μm diameter) N_2-fixing isolates from the ALOHA samples and identified these as *Synechocystis.* These organisms occur in abundance throughout the year to depths of 150 m and actively fix N_2, as shown by an acetylene reduction assay (which is based on the ability of nitrogenase to reduce triple-bonded substrates other than N_2).

Zehr *et al.* (1998) also amplified *nifH* DNA sequences extracted from copepod zooplankton. These sequences were very distinct from those from the picoplankton and clustered with sequences from clostridia and SRB. This suggests that they are derived from anaerobic bacteria that inhabit the gut of the copepods, possibly as symbionts (as occurs in insects and shipworms). Braun *et al.* (1999) amplified *nifH* fragments from anaerobic enrichments of zooplankton and demonstrated that some of the enrichments contained nitrogenase activity using the acetylene reduction assay.

Detection of nitrogen cycling genes in the environment is now an active area of research. The results of Falcon *et al.* (2002), who also studied *nifH* genes, suggest a complex pattern of divergence among N_2-fixing *Cyanobacteria* within and between the Atlantic and Pacific oceans. Mehta *et al.* (2003) have successfully amplified *nifH* genes from hydrothermal vent fluids; these appear to be from a diverse range of prokaryotes, including anaerobic clostridia and SRB, as well as *Proteobacteria* and *Archaea.* The use of new techniques for measuring gene expression, including RT-PCR and oligonucleotide microarrays (Taroncher-Oldenburg *et al.* 2003) is leading to great advances in the detection and quantification of functional N_2-fixing genes in the environment.

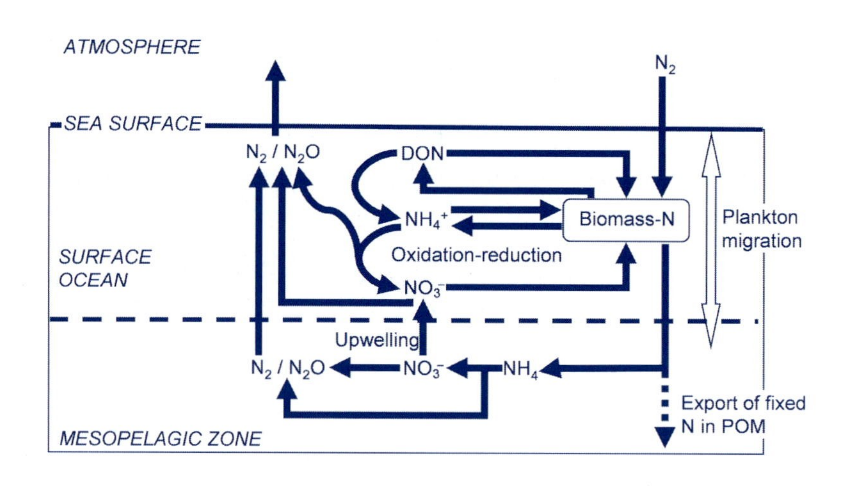

Figure 9.2

Nitrogen cycling in the oceans. Arrows show the transfer of inorganic and organic nitrogen compounds. DON = dissolved organic nitrogen. POM = particulate organic matter. The dotted line represents a depth of about 150–200 m. Adapted from Karl (2002).

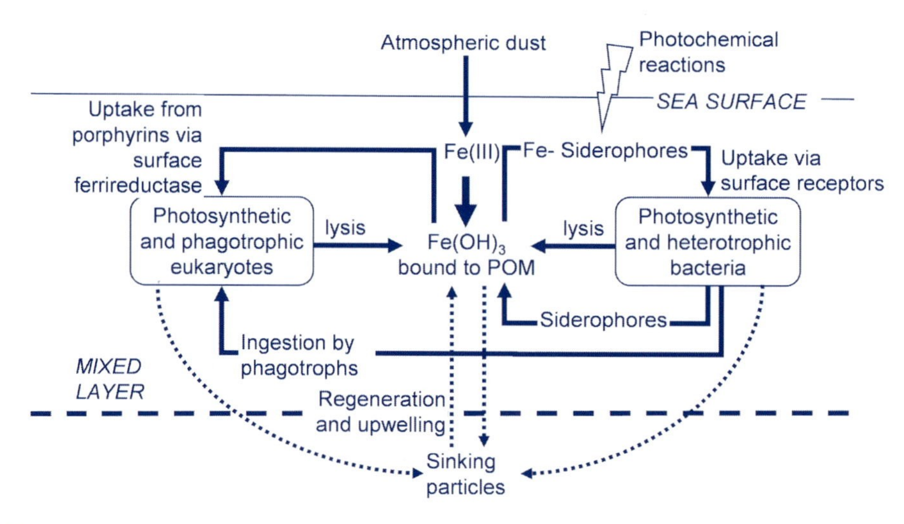

Figure 9.3

Iron cycling in the oceans. Arrows show the transfer of iron and the acquisition mechanisms employed by marine microbes. POM = particulate organic matter.

continents and will only receive iron inputs via windblown dust or from upwelling of deep waters. Iron concentrations in surface water (<200 m) decrease with distance from land, although deep water (>500 m) in open oceans contains about 10 times the level near the surface. This profile resembles that of major nutrients and occurs because biological uptake in the upper (photic) water column is followed by heterotrophic microbial processes of regeneration of DOM as it settles into deeper water. The experimental evidence in support of the iron-limitation theory is discussed in *Box 9.2*, together with the controversial proposals to use iron fertilization of the oceans as a way of removing excess CO_2 from the atmosphere as a countermeasure for global warming.

Box 9.2 RESEARCH FOCUS

Fertilizing the oceans

Controversy surrounds plans to boost phytoplankton with iron

The idea that iron deficiency limits phytoplankton growth in many parts of the ocean was developed by the late John Martin of the Moss Landing Marine Laboratory, California. Martin proposed that climatic changes in the Earth's history (glacial and interglacial periods) were caused by changing patterns of deposition of iron-rich dust, resulting in modification of CO_2 exchange between the atmosphere and oceans. He developed the ultraclean analytical techniques necessary to study trace metals and used bottle experiments to show that iron addition caused rapid growth of phytoplankton in HNLC waters. To prove his hypothesis, Martin conceived the idea of conducting iron-enrichment experiments in the open ocean, whilst monitoring biological and geochemical parameters.

The first experiments (IronEx I and II) were conducted in 1993 and 1995 in the equatorial Pacific off the Galapagos Islands. In IronEx II (Coale *et al.* 1996), a 100 km^2 area was fertilized by the addition of iron sulfate solution dispersed in the wake of a ship's propellor as it sailed over the area. An inert chemical tracer (sulfur hexafluoride, SF_6) was used to track the enriched patch in conjunction with global positioning satellite (GPS) data. Within 4 days, there was a 30-fold increase in phytoplankton biomass, causing a marked change in color of the surface water from blue to green. Uptake of nitrate accelerated and the center of the patch showed a significant transient drawdown of CO_2. From this study, Cooper *et al.* (1996) concluded that iron supply to the equatorial Pacific could strongly affect the atmosphere:ocean CO_2 balance in the short term, but has little long-term influence. In contrast, the Southern Ocean is thought to play an especially important role in the global carbon cycle due to large-scale export of fixed carbon to deep waters. Due to its distance from continental landmasses, the Southern Ocean receives little dust and hence has very low iron levels. Consequently, it has large reservoirs of unused macronutrients in the upper and deep mixed layers. An iron-enrichment experiment was therefore planned for the Southern Ocean and a consortium of international researchers conducted the SOIREE experiment over 13 days in 1999 in which they distributed 8663 kg of iron sulfate, plus the SF_6 tracer, over an 8 km diameter patch of ocean. Boyd *et al.* (2000) reported the results, which depended on interpretation of a wide range of biochemical, geochemical, hydrographic and physical parameters. Phytoplankton biomass (measured by chlorophyll *a*) and primary production (CO_2 fixation) increased within a few days, accompanied by increased utilization of macronutrients. The tripling of phytoplankton biomass was due largely to a shift from small to large-celled diatoms, which grew more rapidly in response to the relief of iron limitation. Abraham *et al.* (2000) reported that phytoplankton biomass in the patch, measured by remote sensing, was still three times the normal level a month later. They emphasize the importance of stirring, which mixes phytoplankton and iron out of the patch, but also entrains silicate. This may have allowed the bloom to continue for as long as there was sufficient iron, without the complication of silicate limitation. An estimated 600 to 3000 tonnes of carbon was accumulated in the bloom, but there was no evidence that any of this carbon was exported to deep ocean waters. Indeed, because the faster-growing cells were lighter, they actually settled more slowly within the enriched patch than in surrounding waters. Thus, SOIREE failed to confirm one of the key tenets of Martin's hypothesis, i.e. that iron fertilization would lead to increased export of carbon.

Watson *et al.* (2000) combined the experimental results with a model based on data from dust deposition in ancient Antarctic ice cores. They concluded that modest sequestration of atmospheric CO_2 by artificial additions of iron to the Southern Ocean is possible, although knowledge of the period and geographical extent over which sequestration would be effective remains poor. The most recent investigation (SOFeX) was conducted at the beginning of 2002 and involved three ships and 76 scientists from 17 institutions. At the time of writing, analysis of the results has not been published, but preliminary data suggest that, as in SOIREE, a significant increase in primary productivity (in this case, about a 10-fold increase) occurred, but with no clear evidence of carbon export (Falkowski 2002). Although most attention is focused on the Southern Ocean, Hutchins and Bruland (1998) have shown that iron limitation may not be restricted to open ocean environments, as enrichment led to increased phytoplankton production in HNLC waters off the coast of California. They suggest that iron limitation may affect the silica content of diatoms, which would affect the rate of sinking and hence carbon export (see also *Box 7.1*).

One of the drivers for these experiments is the idea that ocean fertilization could remove CO_2 from the atmosphere and hence help to control global warming. Following the Kyoto conference on reduction of CO_2 emissions and global warming, the concept of a global market in the trading of 'carbon offset credits' has emerged. Entrepreneur Michael Markels has set up the commercial organization GreenSea Venture Inc., with the aim of profiting from ocean fertilization whilst mitigating the consistent rise in atmospheric CO_2 levels. The company has already patented certain aspects of the technology needed to deliver iron on a large scale. Some scientists and entrepreneurs also envisage ocean fertilization as having potential benefits for enhancement of fisheries. GreenSea's immediate scientific mission is to develop analytical models that describe the long-term flux of carbon from the atmosphere into the marine environment in response to iron fertilization and to determine the consequences of large-scale iron fertilization of a sufficiently large ocean area over a long period. If warranted by the modeled predictions, they plan to work with oceanographers to conduct large-scale experiments to measure carbon fluxes and other consequences. 'Planetary engineering' on such a scale is clearly a matter that will require careful evaluation. Some of the world's leading oceanographers are divided in their opinions about whether we should even contemplate such a course of action. Chisholm *et al.* (2001) argue that the long-term consequences of such deliberate eutrophication are unknown, and just not worth the risk. They argue that carbon sequestration cannot be easily verified, because of the complexities of ocean physics and biogeochemistry, and that changes to composition of the phytoplankton community will have unforeseen consequences. Lam and Chisholm (2002) provide a detailed analysis, based on existing models, which they use to claim that Markel's predictions for carbon sequestration are seriously flawed; 'even if iron fertilization is fully optimized to sequester the maximum amount of carbon, it will make only the smallest of dents in atmospheric CO_2 if fossil fuel burning continues to grow exponentially'. In response to these concerns, Johnson and Karl (2002) argue that Chisholm *et al.* 'have greatly overstated the current knowledge of ocean processes in reaching their opinion that iron fertilization is not a viable option for CO_2 management'. They counter the claim by Chisholm *et al.* that ocean fertilization is not easily controlled, by pointing out that the rapid turnover of phytoplankton and regular mixing of the oceans would quickly restore the status quo, and argue that there is currently no evidence to support claims that hypoxia or anoxia would result in the deep oceans. Johnson and Karl state that we can only make the decision to initiate or abandon ocean fertilization when we know a lot more about carbon cycling processes, and risks have to be weighed against the inescapable fact that global warming is occurring now and causing large changes in ocean processes anyway. Based on past experience (e.g. how to *reduce* CO_2 emissions), getting a clear consensus about such a major issue among commercial interests, scientific experts, public opinion and governments is going to be very difficult!

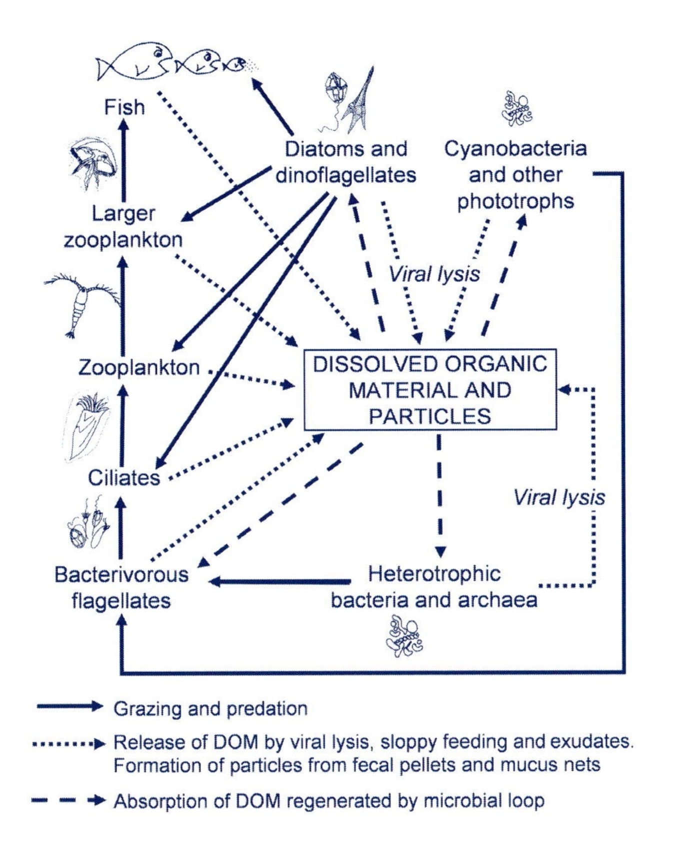

Figure 9.4

Simplified representation of food webs in the sea. Arrows show the transfer of organic material by various routes. For clarity, not all trophic connections are shown and regeneration of inorganic nutrients by heterotrophic bacterial activity is omitted.

9.5 The microbial loop in ocean food webs

9.5.1 Classic and modern food webs compared

As noted earlier, the classic view of marine trophic interactions was a simple pyramidal food chain, summarized as microalgae (mainly diatoms) copepods, fish and whales. A schematic representation of our modern view of ocean food webs is shown in *Figure 9.4*. This shows that there are trophic interactions at multiple levels with microbial processes occupying center stage. However, it is important to recognize that elements of the classic model are still valid and the relative importance of the various components shown in this diagram varies according to the circumstances. In weakly stratified water masses, with a high degree of mixing and turbulence (such as polar and coastal temperate seas during the spring bloom) food webs are often based strongly on the dominance of organisms in the microplankton size class or greater (i.e. microalgae such as diatoms, dinoflagellates or prymnesiophytes and copepod grazers). By contrast, the large areas of the ocean (especially in the tropics and subtropics) that have highly stratified waters with constant low nutrient levels have food webs dominated by microbial loop processes (in this context, interpreting 'microbial' as referring to organisms in the picoplankton or nanoplankton size classes). As a broad generalization, the relative influence of the pico- or nanoplankton size classes of microbes is greatest in oligotrophic waters, because conditions of low nutrient flux exert positive selection for small size (*Section 1.4*). The relative importance of activity of larger particle-feeding protozoa, zooplankton and fish becomes progressively greater as nutrients increases. Of course, it must be remembered that many anaerobic geochemical processes such as the production/oxidation of methane, sulfide oxidation and sulfate reduction are entirely due to microbes, and these are accentuated in eutrophic systems. A useful analogy may be made between the behavior of food webs in nature and the behavior of laboratory cultures. Food webs dominated by microbial loops tend to behave like chemostat (steady state) cultures. Despite rapid multiplication rates of many of the component organisms, species abundance (and the concomitant geochemical processes) in the oligotrophic ocean gyres seem remarkably stable if viewed over fairly long timescales. By contrast, microalgae/copepod-dominated food chains typical of spring bloom events have dynamics that resemble batch cultures. Biomass and photosynthetic activity increase rapidly and then decline due to nutrient depletion and the pressure of copepod grazing, leading to a greater export of photosynthetically fixed carbon to deeper waters through the sedimentation of zooplankton fecal pellets. Certain hydrographic, nutritional and climatic conditions can lead to the very sudden development of massive, dense blooms, such as dinoflagellates or diatoms (*Section 12.6*) and prymnesiophytes (*Section 7.7*), which then disappear equally suddenly. The role of viruses in such 'bloom and crash' dynamics is an area of active research (see *Box 8.1*).

9.5.2 The formation and fate of DOM and POM

About 50% of the daily net production from photosynthesis enters the ocean system as DOM, which supports the growth of heterotrophic bacteria; hence, the microbial loop results in greater retention of dissolved nutrients in the upper layers of the ocean. Some DOM is formed from direct extracellular release of carbohydrates, amino acids, lipids and organic acids from bacteria and phytoplankton cells as they grow. The amounts released by this route are very variable, but are generally highest in the most photosynthetically active region in high light intensities. Extracellular release might result from overproduction of photosynthate when CO_2 fixation provides more organic material than can be incorporated into growing cells due to nutrient limitation. There is probably also a constant leakage of low-molecular-weight compounds across cell membranes due to the steep concentration gradient caused by very low solute concentrations in seawater. Some calculations suggest that phytoplankton cells can lose 5–10% of cell mass per day. A large amount of DOM/POM is released by protistan grazing on picoplankton. Partially digested particles of prey, unabsorbed small molecules and digestive enzymes are released as colloidal material during the egestion process, when food vacuoles fuse with the external membrane. DOM/POM is also released during grazing by zooplankton such as copepods as a result of 'sloppy feeding', when phytoplankton cells are broken apart by the mouthparts of zooplankton, and more DOM/POM is released in fecal pellets. The remaining DOM results from viral lysis of phytoplankton, bacteria and protists. Since the importance of viruses in marine systems has only recently been recognized, their exact contribution to the release of DOM is unclear, but it is undoubtedly very large and probably exceeds that of grazing protists. Up to 80% of the total photosynthetic production can be released as DOM during the collapse of phytoplankton blooms.

Some DOM is converted by purely chemical mechanisms into humic substances, complex polymers that resist biodegradation and form a reservoir of refractory organic matter, which eventually sinks to the ocean floor. DOM and POM aggregate into larger particles containing large numbers of microbes and these settle to deeper waters as marine snow. Thus, carbon fixed in the upper layers is removed to deeper waters. As it sinks, it is remineralized to CO_2 by the respiratory activities of microbes or consumption by zooplankton and fish. The activity of viruses is important here also. As discussed in *Section 1.7.1*, there is substantial release of DOM from particles as they sink and some of this is caused by viral lysis of microbes within the aggregates. Microorganisms in sediments will break down that fraction of DOM that reaches the ocean floor. Large amounts of carbon (possibly about one quarter of net production) are also removed to the ocean floor as $CaCO_3$ in the skeletons of plankton such as coccolithophores and foraminifera (*Section 1.7.2*). In very deep waters, some of this $CaCO_3$ redissolves to HCO_3^-, but in shallower waters the skeletons of these planktonic forms lead to the formation of calcareous sediments. The vertical movement of organic matter is accentuated by the daily migration of plankton over hundreds of meters in the water column. Together, these processes constitute the biological pump, which results in the transport of CO_2 from the atmosphere to deep ocean waters. Despite many advances in recent years, we do not yet have reliable estimates of the amounts of carbon exported from the upper ocean and the effects of mixing and water flow in its redistribution. Nor do we have a clear idea of the quantitative role of bacteria in carbon flux in the oceans. Because bacterial growth efficiency is generally low (i.e. large nutrient and energy inputs result in low yield of biomass, see *Section 4.5.3*) their overall role in carbon flux may be as a sink rather than a link to higher trophic levels.

9.5.3 **Protistan grazing**

A key aspect of the microbial loop concept is the process of bacterivory, i.e. the consumption of heterotrophic bacteria by protistan microbes (*Chapter 7*), which helps to explain the regulation of bacterial biomass at near-constant levels. Bacterivory constitutes a mechanism by which very small bacterial cells are made available to larger planktonic organisms and is critical to the ocean food web. In addition to its importance as a 'top-down' control on bacterial production involved in heterotrophic recycling, grazing of *Cyanobacteria* by protists will influence primary productivity. It should be noted that, for many species of bacterivorous protists, bacteria are not the sole diet. Feeding experiments and analysis of the food vacuole contents of protists shows that many can feed on larger photosynthetic and heterotrophic protists as well as inanimate organic particles (by phagotrophy) or dissolved compounds (by absorption). The most active bacterivorous protists are flagellates in the nanoplankton size class (2–20 μm), and most are less than 5 μm (*Section 7.3*). Flagellated protists, including dinoflagellates, cryptomonads, euglenoids and ciliates in the microplankton (20–200 μm) class are also active bacterivores. These organisms feed by the generation of water currents with cilia and flagella and they can process hundreds of thousands of body volumes per hour. The importance of larger protists such as radiolarians and foraminifera is less well known.

Some metazoan zooplankton can also consume bacteria directly. The larvae and juvenile stages of copepods have been shown to consume labeled bacteria (1–5 μm) in experimental studies, but they are possibly less efficient grazers of the very small bacteria that dominate pelagic systems. As noted in *Section 1.7.1*, larvaceans trap bacteria in fine (<1 μm) mesh structures constructed of gelatinous mucus and the discarded structures make a major contribution to the formation of marine snow. Indeed, since many pelagic bacteria associate with particles of marine snow rather than exist as freely suspended forms, they may be consumed by a wide range of larger metazoan zooplankton and small fish, which would not be able to feed directly on individual bacteria.

A further key component of the microbial loop is the release of DOM through protistan grazing. Large amounts of DOM (up to 25% of ingested prey carbon) are egested by protists in the form of fecal pellets and some of this is readily metabolizable, entering the DOM pool for further recycling by bacteria, whilst the remainder enters the 'sink' of long-term refractory material. The relatively high respiration and excretion rates of protistan grazers mean that the cycling of 'lost' photosynthetic products through the microbial loop is rather inefficient.

Detailed measurements of grazing rates are technically difficult. In one type of experimental study, the uptake of labeled bacteria and their accumulation in food vacuoles is followed using microscopy (fluorescently labeled prey) or radioactive tracer methods ([^{3}H]-tritium-labeled prey). Another approach is the dilution method, in which a dilution series of seawater is prepared, so that the natural growth rate of the prey bacteria is unaltered whilst the predator:prey ratio is reduced. The rate of change in bacterial density is then monitored during incubation at different dilution levels. Separation of bacteria and their grazers in different size classes may also be achieved by filtration. Using such techniques, many studies have attempted to evaluate the impact of protist grazing on the density, dynamics and structure of bacterial populations in the oceans. In low-productivity waters, grazing seems to be the dominant factor in balancing bacterial production. However, in higher-productivity waters (such as coastal regions and estuaries), other factors may be important in controlling bacterial numbers. These include removal by benthic filter-feeding animals, but of greatest significance is the role of viral lysis, which will be favored because of the higher bacterial population densities, as discussed below.

Up to a certain limit, which depends on their own size, protistan grazers show a preference for larger prey as food. Therefore, one consequence of grazing pressure is to encourage the domination of microbial assemblages by small cells. As discussed in *Section 1.4,* most bacteria in oligotrophic pelagic environments are less than 0.6 μm, with many less than 0.3 μm, and the smallest bacteria may be in a state of metabolic maintenance rather than active cell division (*Section 4.5*). Thus, the interesting concept arises that protistan grazers are preferentially removing the larger, actively growing/dividing bacteria, leaving a large stock of bacteria that are growing slowly, if at all. However, selective bacterivory of larger, active bacteria seems to stimulate the growth of other bacteria by the release of regenerated nutrients such as ammonium and phosphate. Use of high-resolution video techniques shows that, as well as chance encounters, flagellates may actively select their prey. Species-specific differences in the processing of food particles explain the coexistence of various bacterivorous nanoflagellates in the 3–5 μm size range and indicate the existence of specific predation pressure on different bacteria. Motility, surface characteristics and toxicity can all affect the outcome of bacterium–protist interactions at the various stages of capture, ingestion and digestion.

9.5.4 Viral lysis

The increasing recognition of the abundance of viruses that infect planktonic microbes has led to evaluation of their role in ocean processes. As discussed in *Chapter 8,* both lytic and lysogenic cycles of infection occur when 'phages infect marine bacteria. Although the lysogenic state is common and has important consequences for genetic transfer, induction in natural marine systems seems to be relatively low, so the majority of viral infection appears to occur via encounter of bacteria with active virions that initiate the lytic cycle. This process is highly dynamic and is affected by the virus:host ratio, the rate of viral replication, the burst size of virus progeny and the rates of viral decay. Trying to estimate the impact of 'phage lysis on bacterial populations produces very variable results, depending on the techniques used. These include measuring the net viral DNA synthesis using radioactive precursors, measuring the decline of labeled DNA from bacteria in the absence of protists (removed by filtration), or monitoring the relative proportions of fluorescent-labeled and -unlabeled viruses with time. Results from a range of studies indicate that viruses are responsible for a significant proportion of mortality of marine planktonic bacteria, ranging from about 10–50%. As noted above, it seems that viral lysis is relatively more important in eutrophic coastal waters, whereas protist grazing is more important in oligotrophic ocean waters.

As shown in *Figure 9.4,* viruses play a key role in the release of DOM from bacteria. Viral lysis leads to the release of bacterial cell contents, much of which enters the DOM pool and is readily recycled by heterotrophs. However, cell fragments and high-molecular-weight components may be more recalcitrant to breakdown. For example, bacterial porins (outer membrane proteins) are released as embedded components of membrane fragments and are relatively resistant to proteolytic degradation. Some components of algal cells lysed by viruses are also very refractory. The quantity and dynamics of these less labile components resulting from viral lysis are currently unknown. Mathematical models can be constructed to compare the effects of different levels of viral mortality on nutrient budgets. At a level of 50% bacterial mortality from viruses, the overall level of bacterial production and respiration rate is increased by about a third (compared to zero mortality due to viruses). In these models, a high level of protist grazing leads to carbon input to the higher trophic levels of the food chain (animals), whereas a high level of viral lysis diverts the flow of carbon from the food chain into a semiclosed cycle of bacterial uptake and release of organic matter. Because cell fragments, viruses and dissolved substances do not sink (unless they aggregate into larger particles), viral lysis has the effect of maintaining carbon and inorganic nutrients such as nitrogen, phosphorus (and possibly iron) in the upper levels of the ocean. Viral lysis also contributes to the microscale heterogeneity of seawater through the release of polymeric substances and the dissolution of material from marine snow (*Section 1.7.1*).

Bacteria are also susceptible to predation by other bacteria. The only well-studied example is *Bdellovibrio* (*Section 5.14.3*), but it is likely that many other types remain to be discovered; these could have considerable ecological importance in nutrient dynamics in marine systems.

9.6 Microbial processes in eutrophication of coastal waters

Nutrient enrichment of estuaries and coastal waters is a growing problem, due to anthropogenic sources of pollutants such as sewage, animal wastes and terrestrial run-off from heavily fertilized land. The impact of eutrophication depends on the source, nature and level of nutrient inputs, as well as hydrographic (especially tidal flushing and mixing) and other physical factors (especially light and temperature). Increased nutrient loading stimulates growth of phytoplankton (especially by input of nitrates), beyond the point where grazing by zooplankton can control it. For example, massive cyanobacterial blooms (e.g. *Nodularia, Microcystis* and *Oscillatoria*) occur regularly in the Baltic Sea, and eutrophication is probably a major factor in the increased occurrence of toxic dinoflagellate blooms (*Section 12.6*). Growth of macroalgae frequently occurs. Active microbial loop processes convert excess primary production, but these too may be overwhelmed and large amounts of decaying detritus and particles of organic material sink to the seafloor. There, bacterial decomposition leads to a heavy demand for O_2 and the overlying water column may become hypoxic or anoxic, resulting in mass mortality of benthic animals and fish. Despite the fact that nitrogen enrichment is responsible, N_2-fixing bacteria (especially *Cyanobacteria*) often increase in abundance and activity. This apparent paradox occurs because of the high activity of denitrifying bacteria, which convert nitrate to N_2 in estuaries and coastal seas. If nitrifying bacteria are inhibited by the development of anoxia (the process is O_2-dependent), the conversion of ammonia to nitrate will decrease and the nitrification–denitrification processes become uncoupled; N_2-fixing organisms may then be at an advantage.

9.7 Microbial processes and climate

This chapter concludes with a brief commentary on the importance of microbial processes on the planetary climate. It should be clear from the earlier sections that the activities of photosynthetic and heterotrophic microbes have profound effects on geochemical cycles and ocean–atmosphere interactions. The solubility and biological carbon pumps are of paramount importance and there is much current concern about changes to these established ocean circulation systems. Natural sources of CO_2 to the atmosphere are approximately equally divided between oceanic and terrestrial sources. In the past 200 years, the amount of CO_2 in the atmosphere has increased by 25%, as a result of the burning of fossil fuels and other anthropogenic effects since the Industrial Revolution. The rate of addition is accelerating and levels are projected to double again by 2100. Increased accumulation of other gases (especially methane and nitrous oxide) accentuates the 'greenhouse effect'. It is indisputable that global warming and climate change are happening now, even though experts may disagree about the rate of change and its biological effects. Understanding ocean processes in the carbon cycle is therefore vital and such knowledge may offer opportunities for mitigating some effects of elevated levels of CO_2 in the atmosphere, using the oceans as a 'sink' for excess carbon. Much of the current research in microbial oceanography is funded by this imperative. The most important potential application of this knowledge is the suggested use of iron fertilization to increase the uptake of CO_2 from the atmosphere to long-term storage in the oceans. Injection of CO_2 into the deep sea has also been proposed. As discussed in *Box 9.2,* we do not know enough about effects on nutrient cycles, phytoplankton and microbial loop community structure and carbon export processes to decide whether or not these are viable or sensible approaches to the problem. There are probably significant geographical and seasonal differences in the responses of the diverse ecological regions of the oceans. *Boxes 3.1* and *7.1* describe other research findings that demonstrate the importance of marine microbes in microscale nutrient cycling processes that have global effects.

The volatile compound DMS is another factor of immense importance in the global climate. The precursor of DMS is DMSP, which is produced by many marine microalgae because of its osmotic protective function. Zooplankton grazing leads to DMS production by bringing DMSP together with the enzyme DMSP lyase, as these appear to be in separate compartments in intact cells. Interestingly, release of DMS can act as a defense mechanism against further grazing. In the past few years, viruses have also emerged as a major cause of DMS release during lysis of phytoplankton cells (*Box 8.1*). Eutrophication may be responsible for the increasing blooms of coccolithophorids, dinoflagellates and other algae in neritic zones, leading to increased DMS flux to the atmosphere where it produces aerosols that absorb and reflect solar radiation and act as cloud condensation nuclei. These properties have an albedo effect that may act as a climate-regulating mechanism (see figure in *Box 8.1*). Anthropogenic changes to atmospheric composition could disturb the balanced feedback processes by altering the complex ecology of phytoplankton species and their predators.

International cooperation in research will further our knowledge of microbial processes in the oceans and may well produce some solutions to the problems of climate change. Sadly, prospects for intergovernmental agreements to do something about the *cause* of the problem (such as implementation of the Kyoto protocol to limit greenhouse gas emissions) show little sign of success. We will probably be unable to slow significantly, let alone halt, the predicted changes to our atmosphere and oceans during this century.

References and further reading

Abraham, E.R., Law, C.S., Boyd, P.W., Lavender, S.J., Maldonado, M.T., and Bowie, A.R. (2000) Importance of stirring in the development of an iron-fertilized phytoplankton bloom. *Nature* **407**: 727–730.

Azam, F., Fenchel, T., Field, J.G., Meyer-Reil, R.A., and Thingstad, F. (1983) The ecological role of water column microbes in the sea. *Mar Ecol Prog Ser* **10**: 257–263.

Barbeau, K., Rue, E.L., Bruland, K.W., and Butler, A. (2001) Photochemical cycling of iron in the surface ocean mediated by microbial iron(III)-binding ligands. *Nature* **413**: 409–413.

Behrenfeld, M.J., and Falkowski, P.G. (1997) Photosynthetic rates derived from satellite-based chlorophyll concentration. *Limnol Oceanog* **42**: 1–20.

Boenigk, J., and Arndt, H. (2002) Bacterivory by heterotrophic flagellates: community structure and feeding strategies. *Ant Van Leeuw Int J Gen Microbiol* **81**: 465–480.

Boyd, P.W., Watson, A.J., Law, C.S., *et al.* (1999) Molecular evidence for zooplankton-associated nitrogen-fixing anaerobes based on amplification of the *nifH* gene. *FEMS Microbiol Ecol* **28**: 273–279.

Buesseler, K.O., Chang, H., Charette, M., *et al.* (2000) A mesoscale phytoplankton bloom in the polar Southern Ocean stimulated by iron fertilization. *Nature* **407**: 695–702.

Coale, K.H., Johnson, K.S., Fitzwater, S.E., *et al.* (1996) A massive phytoplankton bloom induced by an ecosystem-scale iron fertilization experiment in the equatorial Pacific Ocean. *Nature* **383**: 495–501.

Cooper, D.J., Watson, A.J., and Nightingale P.D. (1996) Large decrease in ocean-surface CO_2 fugacity in response to *in situ* iron fertilization. *Nature* **383**: 511–513.

Copley, J. (2002) All at sea. News feature. *Nature* **415**: 572–574.

Cotner, J.B., and Biddanda, B.A. (2002) Small players, large role: microbial influence on biogeochemical processes in pelagic aquatic ecosytems. *Ecosys* **5**: 105–121.

Cullen, J.J. (2001) Primary production methods. In: Steele, J.H., Turekian, K.K., and Thorpe, S.A. (eds) *Encyclopedia of Ocean Sciences*, pp. 2277–2284. Academic Press, New York.

Falcon, L.I., Cipriano, F., Christoserdov, A.Y., and Carpenter, E.J. (2002) Diversity of diazotrophic unicellular cyanobacteria in the tropical North Atlantic Ocean. *Appl Environ Mcrobiol* **68**: 5760–5764.

Falkowski, P.G. (2002) The ocean's invisible forest. *Sci Am* **287**: 54–61.

Fuhrman, J.A., and Noble, R.T. (1999) *Causative agents of bacterial mortality and the consequences to marine food webs*. Proceedings of the 8th International Symposium of Microbial Ecology. Atlantic Canada Society for Microbial Ecology, Halifax, Canada. http://plato.acadiau.ca/isme/Symposium06/fuhrman.PDF (accessed March 31 2003).

GreenSea Venture Inc. (2002) *Mission to develop iron fertilization of marine phytoplankton as a means of managing atmospheric carbon dioxide*. http://www. greenseaventure.com/index.html (accessed March 31 2003).

Herbert, R.A. (1999) Nitrogen cycling in coastal marine ecosystems. *FEMS Microbiol Rev* **23**: 563–590.

Hutchins, D.A., and Bruland, K.W. (1998) Iron-limited diatom growth and Si : N uptake ratios in a coastal upwelling regime. *Nature* **393**: 561–564.

Hutchins, D.A., Witter, A.E., Butler, A., and Luther, G.W. (1999) Competition among marine phytoplankton for different chelated iron species. *Nature* **400**: 858–861.

IEA Greenhouse Gas R&D Programme (2002) *Ocean storage of CO_2*. http://www.ieagreen.org.uk/ocean.htm (accessed March 31 2003).

Johnson, K.S., and Karl, D.M. (2002) Is ocean fertilization credible and creditable? *Science* **296**: 467–467.

Jurgens, K., and Matz, C. (2002) Predation as a shaping force for the phenotypic and genotypic composition of planktonic blooms. *Ant Van Leeuw Int J Gen Microbiol* **81**: 413–434.

Karl, D.M. (2002) Nutrient dynamics in the deep blue sea. *Trends Microbiol* **10**: 410–418.

Kirchman, D.L. (ed.) (2000) *Microbial Ecology of the Oceans*. Wiley-Liss, New York.

Lam, P. and Chisholm, S.W. (2002) *Iron fertilization of the oceans: reconciling commercial claims with published models*. http://web.mit.edu/chisholm/www/ Fefert.pdf (accessed March 31 2003).

Landry, M. (2001) Microbial loops. In: Steele, J.H., Turekian, K.K., and Thorpe, S.A. (eds) *Encyclopedia of Ocean Sciences*, 1763–1770. Academic Press, New York.

Malin, G. (1997) Sulphur, climate and the microbial maze. *Nature* **387**: 857–859.

Maranger, R., Bird, D.F., and Price, N.M. (1998) Iron acquisition by photosynthetic marine phytoplankton from ingested bacteria. *Nature* **396**: 248–251.

Martin, M.O. (2002) Predatory prokaryotes: an emerging research opportunity. *J Molec Microbiol Biotechnol* **4**: 467–477.

Mehta, M.P., Butterfield, D.A., and Baross, J.A. (2003) Phylogenetic diversity of nitrogenase (*nifH*) genes in deep-sea and hydrothermal vent environments of the Juan de Fuca ridge. *Appl Environ Microbiol* **69**: 960–970.

Morel, A., and Antoine, D. (2002) Perspectives: oceanography. Small critters – big effects. *Science* **296**: 1980–1982.

Ocean Primary Productivity Study (2000) Rutgers, State University of New Jersey Institute of Marine and Coastal Sciences. http://marine.rutgers.edu/opp/ (accessed March 31 2003).

Paerl, H.W. (1998) Structure and function of anthropogenically altered microbial communities in coastal waters. *Curr Opin Microbiol* **1**: 296–302.

Pomeroy, L.R. (1974) The ocean's food web: a changing paradigm. *BioScience* **24**: 409–504.

Pomeroy, L.R. (2001) Caught in the food web: complexity made simple? *Sci Mar* **65**: 31–40.

Samuelsson, K., and Anderson, A. (2003) Predation limitation in the pelagic microbial food web in an oligotrophic aquatic system. *Aquat Microb Ecol* **30**: 239–250.

Taroncher-Oldenburg, G., Griner, E.M., Francis, C.A., and Ward, B.B. (2003) Oligonucleotide microarray for the study of functional gene diversity in the nitrogen cycle in the environment. *Appl Environ Microbiol* **69**: 960–970.

Tortell, P.D., Maldonado, M.T., Granger, J., and Price, N.M. (1999) Marine bacteria and biogeochemical cycling of iron in the oceans. *FEMS Microbiol Ecol* **29**: 1–11.

Vaulot, D.A. (2001) Phytoplankton. In: *Encyclopedia of Life Sciences.* Macmillan. http://www.els.net (accessed March 31 2003).

Watson, A.J., Bakker, D.C.E., Ridgwell, A.J., Boyd, P.W., and Law, C.S. (2000) Effect of iron supply on Southern Ocean CO_2 uptake and implications for glacial atmospheric CO_2. *Nature* **407**: 730–733.

Zehr, J.P., and Ward, B.B. (2002) Nitrogen cycling in the oceans: new perspectives on processes and paradigms. *Appl Environ Microbiol* **68**: 1015–1024.

Zehr, J.P., Mellon, M.T., and Zani, S. (1998) New nitrogen-fixing microorganisms detected in oligotrophic oceans by amplification of nitrogenase (*nifH*) genes. *Appl Environ Microbiol* **64**: 3444–3450.

Zehr, J.P., Carpenter, E.J., and Villareal, T.A. (2000) New perspectives on nitrogen-fixing microorganisms in tropical and subtropical oceans. *Trends Microbiol* **8**: 68–73.

Zehr, J.P., Waterbury, J.B., Turner, P.J., *et al.* (2001) Unicellular cyanobacteria fix N_2 in the subtropical North Pacific Ocean. *Nature* **412**: 635–638.

Symbiotic associations 10

10.1 What is symbiosis?

In the broadest definition, the term symbiosis (literally 'living together') is used to describe any close, permanent relationship between two different organisms, which can range from commensalism (a loose association in which one partner gains benefit but does no harm to the host) to parasitism (in which the partner benefits at the expense of the host). In common usage, the term symbiosis usually refers to a mutualistic relationship, in which both partners benefit, and this definition is used in this chapter. We can further classify the relationship in terms of the degree of 'intimacy' of the association. Ectosymbionts (or episymbionts) are microorganisms that colonize the external surfaces (including infoldings or chambers such as the gut cavity or exoskeleton of crustaceans). Endosymbionts are those that are inside the cells of the host; these often show special adaptations to intracellular existence. In many cases, microorganisms involved in symbioses are also capable of life outside the host (facultative), but in other situations they may also have lost the ability to exist independently (*obligate* relationship). Interactions (parasitic, commensal and mutualistic) between different types of microorganism (bacteria with protists, protists with protists); and between microorganisms and higher forms of life (animals and plants) are of fundamental importance in the ecology of the marine environment. This chapter provides some examples of mutualisms that illustrate the diversity and importance of such interactions.

10.2 Symbioses of microalgae with animals

10.2.1 Types of association

Biologists first recognized associations between microalgae and a wide range of invertebrate animals over a century ago. A number of terms were introduced at this time; these distinctions were based largely on the coloration of the tissue often seen due to pigments from the algal symbiont. Thus, zooxanthellae (the term refers to the golden-brown color) occur in a wide range of corals, anemones and some molluscs and sponges. Zoochlorellae (green color) occur mainly in sponges, coelenterates and flatworms. Finally, the term zoocyanellae (blue-green color) is applied to the cyanobacteria (including *Prochloron*) found in some seasquirts and molluscs. The zooxanthellae, most of which are dinoflagellates, are of particular significance and will be considered in detail.

10.2.2 Nature of dinoflagellate endosymbionts (zooxanthellae)

A variety of dinoflagellate types have been recognized in association with a wide range of host species. Classification based on morphology and physiological characteristics is very difficult, but molecular studies reveal a very high diversity of the dinoflagellates, leading to revision of earlier ideas. Previously, it was thought that all symbiotic dinoflagellates belong to a single genus, *Symbiodinium*. It is now clear that there is a degree of diversity approaching that seen in different orders of free-living dinoflagellates. The traditional view has been that each host harbors one species of symbiont. Recently, RFLP analysis, in conjunction with PCR of 18S rRNA genes (see *Section 2.6*), has shown that some corals and clams can associate with genetically different types (clades) of dinoflagellate and may host one or two distinct populations at a time.

Box 10.1 RESEARCH FOCUS

Can coral reefs survive?

New insights into the coral-zooxanthellae symbiosis

Coral reefs are of major importance in marine ecosystems because of the huge biodiversity that they include. Reefs are fragile structures and they are easily damaged in shallow waters by natural phenomena such as storms and wave action. Such cycles of damage and rebuilding have been going on for millennia, but many scientists fear that the recent impact of human activities poses a real threat to their future survival. Local direct anthropogenic effects are common. For example, damage to reefs by tourist boats, divers, and hotel and marina construction has become a serious problem. Dynamite fishing and mining of reefs for building materials is common in some island communities.

In addition to this direct damage, one of the areas of greatest concern is the worldwide spread of coral bleaching, in which the symbiosis between the coral and its resident zooxanthellae is disturbed. Bleaching involves the loss of photosynthetic pigments or the expulsion of the zooxanthellae from the coral tissue. If severe or prolonged, growth and reproductive ability is impaired and the colony can become susceptible to disease or overgrowth by algae. Bleaching is not a new phenomenon, but the frequency and scale of mass bleaching episodes, which can affect thousands of km^2, seem to have increased dramatically in the last two decades. Many have expressed fears for the survival of coral reefs in the near future and this could have major effects on marine ecology and cause devastating economic effects on fisheries and tourism. There are many alternative theories about the causes of large-scale coral bleaching, but it is generally agreed that it is initiated by some kind of environmental stress, in particular elevated sea temperatures and high solar irradiance (especially UV light). Rapid global climate change and ozone layer depletion may be responsible for these effects. Elevation of surface water temperatures by just 1 or 2°C above normal for a few weeks often leads to bleaching. Loss of the symbiotic zooxanthellae can also be induced in aquarium experiments by small temperature shifts. However, it is not yet clear whether the dinoflagellate or the coral host detects the change, or whether they both do. Areas of active research include the monitoring of the induction of heat-shock proteins following temperature shifts (Fang *et al.* 1997) and the early effects of temperature stress on electron flow in the zooxanthellae photosytem (Jones *et al.* 1998).

Although the coral–algal symbiosis is described as a mutualistic relationship, Knowlton (2001) explains the great sensitivity of the ecological balance of the partners in terms of a 'reciprocally selfish' relationship, in which each partner tries to minimize costs and maximize benefits. The relationship could be terminated if stress means that one partner does not get its 'fair share' of benefits. The establishment and maintenance of symbioses between microalgae and their host depends on the continuous movement of molecular signals between the partners. The specificity of host–symbiont interactions is caused by subtle changes in the molecular structure of the signal molecules and their receptors. Some corals can act as a host for different species of zooxanthellae, and an area of current research is the possibility that different genotypes with a greater tolerance to 'stress' may recolonize bleached tissue. This has been termed the 'adaptive bleaching hypothesis'. A key discovery was made by Rowan *et al.* (1997) following study of a bleaching episode in two Caribbean corals, *Montastraea annularis* and *M. faveolata*. They showed that these corals each host three phylogenetic types (clades) of *Symbiodinium* dinoflagellate (denoted A, B and C) which can be distinguished using RFLPs in their rRNA. The proportion of A:B:C found in the coral tissue depends on the depth and shading of the coral colony because the three types are adapted to different levels of irradiance. Careful analysis of the pattern of bleaching in colonies during a natural episode of bleaching associated with increased water temperature and higher irradiance, showed selective expulsion of

Symbiodinium type C. Fagoonee *et al.* (1999) carried out a long-term field study which showed large variability in the zooxanthellae population with regular episodes of very low densities. They concluded that bleaching events are part of a constant variability in zooxanthellae density caused by environmental fluctuations superimposed on a strong seasonal cycle in abundance. It now appears that there are at least five clades of *Symbiodinium* associated with corals of which type D seems to be the most heat-tolerant.

The complexity of the host environment is further increased by the fact that corals may contain many different types of bacteria. A recent 16S rRNA-based study by Rowher *et al.* (2001) showed a wide diversity of bacterial types in the Caribbean coral *Montastraea franski*, and one type of α-*Proteobacteria* appeared to have a specific association with the coral. Using a similar approach, Munn and Bourne are currently analyzing various species of coral from the Great Barrier Reef, Australia; it seems that specific bacteria may associate with particular corals. It is possible that previously unidentified symbiotic bacteria contribute to the nutrition of corals, for example through N_2 fixation or photoautotrophy. An understanding of the multifactorial interactions between the host and its complex microbial community is essential if we are to fully explain the phenomenon of coral bleaching. This point will be considered further in *Chapter 15*, where recent evidence that bacterial or viral infections might be directly implicated in coral bleaching is discussed.

This may have advantages for the host, depending on the extent to which the algal symbionts vary physiologically. For example, hosts could 'select' symbionts that are best adapted to tolerance of environmental conditions or photosynthesis at different depths. Hosts that are able to associate with several types of symbiont could possibly be at an advantage during periods of environmental disturbance (*Box 10.1*).

In the host, the dinoflagellates form coccoid cells surrounded by a cellulose cell wall, within a host vacuole called a symbiosome. In true endosymbioses, the algae are located within the tissue cells of the host, but some borderline cases exist in which the algal cells lie extracellularly in the body cavities. Many symbiotic dinoflagellates can be isolated and maintained in culture given suitable conditions. Most investigators have concluded that the morphology (e.g. the loss of flagella) and life cycle of the free-living and symbiotic forms are fundamentally different, but this is probably an oversimplification.

The dinoflagellates carry out photosynthesis, harvesting light energy via a complex of chlorophyll *a*, chlorophyll c_2 and the protein peridinin. Light energy is used to fix CO_2 which occurs mainly via the C3 pathway (Calvin–Benson cycle) using RubisCO (*Section 4.4.3*). This can be shown by measuring the incorporation of radiolabeled $^{14}CO_2$ into the tissues. Some species also use a C4 route employing phospho-enol pyruvate carboxylase as the key enzyme. The animal host clearly derives nutritional benefit from the relationship in the form of photosynthetically fixed carbon compounds, although views differ on the exact contribution which each partner makes to the acquisition of other nutrients (especially nitrogen and phosphorus). Radiolabeling experiments show that the zooxanthellae release a high proportion of photosynthetically fixed carbon as small molecules including glycerol, glucose and organic acids. It is possible that all the essential amino acids are provided by the zooxanthellae. In return, the host provides certain key nutrients and suitable environmental conditions for the dinoflagellate symbionts, including protection from predators. The density of zooxanthellae varies widely in different hosts; high numbers are often associated with a reduced dependence on feeding by capture of plankton, manifested by smaller tentacles or reduced digestive systems.

We know little about how these intimate relationships came about, but available evidence suggests that they evolved in the mid-Triassic period (about 250 million years ago). In all of the animal phyla that harbor dinoflagellates, the final step in digestion is intracellular. Cells of the animal's

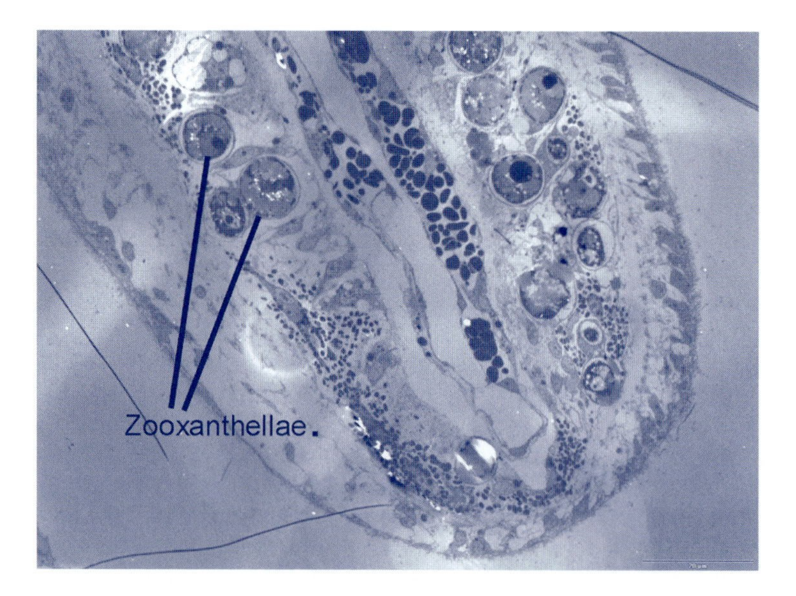

Figure 10.1

Zooxanthellae in corals. Transmission electron micrograph of section of a coral tentacle. Image courtesy of Simon Davy, University of Plymouth and Willie Wilson, Marine Biological Association UK.

digestive tract may have retained algal cells that were resistant to digestion, and subsequent evolution 'refined' this association. The importance of the association to the host is emphasized by anatomical and behavioral adaptations during evolution in the various animal groups. In some cases, free-living dinoflagellates infect hosts directly, but it is likely that most symbionts are usually passed vertically during reproduction of the host. Further use of molecular phylogeny may shed further light on the origins of these associations.

10.2.3 Corals

Zooxanthellae are very widespread and of major importance in the nutrition of corals, especially in tropical waters. In the group known as hermatypic corals, the individual animals (polyps) usually associate into colonies and secrete a skeleton, which leads to the development of coral reefs. *Figure 10.1* shows the ultrastructure of a coral tentacle containing zooxanthellae. Usually, the symbionts occur in the gastroderm (innermost cell layer) of the coral tissue. Reef-building corals lay down the calcium carbonate skeleton over a very fine organic matrix and field observations show that a high content and activity of zooxanthellae is essential for the rapid build up of the reef structure. Under optimum growth conditions, zooxanthellae may produce up to 100 times more carbon than they need for their own growth and reproduction, and most of this excess is transferred to the coral, where it is mainly respired. The coral may stimulate the release of compounds from the zooxanthellae by production of signal molecules which alter membrane permeability. When photosynthesis is inhibited, for example by restriction of light, the uptake of calcium and subsequent secretion of calcium carbonate is reduced. Excretion of lipids by the zooxanthellae is also important in construction of the skeleton. Most corals supplement their nutrition by feeding on zooplankton via their tentacles, but some groups (especially soft corals) appear to rely entirely on the zooxanthellae.

The importance of zooxanthellae to corals is shown dramatically by the phenomenon of coral bleaching. Bleaching occurs either when the host loses its symbiotic zooxanthellae, or when their photosynthetic ability is reduced through loss of pigment. Loss of zooxanthellae does not necessarily lead to immediate death of the corals, but the health of the colony is severely impaired and the corals may not recover. Although bleaching or whitening of corals has been known since the early part of the twentieth century, it has been the subject of intense study since the 1980s and there is currently great concern about the future of coral reefs (*Box 10.1*).

10.2.4 Tridacnid clams

These huge bivalve molluscs, especially *Tridacna gigas* that grows on Pacific reefs, can reach enormous sizes (up to 300 kg). Anyone who has dived or snorkelled on the Great Barrier Reef or Polynesian islands will have marveled at their size and the beautiful colors of their shell mantles. How can these animals grow so rapidly when the waters they inhabit are very poor in nutrients? The siphon tissue around the mantle is packed with endosymbiotic zooxanthellae. In this case, there are additional pigments of many different colors that provide protection to the zooxanthellae from the high UV irradiation in the clear waters that the clams inhabit. Recent studies of the importance of light in the symbiotic relationship have shown that, under the right experimental conditions, the zooxanthellae can provide all of the clam's carbon requirements through the release of small molecules such as glycerol and organic acids. However, as in many corals, acquisition of nutrients by heterotrophic feeding is also important. Tridacnid clams have particularly efficient filter-feeding mechanisms and can extract significant quantities of plankton from the water, even though the plankton is in low concentrations because of the oligotrophic nature of the environment. Thus it appears that, under natural conditions, the clams rely on both autotrophic and heterotrophic sources for their nutrition, with between 35% and 70% of the carbon requirements coming from photosynthetic activity of the zooxanthellae. These animals have evolved remarkable anatomical and behavioral modifications in order to optimize the benefits from their symbiotic partners. By comparison of *Tridacna* with known relatives it is clear that, during evolution, the orientation of the body has changed and the siphon tissue has expanded so that the maximum surface area colonized by the zooanthellae is exposed to the light.

10.3 Symbioses of chemoautotrophic prokaryotes with animals

Prokaryotes form a wide range of associations with marine animals and there have been some very exciting discoveries in the last few years. The details of only a fraction of these are currently known; for example, current research is revealing the importance of *Archaea* in symbioses with sponges and other animals. Study of these relationships is very exciting not just for the interest in gaining fundamental biological knowledge, but also because they have significant biotechnological potential.

10.3.1 Chemoautotrophic endosymbionts in hydrothermal vent animals

Section 1.7.6 describes the nature of the hydrothermal vents found in the deep sea, and the discovery of dense undersea communities of giant tubeworms and bivalve molluscs growing around the hydrothermal vents. As more was learnt about the vent communities, a key question arose – how could such a highly productive ecosystem be sustained? Vent habitats are completely dark, under enormous pressure and at such great depth that organic material from the upper layers of water settles too slowly to support this level of growth. The idea grew up that chemoautotrophic bacteria might somehow be involved in the nutrition of the dense communities of animals and evidence in support of this came from a number of sources.

By measuring the ratios of different isotopes of carbon, it was found that the cellular material of the bivalves does not have its ultimate origin in photosynthetically fixed carbon. The ratios of ^{13}C to ^{12}C reflect the efficiency with which different enzymes deal with the different isotopes (*Section 2.7.3*). The ratio found in these animals is well outside this value, proving that they are not feeding on material derived from carbon fixed in the upper photic zone. It was thought at first that the vent

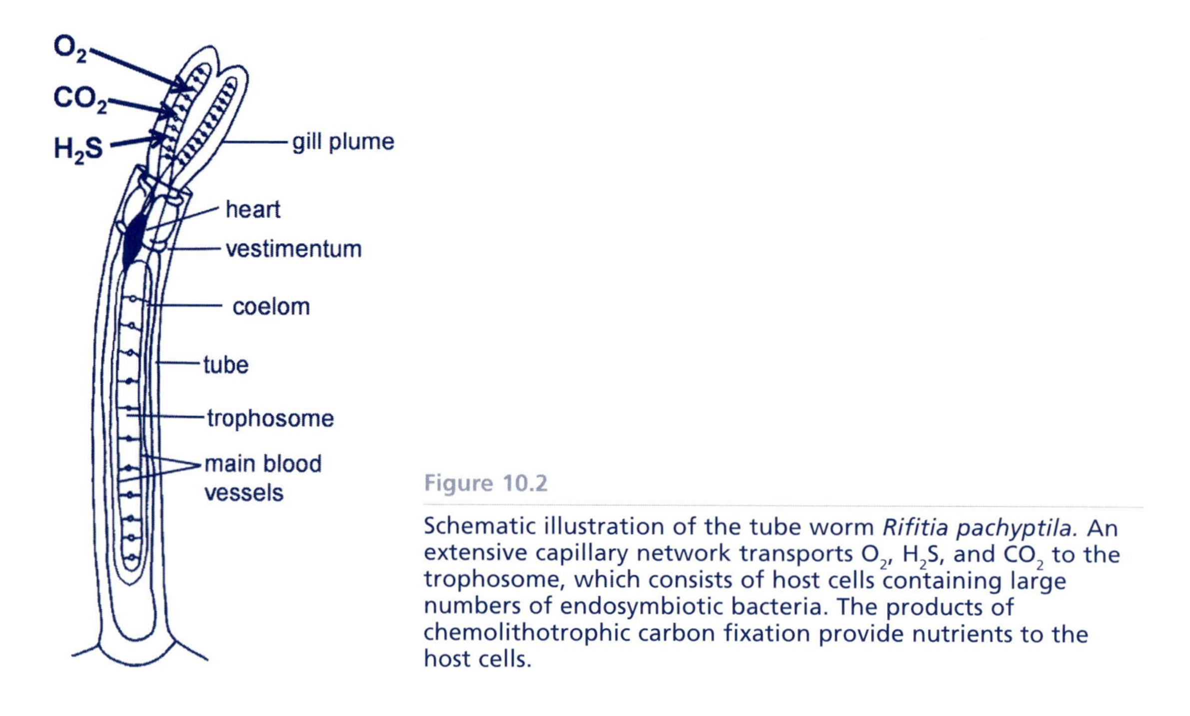

Figure 10.2

Schematic illustration of the tube worm *Rifitia pachyptila.* An extensive capillary network transports O_2, H_2S, and CO_2 to the trophosome, which consists of host cells containing large numbers of endosymbiotic bacteria. The products of chemolithotrophic carbon fixation provide nutrients to the host cells.

animals feed by filtration of chemoautotrophic bacteria that exist in the water around the vents, concentrated by local warm-water currents created by the vent activity. This idea became unlikely when it was discovered that the filter-feeding mechanism and digestive tract of the clams (*Calyptogena magnifica*) found at the vents was greatly reduced.

With investigation of the giant tubeworm *Riftia pachyptila,* new questions arose about how this dense animal community obtains its nutrients. *Riftia* has no gut, so it was assumed at first that it must feed by simply absorbing nutrients from the surrounding water. However, the rate of uptake of organic material is insufficient to explain the high metabolic rates of these animals. The cavity of the worm is filled by an organ (the trophosome) supplied with many blood vessels, and this contains granules of elemental sulfur. Electron microscopy reveals the presence of large numbers of structures strongly resembling prokaryotic cells. The trophosome tissue contains a large amount of LPS and enzymes associated with sulfur metabolism, including ATP-reductase and ATP-sulfurylase. Further proof of endogenous autotrophic metabolism is provided by the presence of high levels of RubisCO. Therefore, it was concluded that the trophosome is packed with Gram-negative bacteria as endosymbionts. This is a truly remarkable animal–bacterium symbiosis, as up to 50% of the mass of the tubeworm can be bacteria. The endosymbiotic bacteria use the oxidation of reduced sulfur compounds as a source of energy for the fixation of CO_2 into organic material, which is transferred to the animal tissue for growth. *Riftia* has evolved into a truly autotrophic animal with sophisticated mechanisms to ensure the optimum conditions for the maintenance of its endosymbiont population and the acquisition of nutrients, as shown in *Figure 10.2.*

In the vent clam *Calyptogena magnifica,* microscopy reveals that the symbiotic bacteria are contained in large cells which are exposed to the seawater on one side and the blood supply on the other, although the anatomical adaptations to ensure efficient transport of nutrients to the symbionts are not as sophisticated as those of *Riftia*. Once again, enzyme assays show the thioautotrophic nature of the bacteria. The vent mussel *Bathymodiolus thermophilus* seems to be the least specialized of the vent animals with respect to the symbiotic interaction and there is little control of the nutrient environment of the symbiont. As discussed below, mussels of this group can associate with both sulfide-oxidizing and methanotrophic bacteria. *Bathymodiolus* also has a highly efficient filter-feeding apparatus and it probably relies on a mixture of autotrophic and heterotrophic feeding. This, combined with its greater mobility, explains the fact that the mussels can tolerate a wider range of habitats at the hydrothermal vent site.

10.3.2 Episymbiotic bacteria on vent animals

When hydrothermal vents on the mid-Atlantic Ridge were investigated in 1985, it was found that large populations of a type of shrimp (*Rimicarius exoculata*) dominated the vent chimneys. Microscopy and enzyme assays similar to those described above showed that these animals do not contain endosymbiotic bacteria. Instead, they feed mainly by grazing on microbial mats of chemoautotrophs on the vent chimney walls. However, there also seems to be a resident population of episymbiotic bacteria on the mouthparts and internal exoskeleton and the shrimps derive much of their nutrition by 'farming' the bacteria; seeking out the right concentrations of sulfide and O_2 and fanning water currents over the bacteria to ensure optimum gas exchange.

Another vent animal that is the subject of intensive research is *Alvinella pompejana,* an oligochaete worm about 9 cm long and 2 cm wide, which lives in dense masses in the walls of black smoker chimneys on the East Pacific vents. This animal is known as the 'Pompeii worm' in recognition of the association with volcanic activity and it is remarkable because of the temperature gradient across its body. The anterior end is in galleries within the chimney wall (approximately 85°C), while the posterior end projects into the surrounding water (about 20°C). Hair-like structures secreted from pores on the dorsal surface cover the body; these have large colonies of episymbiotic filamentous bacteria over 100 μm in length and are visible to the naked eye. Use of different FISH probes (*Section 2.6.10*) shows that there are two distinct bacterial types, one of which is a member of the γ-*Proteobacteria,* whilst the other is a member of the δ-*Proteobacteria* (SRB of the *Desulfosarcina/Desulfococcus* group). Some of the bacteria are clearly active in chemoautotrophic metabolism, but the functions that they perform for the host are not yet known.

10.3.3 Chemoautotrophic endosymbionts in non-vent animals

Chemoautotrophic symbioses do not exist solely in hydrothermal vent communities. There seems to be a very wide diversity of animal hosts, some (but not all) taxonomically related to the vent animals, in many habitats where the correct balance of reducing conditions and sulfide concentration exists. Methanotrophic symbioses in mussels have been investigated extensively following the discovery of huge communities of mussels and tubeworms in methane-rich 'cold seeps' in the Gulf of Mexico. The bacteria contain densely packed lamellae and utilize methane as a source of both carbon and energy. Using enzyme assays and oligonucleotide probes, it can be shown that some mussels host two distinct populations of chemoautotrophic bacteria, one of which oxidizes sulfur and one that oxidizes methane. The ratio of the two types of endosymbiont within the host reflects the relative abundance of sulfide and methane in the site at which the animal is growing.

A number of examples have now been described that negate the common misconception that animal hosts harbor only one type of symbiont. Indeed, multiple symbionts may confer ecological or physiological flexibility to the host. Usually, we would consider that the symbionts are competing for resources within the host, but what if the different microbial symbionts could themselves gain some mutual direct benefit? Perhaps the most remarkable example of such a dual symbiosis is the recent description of a small oligochaete worm (*Olavius algarvensis*) which contains both sulfate-reducing and sulfur-oxidizing bacteria, which cooperate through syntrophy to provide mutual advantage to themselves and their host (*Box 10.2*).

10.3.4 Phylogeny and acquisition of symbiotic bacteria

By analysis of 16S rRNA sequences, it can be shown that thioautotrophic endosymbiotic bacteria belong mainly to the γ-*Proteobacteria* (see *Section 5.6.1*). Symbionts isolated from closely related animal hosts show a high degree of homology, but those isolated from different host

Box 10.2 RESEARCH FOCUS

'Self-contained accommodation'

Two types of bacteria cooperate to feed a gutless worm

The oligochaete worm *Olavius algarvensis* is found in sediments in the Mediterranean Sea. Recent studies by Nicole Dubilier of the Max-Planck Institute of Marine Microbiology have provided a fascinating insight into the mode of nutrition of this animal, which is completely devoid of a gut system. Examination of cross-sections shows that, just below the cuticle of the worm, there are large numbers of bacteria (of two different sizes) and deposits of elemental sulfur. Using FISH, Dubilier *et al.* (2001) showed that the larger symbiont is a member of the γ-*Proteobacteria*, which includes known sulfide-oxidizing bacteria, whilst the smaller type is a member of the δ-*Proteobacteria*, which is known to contain SRB. The closest match of sequence is to the free-living SRB *Desulfosarcina.* The autotrophic nature of the sulfide-oxidizing bacteria (γ-type) was shown using an immunocytochemical stain directed against RubisCO. The activity of the SRB (δ-type) was shown by implanting living worms with tiny silver needles impregnated with radioactive sulfate; this was shown by autoradiography to be converted to sulfide. The two bacteria show cooperative metabolism (syntrophy) which could be regarded as a bacterium–bacterium symbiosis and together they provide nutrients for the host worm. Sulfide produced within the worm by the δ-type is used by the γ-type as an electron donor for the autotrophic fixation of CO_2. This will sustain a cycle of energy, but in order to obtain net growth and transfer of nutrients to the host there must be an external source of electron donors. The SRBs are probably capable of breaking down a variety of organic compounds. By movement through the sediment, the worms maintain the correct balance of O_2 (for its own respiration and that of the sulfur-oxidizing γ-type) and reducing conditions for the SRB δ-type. Possibly, the association with the sulfide-oxidizing γ-type evolved first and the worm may then have acquired the SRB δ-type. Dubilier *et al.* suggest that this may have enabled the worm to colonize new habitats, because it is not dependent on a continual supply of sulfide.

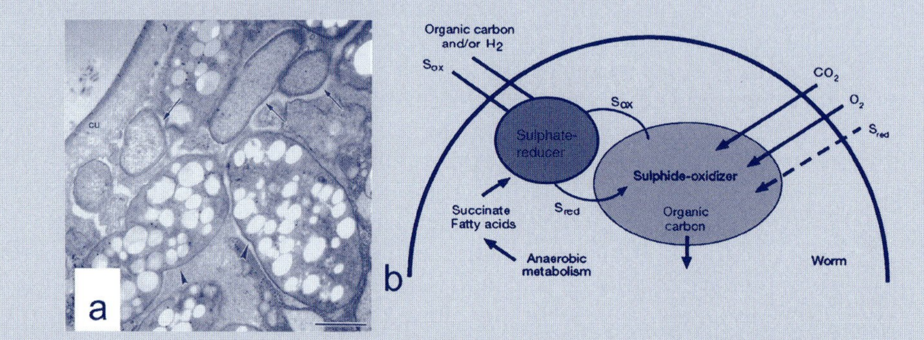

Bacterial endosymbiosis in *O. algarvensis*. (a) Electron micrograph of tissue section showing small (arrows) and large (arrowheads) endosymbiotic bacteria. (b) Model of nutrient and gas exchange. Reproduced from Dubilier *et al.* (2001), with permission from Nature-Macmillan Ltd.

families do not. There appears to have been a remarkable coevolution of the hosts and their symbionts. How animals acquire their symbionts is not entirely certain. Use of 16S rRNA gene probes shows that mussels and clams found at vents and cold seeps probably transmit the bacteria from one generation to the next via the ova, whilst tubeworms acquire their symbionts at the larval stage from the surrounding water.

10.4 Light organ symbioses in fish and invertebrates

Another type of cooperative association exists between bacteria and animals, in which the activities of the bacteria confer a behavioral and ecological benefit to the host, rather than a nutritional one. Bioluminescence is the emission of blue or green light from O_2-utilizing reactions via the luciferase enzymes (*Section 5.12.2*). Bioluminescence occurs very widely in the oceans, particularly in animals that inhabit deep waters where no sunlight penetrates or those which are active in shallow waters at night. Bioluminescent animals use light emission for three main functions: (a) avoidance or escape from predators; (b) attracting prey; or (c) as a means of communication, such as mating recognition. Animals often possess complex structures such as lenses, filters and shutters to control or modify the light emitted. In many animals, the source of the light is the presence of symbiotic bacteria that inhabit the surfaces of specialized glands or crypts. However, most bioluminescent animals appear to be 'self-luminescent' and do not rely on bacterial symbionts. Bioluminescence appears to have multiple evolutionary origins, as there is a very wide diversity of enzymes and substrates for the bioluminescent reaction mechanism. Apart from light emission, the only common feature is the requirement for O_2 and it is possible that bioluminescence evolved originally as an antioxidative mechanism (*Box 5.2*).

Several groups of fish and invertebrates have specialized light organs containing bioluminescent bacteria. In most cases, these bacteria can be isolated and will emit light in culture. Another clue to the bacterial origin is that bacteria are usually continuously luminescent, whereas light produced by eukaryotic enzymes tends to occur as brief flashes. Final proof of the role of bacteria in light emission is the use of an assay for luciferase in the presence of reduced flavin mononucleotide, as this reaction is unique to bacterial bioluminescence. Bioluminescent animals are found in a wide range of phylogenetically distinct groups and the nature of bacterial associations ranges from relatively unspecialized facultative colonization of the intestinal tract or organs derived from it, through to obligate interactions with specialized external light organs. The bacteria associated with bioluminescence have probably evolved via adaptive radiation with fish and invertebrate hosts over the last few hundred million years. Bioluminescent bacteria are often found as members of the commensal gut microbiota in fish and luminescence in the feces may encourage ingestion by other animals and thus facilitate their transmission to new hosts. The commonest culturable types of bacteria are *Photobacterium phosphoreum*, *P. leiognathi*, and *Vibrio* (=*Photobacterium*) *fischeri*. In the light organ, the bacteria grow with much lower generation times than they are capable of in culture and it seems that some mechanism of the animal host controls this. It is in the host's interest to maximize bioluminescence but minimize diversion of nutrients to the bacteria; restriction of O_2 supply or iron limitation may be important factors in this process. In all cases, the light organs contain dense communities of extracellular bacteria in tubules that communicate with the intestine or the external environment. Release of bacteria is an important component in the control of the bacterial population.

10.4.1 Flashlight fishes and anglerfishes

Figure 10.3 illustrates the location of external light organs in the flash-light fishes and deep-sea anglerfishes, which in both cases contain obligate symbionts that have not been isolated in culture. It is likely that the bacteria have undergone reductive evolution through loss of genes to become obligate symbionts. Ribosomal RNA sequencing and RFLP analysis shows that the symbionts associated with specific fish genera show significant genetic variation, as would be expected in obligate symbioses showing parallel divergence. Interestingly, obligate light organ symbionts appear to have a very low number of rRNA genes, which is probably related to their extreme specialization for slow, constant growth in the light organ.

Bacterial light organs in flashlight fishes (members of the family *Anomalopidae*) are the largest in any fish relative to body size. The light organs in these strictly nocturnal tropical reef fish are

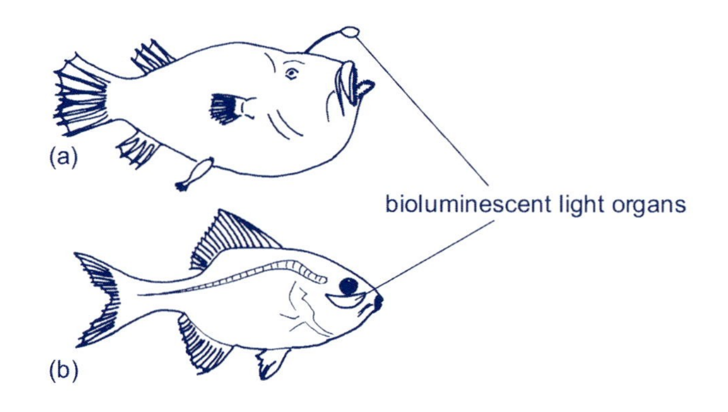

Figure 10.3

Illustration showing location of external light organs in fish harboring bioluminescent bacteria. (a) Anglerfish, (b) Flashlight fish.

located below the eyes. In very dark conditions, the forward illumination is bright enough to allow the fish to seek out prey, and the fish can also control the light emission in order to communicate with other members of the species (possibly for mating).

Anglerfishes, which belong to nine families in the suborder *Ceratoidei*, are usually solitary animals found in the deep sea. They have a small light organ (esca) at the end of a projection from the head. This acts as a lure to attract prey near to the jaws and light emission is controlled by the supply of oxygenated blood to the esca. It seems that only the female possesses this structure. In many anglerfishes the male has only a short lifecycle as free-living form and is attracted to the female and becomes permanently fused to her body. It is not known whether bioluminescence plays a role in location of mates.

10.4.2 **Sepiolids (bobtail squids)**

Hawaiian bobtail squids (*Euprymna scolopes*) inhabit shallow reefs, hiding during the day and feeding at night. They emit light from their ventral surface, adjusted to match the intensity of moonlight. This counter-illumination camouflage helps them to be less visible to predators from below (*Figure 10.4*). A highly specific association with a certain strain of bioluminescent *V. fischeri* is responsible. Newly hatched squid acquire this particular vibrio exclusively from the surrounding seawater within a few hours, even though it is present at very low densities (one to a few hundred cells ml^{-1}) and many other species of bacteria are present. The squid flush seawater in and out of their body during respiration and movement, and this brings bacteria into contact with the nascent light organ. To reach the crypts of the light organ, bacteria must swim through a duct lined with ciliated cells and covered with mucus. In the crypt spaces, the bacteria encounter macrophage-like cells that have an immune surveillance function. The macrophages engulf unwanted bacteria and some of the *V. fischeri*, probably serving to keep the symbionts from overgrowing. Most of the symbiotic bacteria colonize the microvilli of the crypt epithelial cells and this step involves adhesion of the bacteria to cell surface receptors. After adhesion, the bacteria grow rapidly (about three doublings per hour) for the first 10–12 hours until they reach a density of about 10^{11} bacteria ml^{-1} of crypt fluid. Once this critical density is reached, an autoinducer of bioluminescence accumulates and light emission is initiated via QS regulation of the *lux* operon (*Section 5.12.3*). Once established, the bacteria grow more slowly (about 0.2 doublings per hour), and each morning (in response to daylight) the squid squeezes out the contents of the light organ (a thick paste of mucus, bacteria and macrophage-like cells) so that about 90% of the bacteria are expelled. This behavior of the squid ensures that

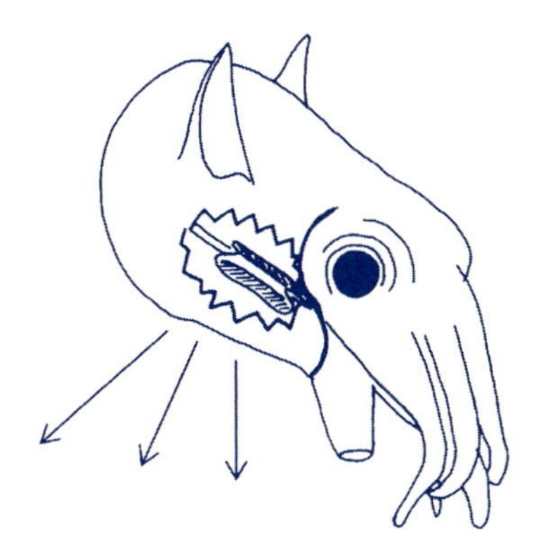

Figure 10.4

Illustration of counterillumination by bioluminescence in *Euprymna scolopes.* The cutaway shows the location of the light organ containing *Vibrio fischeri*, which emit light from the ventral surface. *E. scolopes* swims at night, and it is thought that bioluminescence confuses predators from below because the light masks the shadow created by moonlight or starlight. Adapted from Ruby (1996).

it maintains a 'fresh' active culture of *V. fischeri* that will build up to the high density required for bioluminescence when the squid emerges from its hiding place at night. Study of this remarkable association is revealing many important findings about the 'dialog' between an animal host and its bacterial partner, which has great significance in consideration of the evolution of symbiotic and pathogenic associations (see *Box 10.3*).

10.5 Microbial symbionts of sponges

Sponges (Phylum *Porifera*) have a simple body structure containing a large number of pores and channels. Studies of the microbial content of sponge tissue are largely driven by the interest in natural products such as antibiotics and antitumor compounds. Many sponges contain photosynthetic symbionts, including *Cyanobacteria* (such as *Aphanocapsa* and *Prochloron*) and dinoflagellates (*Symbiodinium*). These are assumed to contribute to the nutrition of sponges but the relationship has been less extensively studied than in corals. In recent years, use of 16S rRNA methods has revealed many types of *Bacteria* and *Archaea* in several species of sponge and they may constitute up to 40% of body mass. This knowledge has important applications in ecotoxicology and biotechnology.

10.6 Symbiosis and mixotrophy in protists

As discussed in *Chapter 1*, the widely accepted evolutionary theory first proposed by Margulis states that eukaryotic organisms evolved as a result of endosymbiosis between early ancestors of prokaryotic cells. Study of marine protists provides powerful evidence in support of this theory. Many protozoa and microalgae form associations with other species of protists, or with *Bacteria* and *Archaea*, ranging from loose associations, through intracellular endosymbionts to fully integrated organelles. This is well illustrated by the phenomenon of mixotrophy, a term used to describe the ability of protists to use both heterotrophic and phototrophic modes of nutrition. Many species of phytoplankton carry out photosynthesis, but can also ingest particulate food

Box 10.3 RESEARCH FOCUS

A perfect partnership?

Molecular interactions in the Euprymna scolopes–Vibrio fischeri *symbiosis*

The pioneering work in the study of this intriguing interaction has been carried out at the University of Hawaii by Ned Ruby who works on the bacterium and Margaret McFall-Ngai who studies the developmental biology of the squid host. With their research colleagues, they have published many papers on the interactions between the symbiont and its host. This interaction lends itself to detailed study because: (a) the animals can be cultured in the laboratory; (b) the infection process can be controlled and monitored; and (c) mutants of the bacterium can be easily generated, permitting detailed molecular analysis.

A colonization assay was developed which allows the symbiont-free newly hatched squid to be inoculated with *V. fischeri,* so that the early events in colonization can be studied (McFall-Ngai & Ruby 1991). The light organ of the newly hatched squid traps the bacteria, which swim using their flagella, probably in response to chemotactic signals. Mutants of the bacteria which have defective motility apparatus cannot initiate the infection (Graf *et al.* 1994). Various classes of mutants are able to initiate infection but do not grow properly and persist in the light organ. Among the factors involved here is the ability to utilize the rich mixture of amino acids provided by the host (Graf & Ruby 1998). One of the most surprising discoveries is that mutants in *luxA, luxi* or *luxR* genes (*see Section 5.12.2*) fail to elicit a programed host cell differentiation resulting in swelling of the epithelial cells in the crypts of the light organ (Visick *et al.* 2000), which normally occurs soon after infection and brings the bacteria into intimate contact with the host

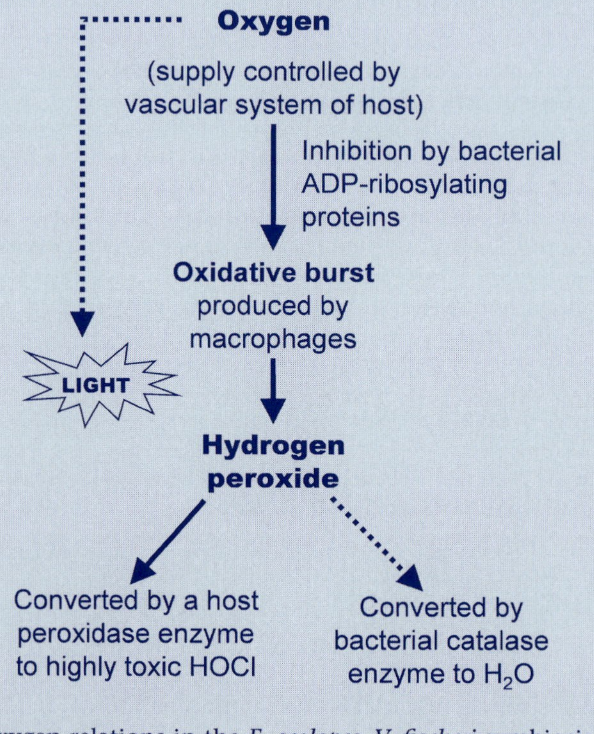

Oxygen relations in the *E. scolopes–V. fischeri* symbiosis.

cells. This finding is the first example of a molecular basis for a mechanism by which bacteria induce changes in the differentiation of host cells. This has far-reaching implications for our understanding of tissue development, since all animals are in contact with bacteria. LPS on the bacterial surface seems to be partly responsible for this effect (Foster *et al.* 2000). In addition, Visick *et al.* note that the association with *V. fisheri* is very specific, and that dark (*lux*$^-$) variants are never found naturally in the light organ. Whilst the selection of luminescent strains is of obvious advantage to the host, it is difficult to explain why luminescent strains out-compete dark strains (which might be expected to grow more quickly) unless luminescence confers a positive advantage to the bacterium also. Visick and Skofos (2001) and Fidiopastis *et al.* (2002) recently showed that additional levels of regulation of bioluminescence exist. The role of luminescence might be related to the control of O_2 levels and avoidance of oxidative stress in the light organ. The luminescence reaction consumes O_2 and possibly creates a hypoxic environment which would stimulate the release of host nutrients. In addition, the reduction in oxygen supply would limit the generation of reactive oxygen species which would be detrimental to the bacteria. Evidence that oxidative stress is an important factor in the light organ is provided by the observation that an inducible periplasmic catalase, which breaks down hydrogen peroxide, is necessary for normal colonization (Visick & Ruby 1998). Ruby and McFall-Ngai (1999) note that *V. fischeri* produces proteins with ADP-ribosylating activity which may also interfere with the oxidative burst as illustrated above. Study of these phenomena is providing major clues to an understanding of the evolution of symbiotic and pathogenic interactions between animals and bacteria.

(phagotrophy). This is particularly common in the dinoflagellates, where over 20% of species have been shown to be mixotrophic. Many small mixotrophic algae can ingest bacteria, whilst larger forms can also ingest other microalgae, ciliates and other prey (*Section 9.5.3*). Different stages in the life cycle of some dinoflagellates may show a spectrum from fully photosynthetic, through mixotrophic to fully phagotrophic lifestyles (this is well illustrated in *Pfiesteria,* as discussed in *Section 12.4*). Phagotrophy clearly provides nutrients in the form of carbon to supplement photosynthesis and other essential substances such as nitrogen, phosphorus and iron. Mixotrophs have sophisticated mechanisms to regulate their metabolism. For example, when high densities of bacteria are present and/or light levels are low, photosynthesis may be switched off completely and the protist relies entirely on ingested bacteria as food. When densities drop below a certain level, the photosynthesis machinery is activated. Protists will grow at different rates depending on the mix of nutrition employed at the time.

There are many examples of the ingestion of photosynthetic microorganisms by 'normally' heterotrophic protozoa, in which the chloroplast organelle is retained by the protozoan and becomes 'enslaved' for its benefit, but has not yet 'progressed' to a fully integrated organelle (in accordance with the Margulis paradigm). Such retention of chloroplasts, also known as kleptoplastidy, is widespread in ciliates, formaniferans and dinoflagellates. Clearly, the integration of the metabolic control of organelles with that of the cytoplasm requires complex adaptations, and many protozoa retain ingested chloroplasts for a few days before they are digested or eliminated.

True endosymbionts must avoid digestion by the host. This can occur either by physical separation of the food vacuoles and symbiont-capturing vacuoles, or by alteration of the phagosome membrane so that lysosomes containing digestive enzymes do not fuse with the phagosome. How kleptoplastidy is achieved is not yet known; here, the protozoan digests the bulk of its digested prey, but not the chloroplasts. The continuum of various modes of protistan nutrition is shown in *Figure 10.5*.

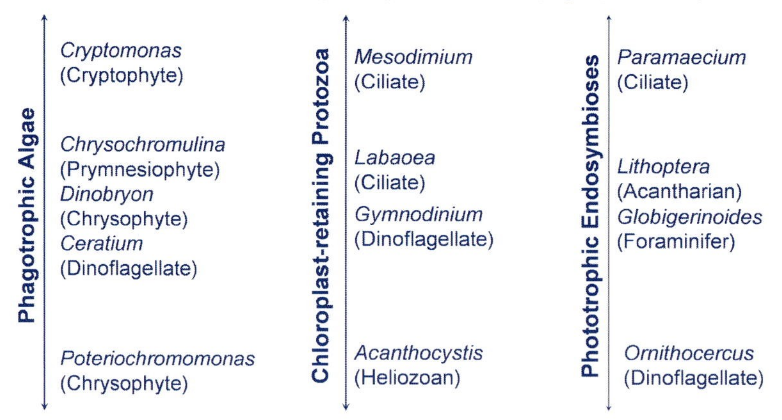

Figure 10.5

Depiction of various modes of protistan nutrition. There is a continuum between absolute autotrophy (phototrophy) and absolute heterotrophy. Representative genera of protists that depict this behavior are indicated on the figure in locations that attempt to approximate their degree of phototrophic/heterotrophic tendency. Reproduced from Caron (2000), with permission from Wiley-Liss Inc.

10.7 Metabolic consortia and mutualism between prokaryotes

There is increasing recognition that many microbes do not exist independently, but instead cooperate to ensure the effective utilization of nutrients (syntrophy). Microbes can work together to carry out particular metabolic transformations that neither organism can carry out alone. Two recently discovered examples discussed in *Research Focus Boxes* are (a) sulfate-reducers and sulfide-oxidizers as cosymbionts in the worm *Olavius* (*Box 10.2*) and (b) the importance of syntrophy in methane oxidation in anaerobic marine sediments by a closely integrated syntrophic consortium of archaeal cells and SRB (*Box 6.1*). It is clear that symbiosis operates within, and between, all major groups of life and has a major impact on marine processes.

References and further reading

Brown, B.E. (1997) Coral bleaching: causes and consequences. *Coral Reefs* **16**: S129–S138.

Buchheim, J. Coral reef bleaching website. http://www.marinebiology.org/coralbleaching.htm (accessed March 31 2003).

Caron, D.A. (2000) Symbiosis and mixotrophy among pelagic microorganisms. In: Kirchmann, D.L. (ed.) *Microbial Ecology of the Oceans*, pp. 495–523. Wiley-Liss, New York.

Dubilier, N., Mulders, C., Ferdelman, T., *et al.* (2001) Endosymbiotic sulphate-reducing and sulphide-oxidizing bacteria in an oligochaete worm. *Nature* **411**: 298–302.

Fagoonee, I., Wilson, H.B., Hassell, M.P., and Turner, J.R. (1999) The dynamics of zooxanthellae populations: a long-term study in the field. *Science* **283**: 843–845.

Fang, L.S., Huang, S.P., and Lin, K.L. (1997). High temperature induces the synthesis of heat-shock proteins and the elevation of intracellular calcium in the coral *Acropora grandis*. *Coral Reefs* **16**: 127–131.

Fidiopastis, P.M., Miyamoto, C.M., Jobling, M.G., Meighen, E.A., and Ruby, E.G. (2002) LitR, a new transcriptional activator in *Vibrio fischeri*, regulates luminescence and symbiotic light organ colonization. *Mol Microbiol* **45**: 131–143.

Foster, J.S., Apicella, M.A., and McFall-Ngai, M.J. (2000) *Vibrio fischeri* lipopolysaccharide induces developmental apoptosis, but not complete morphogenesis, of the *Euprymna scolopes* symbiotic light organ. *Dev Biol* **226**: 242–254.

Graf, J., and Ruby, E.G. (1998) Host-derived amino acids support the proliferation of symbiotic bacteria. *Proc Natl Acad Sci USA* **95**: 1818–1822.

Graf, J., Dunlap, P.V., and Ruby, E.G. (1994) Effect of transposon-induced motility mutations on colonization of the host light organ by *Vibrio fischeri*. *J Bacteriol* **176**: 6986–6991.

Graf, J., and Mirkin, E.S. Symbiosis web site. University of Connecticut. http://www.spuconn.edu/~mcbstaff/graf/Sym.html (accessed March 31 2003).

Harvell, C.D., Kim, K., Burkholder, J.M., *et al.* (1998) Review: marine ecology – emerging marine diseases – climate links and anthropogenic factors. *Science* **285**: 1505–1510.

Haygood, M.C. (1993) Light organ symbioses in fishes. *Crit Rev Microbiol* **19**: 191–216.

Jones, R.J., Hoegh-Guldberg, O., Larkum, A.W.D., and Schreiber, U. (1998) Temperature-induced bleaching of corals begins with impairment of the CO_2 fixation mechanism in zooxanthellae. *Plant Cell Environ* **21**: 1219–1230.

Kinzie, R.A., Takayama, M., Santos, S.R., and Coffroth, M.A. (2001) The adaptive bleaching hypothesis: experimental tests of critical assumptions. *Biol Bull* **200**: 51–58.

Knowlton, N. (2001) The future of coral reefs. *Proc Natl Acad Sci USA* **98**: 5419–5425.

McFall-Ngai, M.J., and Ruby, E.G. (1991) Symbiont recognition and subsequent morphogenesis as early events in an animal–bacterial mutualism. *Science* **254**: 1491–1494.

McFall-Ngai, M.J., and Ruby, E.G. (1998) Sepiolids and vibrios: when first they meet – reciprocal interactions between host and symbiont lead to the creation of a complex light-emitting organ. *Bioscience* **48**: 257–265.

Rohwer, F., Breitbart, M., Jara, J., Azam, F., and Knowlton, N. (2001) Diversity of bacteria associated with the Caribbean coral *Montastraea franksi*. *Coral Reefs* **20**: 85–91.

Rowan, R., Knowlton, N., Baker, A. and Java, J. (1997) Landscape ecology of algal symbionts creates variation in episodes of coral bleaching. *Nature* **388**: 265–269.

Ruby, E.G. (1996) Lessons from a cooperative, bacterial–animal association: the *Vibrio fischeri–Euprymna scolopes* light organ symbiosis. *Ann Rev Microbiol* **50**: 591–624.

Ruby, E.G., and McFall-Ngai, M.J. (1999) Oxygen-utilizing reactions and symbiotic colonization of the squid light organ by *Vibrio fischeri*. *Trends Microbiol* **7**: 414–419.

Stoecker, D.K. (1999) Mixotrophy among dinoflagellates. *J Euk Microbiol* **46**: 397–401.

Van Dover, C. (2000) *Ecology of Hydrothermal Vents*. Princeton University Press, New Jersey.

Visick, K.L., and Ruby, E.G. (1998) The periplasmic group III catalase of *Vibrio fischeri* is required for normal symbiotic competence and is induced both by oxidative stress and by approach to stationary phase. *J Bacteriol* **180**: 2087–2092.

Visick, K.L., and McFall-Ngai, M.J. (2000) An exclusive contract: specificity in the *Vibrio fischeri–Euprymna scolopes* partnership. *J Bacteriol* **182**: 1779–1787.

Visick, K.L., and Skoufos, L.M. (2001) Two-component sensor required for normal symbiotic colonization of *Euprymna scolopes* by *Vibrio fischeri*. *J Bacteriol* **183**: 835–842

Visick, K.L., Foster, J., Doino, J., McFall-Ngai, M.J., and Ruby, E.G. (2000) *Vibrio fischeri lux* genes play an important role in colonization and development of the host light organ. *J Bacteriol* **182**: 4578–4586.

Human disease – bacteria and viruses

11

11.1 Mechanisms of pathogenicity

This chapter considers the importance of bacteria and viruses associated with marine and coastal environments as a source of human disease. Disease can be caused either by multiplication of a bacterial or viral pathogen within the human body (infection), or by consumption of fish or shellfish containing a toxin produced by bacteria when they grow in these animals (foodborne intoxication), as shown in *Figure 11.1*. It is important to emphasize that one of the principal mechanisms of pathogenicity of many bacteria causing infections is the production of a toxin within the body after an 'incubation period' during which the bacteria multiply. This distinction has important consequences for the onset of disease symptoms; as a rule, intoxications caused by the consumption of 'ready-made' toxins cause their effects much more rapidly (minutes or a few

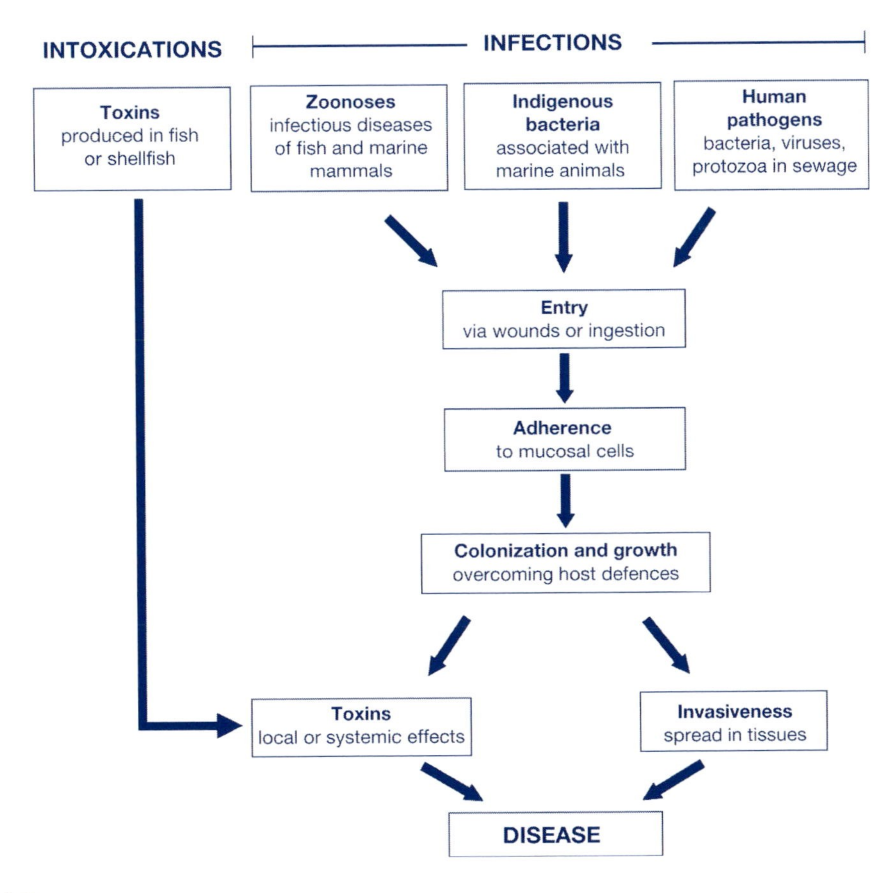

Figure 11.1

Sources of disease in the marine environment.

hours) than infections (which may take several hours or a few days to show symptoms). Toxins are also produced by dinoflagellates and diatoms, and these are considered Chapter 12.

Another important distinction is between those pathogens that are naturally present in the sea (indigenous or autochthonous organisms) and those that are introduced from the human population via the input of sewage containing human waste. In some cases, this distinction is somewhat artificial as pathogens may have their ultimate origin in the marine environment, multiply within the human population and then be returned to estuaries or the sea via fecal contamination. This applies particularly to cholera, as discussed below. However, in most cases, infections acquired from the sea infect individuals and are not normally transmitted from person to person. Many of the indigenous organisms are part of the normal commensal microbiota associated with marine animals, in which they normally cause no obvious effects. By contrast, some pathogens cause infectious diseases of animals which may be transmitted to humans. There are some important examples of such zoonoses acquired from fish and marine mammals, and these are dealt with in *Chapters 13* and *14*. *Table 11.1* shows a summary of the major infections and intoxications caused by various microorganisms associated with the marine environment.

11.2 Indigenous marine bacteria

11.2.1 *Vibrio cholerae*

The disease cholera (caused by *Vibrio cholerae*) was originally confined to parts of India until the early 19th century, when six major pandemics of the classical form of the disease spread to Europe and North America. Throughout the 20th century, cholera spread to become a major cause of mortality throughout the world, aided by increased transport, wars, refugee crises and the like. Since the 1960s, the most dominant form of the organism in the seventh pandemic has been the El Tor variant, which has reduced virulence but is better adapted for transmission. Until quite recently, it is unlikely that cholera would be included in a book about marine microbiology. The classical view of cholera as a disease spread by drinking water contaminated with human excreta stems from John Snow's famous deduction in 1854 that a particular source (the Broad Street water pump) was responsible for clusters of disease outbreaks in London. Cholera is characterized by profuse diarrhea caused by fluid and electrolyte loss from the small intestine and is therefore easily spread from person to person via contamination of water and food. However, the history of cholera through the ages reveals a striking association with the sea, and epidemiologists have long suspected that there might be an aquatic environmental reservoir of *V. cholerae*.

Various serotypes of the bacterium exist and these have different LPS O-antigens on their surface, although they are biochemically similar. Until 1993, all cases of cholera were caused by *V. cholerae* possessing antigen O1 (including the El Tor variant), but in 1993 a new serotype, O139, emerged as the cause of the eighth pandemic which began in the Bay of Bengal. Non-O1 and non-O139 strains are most commonly isolated from estuaries and coastal waters, even in areas where cholera is not endemic. Work by Rita Colwell and colleagues, beginning in the 1970s, produced significant findings which changed our thinking about the ecology of *V. cholerae*. The survival of the pathogen in water is greatly affected by environmental conditions, particularly salinity, temperature and nutrient concentration. *V. cholerae* can survive for long periods in seawater, but the numbers which can be isolated from water are very low and usually thought to be insufficient to initiate infection. Comparison of direct epifluorescence and viable counts shows that a large proportion of the cells enter a type of dormant state termed 'viable but nonculturable' (VBNC). Use of molecular and immunological methods has confirmed the VBNC phenomenon and it has subsequently been shown to occur in many other bacteria, although there is controversy about use of the term VBNC and how the state should be defined (see *Box 4.1*). During transition to the VBNC state, the cells initiate an active program of change resulting in reduced size, alteration of RNA and changes to the cell surface. This adaptation allows bacteria to survive adverse changes in nutrient concentration, salinity, pH and temperature. Colwell's group also showed that *V. cholerae* associates with a range of phytoplankton and zooplankton, especially copepods. It is

Table 11.1 Human infections and intoxications from marine and coastal environments

Disease	Pathogen	Source
	Bacterial infections	
Brucellosis	*Brucella*	Captive marine mammals
Cholera	*Vibrio cholerae*	Drinking water contaminated with *V. cholerae* in estuarine and coastal waters
Dermatitis; wound, ear and eye infections	*Staphylococcus, Pseudomonas* [also *Candida* – a yeast]	Sewage-contaminated water. Swimming, watersports
Fatal soft tissue infection	*Steptococcus iniae*	Handling infected fish
Gastroenteritis	*Salmonella, Shigella* and others	Sewage-contaminated water. Swimming, watersports. Filter-feeding molluscs. Whalemeat
Gastroenteritis	*Vibrio parahaemolyticus, V. vulnificus, V. mimicus, V. hollisae, V. fluvialis,* non-O1 *V. cholerae*	Crustacea and molluscs naturally harboring these bacteria as commensals
Tuberculosis	*Mycobacterium bovis*	Captive marine mammals
Typhoid	*Salmonella typhi*	Sewage-contaminated water. Swimming, watersports. Filter-feeding molluscs
Wound infections	*Erysipilothrix, Clostridium botulinum*	Wounds from handling fish and shellfish (e.g. 'seal finger', 'crayfish-handlers disease'). Seal bites
Wound infections	*Mycobacterium marinum*	Infection when cleaning fish tanks ('aquarist's granuloma')
Wound infections, septicemia	*Vibrio vulnificus, V. alginolyticus,* and others	Wounds from handling fish and shellfish; coral cuts
	Bacterial intoxications	
Botulism (neurotoxic)	*Clostridium botulinum*	Production of botulinum toxin in improperly preserved fish products
Pufferfish poisoning (neurotoxic)	*Vibrio, Photobacterium* (?)	Tetrodotoxin produced by symbiotic bacteria (?)
Scombroid poisoning (neurotoxic)	*Shewanella, Morganella* and others	Production of histamine following growth of commensal bacteria in fish tissue
	Viral infections	
Gastroenteritis	Norwalk virus, SSRVs, rotavirus and others	Fecal contamination of coastal water by sewage
Hepatitis	A, non-A, non-B hepatitis virus	Swimming, watersports
Poliomyelitis	Poliovirus	Filter-feeding molluscs
Influenza	Influenza virus	Captive marine mammals (?)
	Dinoflagellate and diatom intoxications	
Azaspiracid poisoning	*Protoperidinium* sp. (?)	Filter-feeding shellfish accumulating toxic dinoflagellates (low levels)
Ciguatera (neurotoxic)	*Gambierdiscus toxicus*	Accumulation of ciguatoxins in fish at top of food chain
Diarrhetic shellfish poisoning	*Dinophysis*	Filter-feeding shellfish accumulating toxic dinoflagellates (low levels)
Paralytic, neurotoxic, amnesic shellfish poisonings (neurotoxic)	*Alexandrium, Gymnodinium Pyrodinium, Pseudo-nitzschia*	Filter-feeding shellfish following harmful algal bloom
'Pfiesteria-associated syndrome'	*Pfiesteria piscicida*	Laboratory work with pathogen, contact with water or aerosols during fish kills in estuaries

not known if *V. cholerae* is a component of the normal commensal flora or a symbiont of a specific species in the plankton. Like many other vibrios, *V. cholerae* possesses chitinase and mucinase enzymes, which suggest that they are adapted to colonizaton of crustacean zooplankton.

These findings may be better understood if we consider that *V. cholerae* probably evolved in the aquatic environment and at some stage acquired the capacity to infect humans. During infection, the bacterium encounters dramatic changes in its environment, particularly the shock of temperature rise and low pH as it encounters the acid of the stomach. There is a complex process of gene regulation as the bacterium responds to chemotactic signals, swims towards the surface of the gut lumen and attaches to the gut epithelium by the production of pili. It then produces a toxin which stimulates loss of electrolytes and water from the gut cells. Each molecule of the toxin is composed of two different protein subunits. There are five B subunits that bind to ganglioside receptors on the surface of the gut cells, and one A subunit that is inserted into the host cell membrane. The A subunit is an ADP-ribosylating enzyme which modifies a host regulatory protein. This results in continued activation of adenyl cyclase in the gut cells leading to increased levels of the molecule cyclic AMP. This is responsible for the excessive fluid and electrolyte loss from the gut cells. The toxin affects membrane permeability but does not lead to permanent damage to the gut cells. Cholera victims usually die from extreme dehydration as water flows uncontrolled from the body. The symptoms of the disease can be limited by oral rehydration therapy or intravenous drips. In these circumstances, a patient can pass up to 20 liters of fluid a day.

There is evidence that the ancestor of the modern organism has undergone dramatic shifts in its genetic makeup, which may explain its transition from aquatic organism to human pathogen. For many years, we have known that toxin production and the expression of the pili are closely linked (hence, the pili are known as toxin coregulated pili, Tcp). Recently, it has been found that pathogenicity is dependent on a remarkable interaction between different bacteriophages within the *V. cholerae* host cell, as shown in *Figure 11.2*. The origin of the genes encoding the two subunits

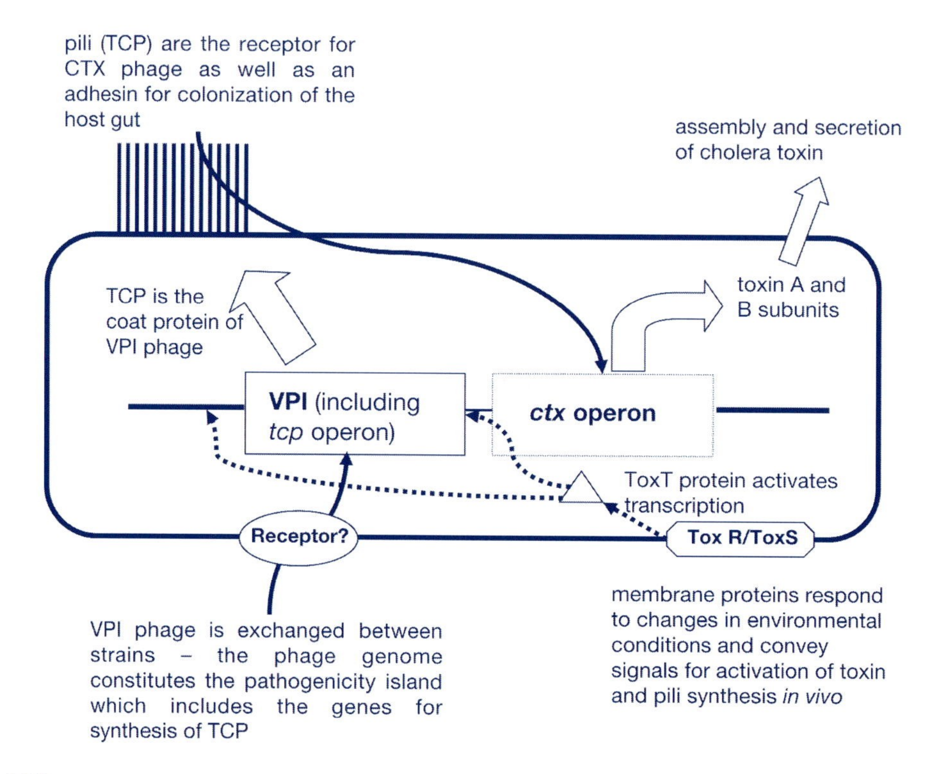

Figure 11.2

Model of the bacteriophage-mediated genetic basis of pathogenicity in *Vibrio cholerae.*

of the toxin (*ctxA* and *ctxB*) is a bacteriophage (CTX-phi). A large pathogenicity island (VPI) is the genome of another bacteriophage (VPI-phi), which encodes the Tcp that promote colonization of the host gut. The pili act as receptors for Ctx-phi bacteriophage. Transcription of ctxAB and the tcp operon are coregulated by environmental signals, involving membrane proteins. Bacteriophages have been shown to mediate virulence in other *Vibrio* species and other pathogens, as discussed in *Box 15.1*. Outside areas contaminated by excreta from infected persons, most environmental isolates do not possess the *ctx* genes. The full genome sequence of *V. cholerae* has recently been determined and this has confirmed that genes for the major pathogenic determinants are clustered in distinct regions of the larger of the two chromosomes (see *Section 3.5*), suggesting that horizontal transmission occurs. Possibly, in the natural aquatic situation, environmental factors induce the lytic cycle in toxigenic *V. cholerae,* resulting in the release of extracellular bacteriophages. Under appropriate conditions, these could infect nontoxigenic bacteria resulting in the emergence of new toxigenic strains. Passage through the human gut results in enrichment of the virulent forms and many genes are expressed *in vivo* which are not expressed outside the host. A model summarizing the ecology of *V. cholerae* is given in *Figure 11.3*. Recent ideas linking the emergence of cholera epidemics with climate are discussed in *Box 11.1*.

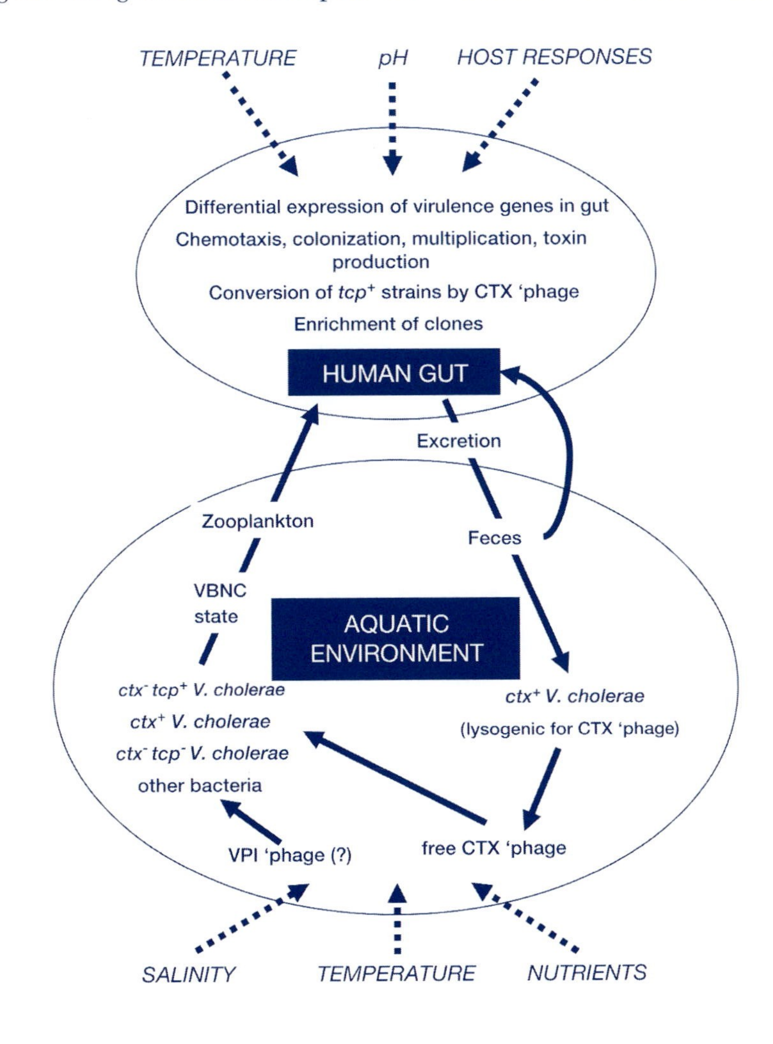

Figure 11.3

Model for the ecological interactions of *Vibrio cholerae.* The ecosystem comprises the human host, the estuarine or coastal environment, copepods and other plankton, *V. cholerae* and other aquatic bacteria, plus bacteriophages mediating gene transfer.

Box 11.1 RESEARCH FOCUS

Under the weather?

Long-term studies of cholera show that climate can affect disease ecology

The sudden explosive outbreak of cholera that occurred in South America in 1991 was unexpected, as epidemics had not occurred there for over a century. From an initial focus near Lima in Peru, the disease spread quickly throughout the continent and by 1992, over 750 000 people were infected and 7500 deaths occurred. One idea for its sudden emergence is that it may have been transported to a harbor near Lima from a ship from a cholera-endemic region and entered the human infectious cycle by persons eating shellfish from the contaminated waters. Ships may pick up ballast water in contaminated areas and transport infectious organisms thousands of miles before depositing them in another country and a variety of pathogens, including *V. cholerae,* can certainly be distributed in this way (Ruiz *et al.* 2000). However, in Peru the disease appeared almost simultaneously along a large distance of coastline, suggesting that other factors were at work. Colwell (1996) hypothesized that a climatic event was responsible, namely the El Nino Southern Oscillation which resulted in warmer coastal waters and influx of nutrient run-off from the land following heavy rains.

Detailed studies of the epidemiology and environmental factors affecting cholera have also been obtained by extensive field studies in Bangladesh. Regular screening of water for *V. cholerae* and monitoring of phytoplankton and zooplankton blooms, together with analysis of climate variations and nutrient status of the water, showed strong seasonal patterns (Colwell & Huq 2001). Mean wave height was also important and was thought to be related to the likely nutrient content of coastal waters. Mathematical models have been developed to study the pattern of epidemics (Pascual *et al.* 2001). Historical mortality data for Bangladesh between 1891 and 1940 shows a biannual seasonal cycle, with a dominant peak in the spring. Spring mortality, particularly in the coastal region, is strongly correlated with sea surface temperatures in the Bay of Bengal. Large epidemics occur in years after El Nino events, when coastal sea temperatures rise, resulting in a shift away from the usual seasonal pattern. However, the winter peak in cholera cases is less consistently associated with water temperatures and suggests the possibility of different aquatic reservoirs for the pathogen. These findings are of particular significance given current concerns about climate change. As discussed in *Chapter 15*, there are similar concerns about the effects of global warming on the emergence of diseases affecting natural marine ecosystems (Harvell *et al.* 1999). Human and animal diseases caused by other *Vibrio* species, such as *V. vulnificus, V. parahaemolyticus, V. shiloi* and *V. harveyi* also show variations with climatic conditions. Elucidation of the complex ecological interactions occurring in these pathogens is vital if we are to predict and manage changes in the pattern of infectious diseases caused by climate change.

Transmission of cholera can be controlled within the human population by good sanitation and clean drinking water, but this is still not achieved in many parts of the world. Cholera still results in about 120 000 deaths per year and untold human misery. Field workers in India and Bangladesh have shown that removal of zooplankton from water used for washing and food preparation by a simple cloth filter can be remarkably effective in preventing cholera. Despite the effectiveness and simplicity of oral rehydration therapy, this is not always easy to deliver in the field. Attempts to develop effective vaccines have had partial success. One approach is to administer purified B-subunits of the toxin, and it may be possible to genetically engineer plants such as bananas to express the immunogenic protein. Another method is to use a live recombinant strain in which the active A-subunit of the toxin has been deleted.

11.2.2 *Vibrio vulnificus*

This pathogen is found in estuarine and coastal water, sediment, plankton and shellfish throughout the world. Human infections are reported mainly from the United States, with the greatest incidence associated with consumption of shellfish in Florida and the Gulf of Mexico (approximately 50 serious cases and an estimated total 41 000 cases occur each year). Infections occur in many other parts of the world, although with lower frequency. In healthy individuals, consumption of contaminated raw shellfish (especially oysters) usually results in an unpleasant but not life-threatening gastroenteritis. However, in individuals with chronic underlying diseases (especially alcohol abuse, liver disease, diabetes or AIDS) a septicemic form of the disease may result, with a mortality rate of over 50%. The infective dose is very low, perhaps as little as 100 cells for susceptible persons. Like *V. cholerae,* the bacterium enters a VBNC state as a survival mechanism in unfavorable conditions. In this case, the trigger is a drop in water temperature below 10°C; isolation from water and incidence of the disease is rare in winter. A number of factors appear to be involved in virulence, including extracellular cytolytic enzymes, resistance to serum bactericidal activity and high-efficiency iron-sequestering systems. In fatal cases, death is primarily due to the effect of endotoxin (LPS) which stimulates over-production of cytokines with consequent host damage. *V. vulnificus* can also cause a very serious infection following contamination of open wounds with seawater or cuts when handling shellfish. *V. vulnificus* (as well as *V. alginolyticus)* is probably also responsible for infection of 'coral cuts' which are a common hazard of SCUBA diving. Unless treated promptly, these wounds may take many weeks to heal, and can sometimes lead to cellulitis which may be fatal.

11.2.3 *Vibrio parahaemolyticus*

This bacterium is isolated frequently from marine and coastal waters throughout the world; it also colonizes the surface of many types of marine animals and plankton. It is responsible for gastroenteritis and the pathogenesis appears to be similar to that of cholera, though not as severe. It is the commonest cause of food poisoning in Japan, due to the popularity of raw seafood. There are many serotypes, but O3:K6 is the most commonly associated with large outbreaks of seafood-related illness which occur in Asia, USA, Europe and elsewhere. Diarrhea, vomiting, cramps and fever are observed, usually about 15 h after ingestion of the bacterium. The disease is associated with raw or incompletely cooked seafood (particularly crabs and bivalve shellfish). It is thought that a large infective dose is required in order for sufficient numbers to survive the acidity of the stomach (use of antacid medicines may reduce this), and outbreaks often follow the cross-contamination of seafood that has been kept in a warm kitchen. The doubling time at 37°C can be as low as 10 minutes, permitting a massive expansion of the population within a few hours.

Isolates from environmental samples are usually much less virulent than those isolated from clinical samples, so it is likely that there are either multiple types of the organism, or that it undergoes some form of genetic changes and selective enrichment in the gut as seen in *V. cholerae*. Bacteriophage or plasmid exchange is probably implicated. *V. parahemolyticus* undergoes a remarkable process of adaptation when it forms biofilms on surfaces, as discussed in *Section 3.9.1*. Strains isolated from clinical samples are 'Kanagawa positive' (a reaction on a specialized blood agar) because they possess a thermostable direct hemolysin (TDH) and/or a related toxin (TRH). The toxins are implicated in the production of diarrhea, but knowledge of the mechanisms of pathogenicity is incomplete; other unknown factors are certainly involved. DNA probes for the identification of TDH and TRH are now routinely used in investigations.

11.2.4 *Clostridium botulinum*

Botulism is one of the most feared foodborne diseases but is fortunately quite rare. *C. botulinum* is a Gram-positive bacterium which forms endospores that can survive high temperatures. The bacterium is a strict anaerobe and under appropriate conditions it produces a powerful neurotoxin in foodstuffs. If ingested, tiny amounts of this toxin (perhaps as little as a few micrograms) cause a severe flaccid paralysis by binding to the membranes of the motor neurons, blocking the release of acetylcholine and therefore preventing muscle contraction. Symptoms usually occur after 12–36 h (unlike other fast-acting neurotoxins, the delay is caused by absorption of the toxin and transport via the central nervous system) and the mortality rate can be over 50%. *C. botulinum* is classified into types A to G based on the serological properties of the toxin produced. Type E is most commonly associated with seafood and can be isolated from marine sediments and the intestinal contents of fish. Commercial canning, the most common method of preserving fish, is designed to ensure destruction of the spores; however, one famous outbreak of botulism in 1979 involved one of the world's most well-known fish-processing companies. In this case, four people were infected (and two died) following consumption of part of a can of Alaskan salmon, which was subsequently found to have a minute hole in the seal. Despite being properly heated, as the can cooled on the production line a small amount of contaminating material from fish viscera was sucked into the can, and this contained enough viable spores to germinate and produce a lethal dose of toxin. Most cases of botulism occur from home-prepared or ethnic foods (e.g. botulism is a serious health problem in Arctic Inuit communities). Several outbreaks have occurred following consumption of smoked fish (smoking kills most but not all the spores) and such products should always be refrigerated to inhibit growth of any survivors. Most salting and drying methods of preservation are safe. *C. botulinum* is normally incapable of growth within the body. However, infection of unprotected wounds can occur in workers handling fish. The bacteria cannot invade healthy tissue, and grow only in necrotic wounds causing local paralysis similar to the closely related tetanus.

11.2.5 Scombroid fish poisoning

This type of foodborne intoxication is associated with eating fish of the family *Scombridae*, which include tuna, mackerel and bonito. The tissue of these fish contains high levels of the amino acid histidine. If there is a delay or breakdown in the refrigeration process between catching the fish and consumption, bacteria from the normal commensal microbiota can multiply and convert histidine to histamine. The spoilage may not be enough to alter the taste or smell of the fish, but levels of histamine can be sufficient to induce an allergic-type response. Reddening of the face and neck, shortness of breath and, in severe cases, respiratory failure can result within a few minutes of eating contaminated fish. Treatment with antihistamine drugs is beneficial and the patient usually recovers fully. In Europe, the incidence of scombroid poisoning is increasing due to the rising popularity of sushi and the import of fresh tuna and swordfish by airfreight over long distances.

11.2.6 Pufferfish (*Fugu*) poisoning

This intoxication is caused by the ingestion of tetrodotoxin, which is found in the intestines, liver and gonads of certain species of pufferfish, especially those of the genus *Fugu*. It is probable that the toxin is synthesized by bacteria which inhabit the intestine of these fish. Various isolates of *Vibrio, Photobacterium* and *Pseudoalteromonas* have been isolated from pufferfish and other tetrodotoxin-containing animals and reported to produce the toxin in culture, although there is some controversy about the bacterial origin of the toxin (see *Box 12.2*). Tetrodotoxin is one of the most active neurotoxins known, and acts by blocking the flow of sodium ions in the nerves. Within a few minutes or hours of eating fish contaminated with the toxin, victims feel tingling sensations in the mouth and a sense of lightness, followed quickly by the onset of total paralysis. The person remains conscious but totally immobilized until the moment of death, which occurs in about 50% of cases. In Japan, fugu is a prized delicacy for which diners are prepared to pay huge amounts for the thrill of eating tiny portions of this risky food. Specially licensed chefs prepare the dish, with the added frisson of allowing a miniscule dose of the toxin to remain in order to give a 'buzz'. There are an estimated 200 cases and 50 deaths per year from fugu poisoning. Considerable media interest was aroused in the 1980s following a report that the zombie culture of Haiti relied on the use by witch-doctors of a powder obtained from pufferfish to induce the 'living dead' state. However, the validity and ethics of the methods used to investigate this phenomenon have been questioned. Tetrodotoxin is now under intensive investigation as a powerful anesthetic and has been used for treatment of heroin addiction. One of the most dangerous venomous marine animals known, the blue-ringed octopus of northern Australia, also contains tetrodotoxin which is used to paralyze its prey. Species of some terrestrial animals such as newts and toads also produce the toxin. It is thought that animals which produce the toxin have single-point mutations in the amino acid sequence of a sodium channel protein, which enables them to be resistant to the toxin's effects. Proof of the role of bacteria in toxin synthesis is needed to understand the evolution of this possible symbiotic association (see *Box 12.2* for discussion of the role of bacteria in the synthesis of marine toxins).

11.3 Health hazards from sewage pollution at sea

11.3.1 Sewage as a source of bacterial and viral infections

A large proportion of the world's population lives near the coast. Many of the largest urban settlements have grown up around river estuaries and natural harbors because of their importance for trade. As these towns and cities developed, it was an easy option to dispose of untreated sewage directly into the rivers and sea; the adage 'the solution for pollution is dilution' applied. In many developed countries, awareness of the potential problems arising from disposal in close proximity to the population led to longer and longer pipelines off the coast, but the grounds for this were usually aesthetic rather than health-related. Vast quantities of raw sewage are still disposed directly to sea. In some communities where disposal of raw sewage has caused particular problems, complex engineering works and disinfection via UV light have been used, but this is a very expensive option.

In the mid 20th century, increased wealth and leisure time (especially in Europe and North America) led to greater use of the sea for recreational use, and a trip to the seaside became an important feature of many peoples' lives. In the 1950s, awareness began to be focused on the potential hazards of swimming in sewage-polluted water. Public awareness of the problem has become more acute because of the growing popularity of sports such as surfing, sailboarding and diving, and the use of wetsuits permits prolonged exposure times. As discussed below, sound epidemiological data on the frequency and incidence of diseases are lacking, but it certainly seems that many people do acquire troublesome (though not usually life-threatening) infections through recreational use of the sea. Coastal waters used for recreation contain a mixture of pathogenic and nonpathogenic microorganisms derived from sewage effluents, bathers themselves, seabirds, run-off from agricultural land contaminated by waste from livestock, as well as indigenous pathogens discussed in the previous section. Most of these pathogens are transmitted via the fecal–oral route and cause disease if the swimmer unwittingly ingests seawater. Infection via the ears, eyes, nose and upper respiratory tract may also occur. The number of organisms required to initiate infection will depend on the specific pathogen, the conditions of exposure and the susceptibility and immune status of the host. Viruses and parasitic protozoa may require only a few viable units (perhaps 100 or less) to initiate infection, whereas most bacteria require large doses. The types and numbers of various pathogens in sewage vary significantly according to the distribution of disease in the population from which sewage is derived, as well as geographic and climatic factors.

The commonest route of infection is via accidental ingestion of water, leading to gastrointestinal illnesses. Over 100 enteric viruses occur in human feces, including picornaviruses, reoviruses, adenoviruses, caliciviruses, astroviruses, Norwalk virus and unclassified small round structured viruses (SRSV). Most cases of swimming-associated gastroenteritis are due to the Norwalk agent (RNA viruses possibly related to the caliciviruses) and SRSV. Disease due to the Norwalk agent is characterized by nausea, vomiting, diarrhea, abdominal pain, headache and slight fever. The infectious dose is thought to be very low (perhaps 1–10 virus particles) as stomach acid does not inactivate the virus. In addition to swimming and seafood-associated infection, hundreds of thousands of cases of Norwalk disease occur each year through contamination of drinking water and foods by infected persons and the disease is a major problem as a cause of local epidemics in hospitals, hotels, holiday camps and cruise ships. Pathogenic enteric bacteria include *Salmonella, Shigella* and virulent strains of *Escherichia coli*. Dermatitis ('swimmers' itch') and infection of cuts and grazes can be caused by bacteria such as *Staphylococcus, Pseudomonas* and *Aeromonas* and the yeast *Candida*. These organisms can also cause ear and eye infections. There is some evidence that aerosols may be a significant route of infection for surfers, causing mild respiratory illness. Of course, there are some more serious diseases transmitted via the fecal–oral route, such as cholera, hepatitis, typhoid and poliomyelitis, but as a general rule these become significant for swimming-associated illness only if there is an epidemic of the infection in the population. In this case, direct person-to-person transmission is likely to be far more important (however, recall the special case of cholera discussed in *Section 11.2.1*).

Sewage pollution of water in which shellfish are harvested for human consumption poses a serious health hazard. In particular, molluscan shellfish (bivalves) such as oysters, mussels, clams and cockles concentrate pathogens from their environment by filter feeding, and there are many documented episodes of infection from this source. Viruses are the most important hazard because they can remain in the shellfish tissue after processing. Hepatitis (of the A or 'non-A, non-B' serotypes) acquired from shellfish is of particular concern; although rarely fatal, it can cause a long, severe, debilitating illness and some large outbreaks involving hundreds of cases have occurred (for example, following the serving of contaminated shellfish at banquets or receptions). Norwalk agent is the most common cause of seafood-associated gastroenteritis but usually only comes to the attention of health authorities when outbreaks involving groups of people are involved. It is likely that thousands of undocumented cases of Norwalk gastroenteritis occur each year, since most victims do not visit their doctor and, in any case, accurate diagnosis is difficult. As well as viral hepatitis and gastroenteritis, outbreaks of serious bacterial infections such as typhoid and cholera have been linked to shellfish. As with the indigenous aquatic organisms such as *Vibrio parahaemolyticus* and *V. vulnificus,* the problem is exacerbated by the consumption of raw or barely cooked shellfish.

11.3.2 Monitoring for potential pathogens – the indicator concept

The testing of marine waters and shellfish to determine risks to public health is based on the use of 'indicator organisms'. This methodology was developed from the system used to assess the safety of drinking water, developed over 100 years ago. The concept of an indicator organism stems from the idea that pathogens may be present in very low numbers in water, such that detection may be too difficult or expensive. An indicator is an easily cultivable organism present in sewage, whose presence indicates the possibility that pathogens may be present. Ideally, there will be a built in 'safety margin' because the indicator should be present in greater numbers and survive longer in the environment than the pathogens. As we shall see, current bacterial indicators for pollution of the marine environment fall far short of this ideal.

11.3.3 Coliforms and *Escherichia coli*

The first organisms used to assess water quality were the coliforms. These are a group of facultatively anaerobic Gram-negative bacilli characterized by sensitivity to bile salts (or similar detergent-like substances) and the ability to ferment lactose at 35°C. The definition of the coliform group is operational rather than taxonomic and there are 80 species in 19 genera, the best known of which are *Escherichia, Citrobacter, Enterobacter, Erwinia, Hafnia, Klebsiella, Serratia* and *Yersinia*. The fecal coliforms (FC, also termed thermotolerant coliforms, depending on the methodology used) are a subset of the coliforms originally distinguished by their ability to grow and ferment lactose at 44°C. The main member of this group is *E. coli*, but confirmatory biochemical tests (especially indole production from tryptophan) are used as a further distinction. Growth on selective media and fermentative properties are used in the traditional methods of identification and enumeration in water (see *Figures 11.4* and *11.5*). However, traditional tests involve several stages and reactions in biochemical tests (including lactose fermentation) are very variable so there has been a move over the last decade towards enzyme- and molecular-based tests. A defining feature of the coliforms is the enzyme β-galactosidase, which can be detected using o-nitrophenol-β-galactopyranoside (ONPG) as a substrate, resulting in a yellow color. *E. coli* can be distinguished by the production of β-glucuronidase, which cleaves methylumbilliferyl-β-glucuronide (MUG), resulting in fluorescence under UV light. These reactions form the basis of several commercial testing methods, which are gaining widespread acceptance because they produce results more rapidly.

Early microbiologists recognized that *E. coli* and other coliform bacteria are present in large numbers (over 10^8 per gram) in the gut of warm-blooded animals and their feces; therefore they became the standard indicator for fecal pollution of water. In fact, obligately anaerobic bacteria such as *Bacteroides, Bifidobacterium, Lactobacillus* and *Clostridium* are about 100 times more abundant, but these have not been used so extensively because they are more difficult to cultivate. It is now generally recognized that, whilst appropriate for testing drinking water, coliform and *E. coli* counts are woefully inadequate indicators of marine water quality. The total coliforms can be derived from a wide variety of sources, including plants and soil, so they often reflect run-off from the land as well as fecal pollution. It is very difficult to distinguish whether *E. coli* in coastal waters originates from human or animal fecal contamination. Most significant of all, as shown in *Table 11.2*, the survival of coliforms and *E. coli* in the environment does not seem to reflect the survival of the pathogens, which are principally viruses. Thus, they fail to meet one of the major requirements of a reliable indicator organism. Many studies have been carried out on the survival of bacteria and viruses in water. The effects of temperature, sunlight, pH, water turbidity, salinity and presence of organic matter are highly complex interacting variables. Of these, the bactericidal effects of UV irradiation are the most important. Bacterial indicators have different survival characteristics in marine and fresh waters, while human viruses are inactivated at similar rates in both. Although many different studies have produced conflicting results, depending on the environmental conditions, some general conclusions can be safely drawn with respect to marine waters: (a) levels of FC (including *E. coli*) do not necessarily indicate good or poor water quality; (b) levels of total coliforms are virtually useless as an indicator; and (c) neither total nor fecal coliform levels are reliable as predictors of health risks, because they do not correlate well with epidemiological findings (see *Box 11.1*).

11.3.4 Fecal streptococci (enterococci)

As early as 1980, it was proposed that the group known as the fecal streptococci (FS) might be a better indicator for monitoring marine water quality. This group includes the genera

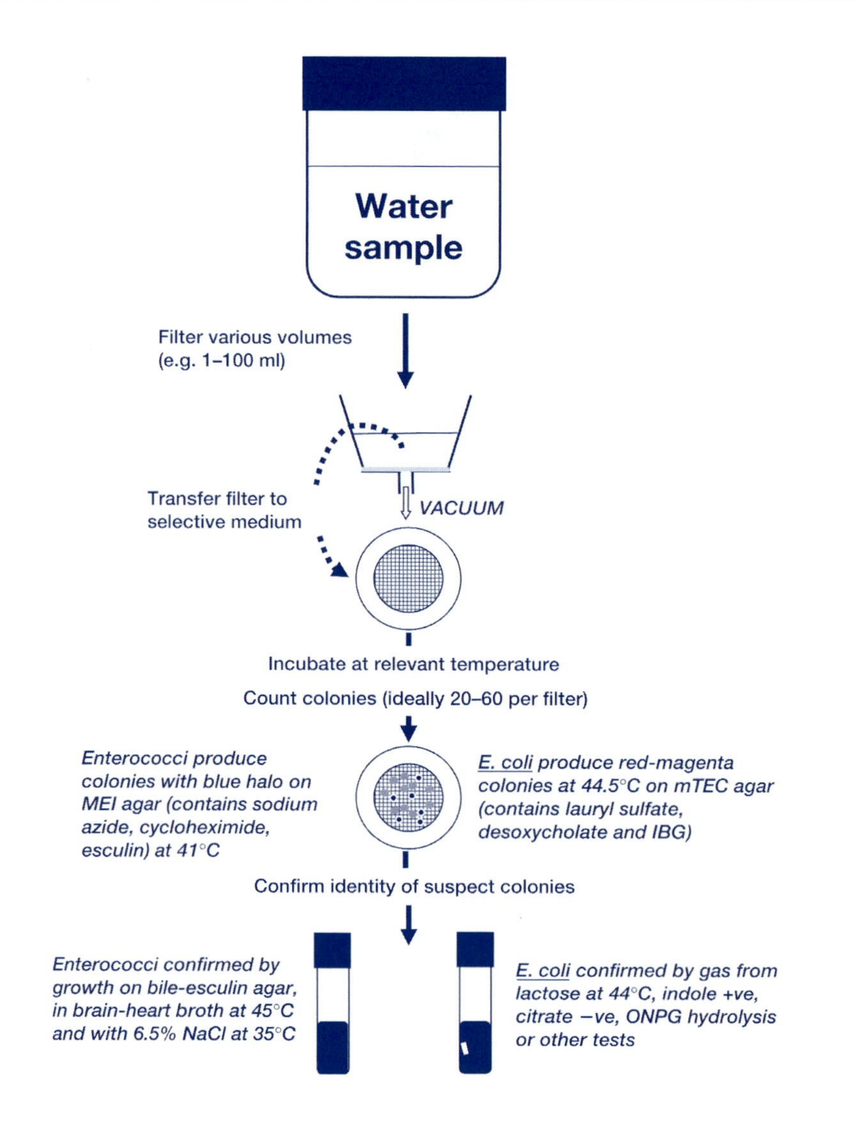

Figure 11.4

Outline of membrane filtration method for enumeration of *E. coli* and enterococci in marine water samples. There is a wide variation in the media used; that shown is based on the US EPA method. With modern selective and chromogenic media, confirmatory tests are usually not necessary for routine analysis and results can be obtained within 24 hours. Filters for *E. coli* are preincubated at 35°C before transfer to 44°C. Abbreviations: IMG = 5-bromo-6-chloro-3-indolyl-β-D-glucuronide; ONPG = o-nitrophenol-β-D-galactopyranoside.

Streptococcus and *Enterococcus,* which are consistently associated with the intestines of humans and warm-blooded animals and, as shown in *Table 11.2,* generally have longer survival times in water than coliforms and *E. coli.* Although the numbers of streptococci excreted in feces are lower than *E. coli,* they are nevertheless much higher than the numbers of pathogenic bacteria and viruses. Methods of enumeration follow the same principles as those for the coliforms (see *Figure 11.4*) and a variety of specialized media incorporating chromogenic substrates, antibiotics and tests for specific biochemical reactions can be used. A fluorigenic substrate, 4-methylumbilliferyl-β-D-glucoside, tests for the distinctive enzyme β-glucuronidase and is the basis of the Enterolert® testing system, which is now being widely adopted as a standard

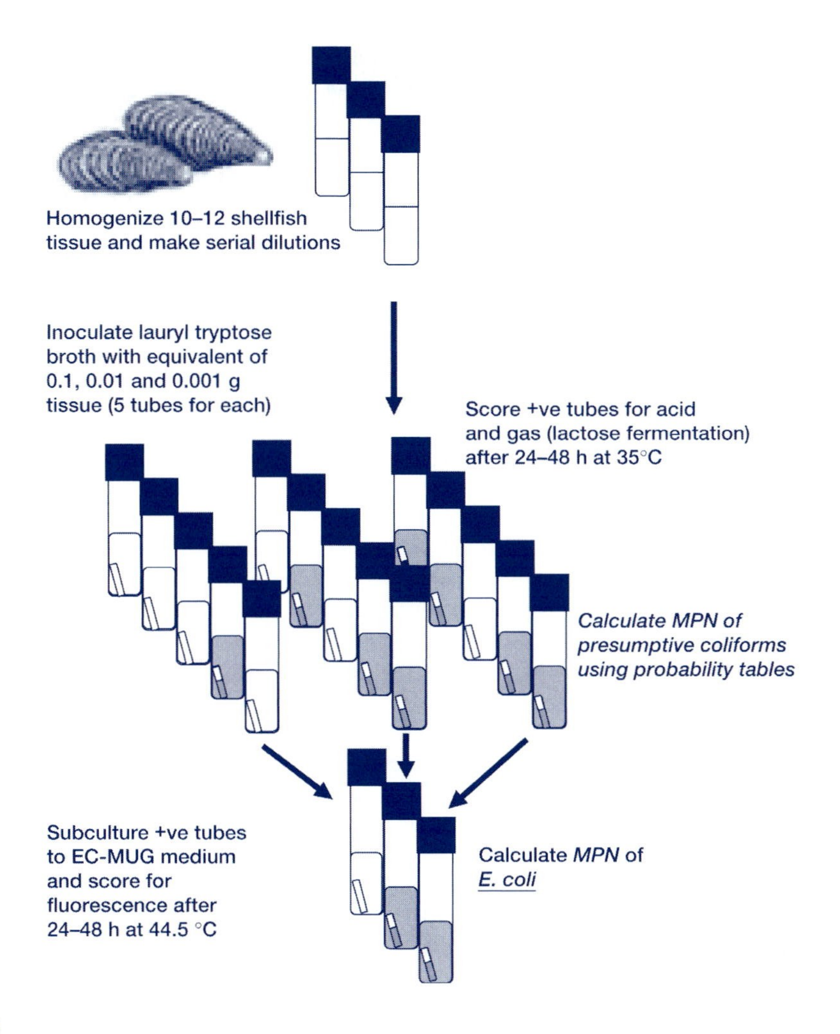

Figure 11.5

Outline of most probable number technique (MPN) for determining coliforms and *E. coli* in shellfish. There is a wide variation in the media used; that shown is based on the US FDA method. The test is done in two stages to ensure recovery of damaged bacteria and reduce interference by substances in shellfish tissue. EC-MUG medium contains 4-methyl-umbelliferyl-β-D-glucuronide, a fluorogenic substrate for glucuronidase (distinctive of *E. coli*). Calculation of MPN is based on the statistical probability that some tubes will be inoculated with a single organism from the dilutions used. MPN methods are also used for quantification of coliforms and *E. coli* in water.

method. Unfortunately, indigenous marine vibrios and aeromonads can interfere with the reactions and lead to false-positive results. Run-off from farmland after heavy rainfall can significantly distort the counts of FS; some studies have attempted to distinguish bacteria of human and animal origin.

11.3.5 Quality standards for recreational marine waters

As discussed in *Box 11.2*, there have been many studies attempting to prove an epidemiological link between the levels of microbial indicators and adverse health risks from recreational use of

Table 11.2 Typical survival times of indicator organisms and viruses in temperate coastal waters. This is a very simplified outline of survival times; results of T_{90} or T_{99} calculations (the time taken to reduce the initial population by 90 or 99%) are highly variable

Type	Conditions[1]		
	Extreme	**Moderate**	**Protected**
Coliforms	1 hour–a few hours	Hours–days	A few days
E. coli	Hours–1 day	A few days	Days–weeks
Fecal streptococci	1 day–a few days	Days–1 week	Weeks
Human viruses	A few days	Days–weeks	Weeks–months

[1]Conditions which affect microbial survival in water are highly complex and interdependent. Extreme conditions would be typified by clear water, bright sunlight, high temperatures; protected conditions would include turbid water, cloud cover, low temperatures. However, some studies show prolonged survival of coliforms under certain conditions, e.g. in tropical waters.

marine waters. The methodological or statistical basis of many studies has been the subject of much controversy. Nevertheless, public health authorities in many countries have developed monitoring programs and quality standards based on microbiological criteria. In Europe, the first definition of microbiological standards was published in 1976 as the European Community Bathing Water Directive. This defined testing protocols based on total and fecal coliform counts and used the concept of guide (which must not be exceeded in 80% of samples) and imperative (which must not be exceeded by 95% of samples) values (*Table 11.3*). The European legislation was used as the basis for improvements to coastal sewage disposal systems, which were designed to comply with the standards for coliform and *E. coli* levels. Interpretation of epidemiological studies led to the realization that although the motive of these expensive schemes was laudable, the scientific basis for evaluating their effectiveness was questionable, since an increased risk of illness seemed to be associated with water quality indicators well within the European limits. Amendments to the European legislation were proposed in 1994 to include an imperative value for FS, but this has not yet been implemented, although a guideline value was introduced in 1998. Much controversy about these standards remains, because the current European guideline level of 100 FS per 100 ml is considerably greater than the level predicted by the epidemiological studies as carrying a significant increased probability of illness and much higher than that in many other developed countries (e.g. the USA and New Zealand both have recommended threshold values of 35 FS per 100 ml). Health authorities will have to make decisions on what constitutes an 'acceptable risk', but public information on the interpretation of the water quality results posted at many seaside beaches is woefully inadequate to enable people to make an informed personal choice about the risks of bathing or undertaking watersports.

There are many different classification systems used in different countries. The nature of the indicators, the frequency of sampling, the methods of quantification and the threshold values for compliance with standards all vary. Within the USA, for example, different states employ different criteria. To some extent, such variations are probably appropriate depending on local conditions (for example, *Clostridium perfringens* may be a more reliable indicator in tropical waters). However, in view of the importance of international tourism, public confidence in monitoring procedures is essential and the current diversity of approaches is confusing.

11.3.6 Shellfish hygiene

The risk of infection from consumption of shellfish that have concentrated pathogens by growing in sewage-contaminated waters is considerable. Many countries have microbiological standards for the classification of waters in which shellfish are cultivated and harvesting is prohibited from areas with very high coliform/*E. coli* counts. For example, in the USA and

Box 11.2 RESEARCH FOCUS

Is it safe to swim?

The difficulties of linking water quality and health risks

Is it possible to establish a link between the health effects of bathing and the levels of indicator organisms? The first attempts to do so were carried out by the US authorities in the 1950s (in freshwater bathing sites). These showed that there was a significant increase in gastrointestinal, ear, nose and throat symptoms in swimmers exposed to waters where the coliform count exceeded 2700 per 100 ml. This was used by the US authorities to set the first standards for recreational waters. They proposed that the FC content should not exceed a log mean of 200 per 100 ml in five samples over a 30-day period, and that 10% of the samples should not exceed 400 per 100 ml. The epidemiological basis for defining these standards has been questioned by many authors; most notably Cabelli *et al.* (1982), who undertook more detailed studies at US beaches. Rigorous attempts to ensure standardization of microbiological methods, reporting of symptoms by participants and the use of suitable control groups were included. Cabelli's group found a statistically significant correlation between the levels of FS and an increased incidence of gastrointestinal symptoms associated with swimming. Many similar surveys have been undertaken, and Prüss (1998) subjected the data from 22 epidemiological studies to meta-analysis and found a consistent correlation between gastrointestinal symptoms and the counts of fecal indicator bacteria (especially FS). Hard evidence linking respiratory, ear and eye infections with indicator numbers was less compelling.

One of the major limitations of the types of study conducted by Cabelli and others is that they rely on *perception* of illness by those taking part, who are usually asked to respond to a questionnaire. It is also very difficult to distinguish the true effect of swimming from other effects (for example food consumed during a trip to the beach). Nevertheless, these data were sufficiently convincing to prompt the United States Environment Protection Agency (USEPA) to define complex standards comprised of a geometric mean standard of 35 FS per 100 ml in five samples over a 30-day period, a single sample density of 10^4 per 100 ml and percentile compliance levels.

As a result of growing public concern and pressure from groups such as Surfers against Sewage and the Marine Conservation Society, the UK Government sponsored new epidemiological research designed to overcome some of the criticisms of earlier studies. Carefully controlled beach surveys were conducted, in which volunteers were given a medical examination before randomized groups undertook supervised swims and water was tested at frequent intervals. One week later, participants received a second medical examination plus testing of throat and ear swabs and fecal specimens. Careful randomization and detailed analysis of questionnaires allowed other risk factors (e.g. food intake) to be taken into account. The UK study (Kay *et al.* 1994, Fleisher *et al.* 1998) confirmed that there was close correlation between levels of FS (measured at chest depth) in bathing water and the increased risk of gastrointestinal illness. A significant increase in gastrointestinal disease occurred when FS exceeded 30–40 per 100 ml and a convincing dose–response curve was observed. Based on these results, new guideline values for the microbiological quality of recreational waters have been proposed (WHO 2001). These use a statistical approach in which, to meet the standard, 95% of the counts must lie below a threshold value. Using this system, the estimated additional risks of bathing (compared to a control group) at less than 40 FS per 100 ml is below 1%. Between 41 and 200 FS per 100 ml, this rises to a 1–5% increased risk, and between 201 and 500 FS per 100 ml the risk is 5–10%. It should be noted that these risk factors apply to healthy adults, as children are usually excluded from this type of epidemiological study for ethical reasons. The risk of infection is probably much higher for lower age groups. Sound epidemiological studies are also needed to evaluate the risks for groups such as surfers and sailboarders, who may have prolonged exposure at times of the year when pathogens survive for longer or are resuspended from sediments by wave action (Bradley & Hanock 2003). It is also important that quality-monitoring programs should test a sufficient number of samples to determine statistically significant 95% compliance levels. After many years of uncertainty, a scientific rationale for microbiological standards and risk assessment is beginning to emerge, which will provide better guidance to policymakers and sanitation engineers.

Table 11.3 Microbiological criteria for quality of recreational marine waters in the European Union

Indicator	Guideline value per 100 ml	Imperative value per 100 ml
Coliforms	500 (80%)	10 000 (95%)
Fecal coliforms	100 (80%)	2000 (5%)
Fecal streptococci	100 (90%)	No mandatory standard

Figures in parentheses are the percentage of samples taking during the monitoring season which must comply with the standard (under current revision, EU Press Communication, October 2002).

Australia, waters used for harvesting shellfish for direct human consumption must not contain more than 70 total or 14 FC per 100 ml. In Europe, testing is carried out on an 'end-product' basis by testing bacterial counts in replicate samples of shellfish flesh, the results of which are used to classify harvesting waters. Only shellfish from category A waters (with maximum levels of 300 FC and 230 *E. coli* per 100 g) may be marketed for direct human consumption. Shellfish from category B areas must not exceed, in 90% of samples, the limit of 4600 *E. coli* (or 6000 FC)/100 g flesh. Such shellfish can only be placed on the market after depuration, relaying in cleaner water or heat treatment in order to meet the category A standards. Depuration is a disinfection procedure which involves transferring the animals to clean seawater which is usually recirculated for 24–48 hours through an UV lamp until the levels of indicator organisms reach acceptable limits. Again, the use of coliforms and *E. coli* as indicators is a very unreliable predictor of health risk and the standards have no sound epidemiological basis. Many outbreaks (sometimes large and severe) of viral illness have been associated with shellfish which appear to be of high standard as judged by the levels of bacterial indicators. The bacteriological quality of shellfish is determined by a lengthy MPN method (*Figure 11.5*). Although there have been proposals to replace coliform/*E. coli* counts with those for FS or other indicators, these have not yet been adopted. More rigorous methods of ensuring seafood safety, such as heat treatment or irradiation, are usually unsatisfactory because they spoil the taste or texture of the product. Shellfish must also be tested for the presence of microbial toxins (*Section 12.2.1*).

11.3.7 **Alternative indicators**

Recognition of the inadequacy of coliforms, *E. coli* and (to a lesser degree) FS as indicators of health risks has led to investigations of a number of alternative indicators. *Clostridium perfringens* has been used as an indicator because it forms resistant endospores which survive for long periods in the environment. This is particularly valuable in monitoring long-term effects of sewage disposal, for example in sediments and offshore sewage sludge dumps. Bacteriophages have been studied extensively; the rationale being that these viruses could be expected to have similar survival characteristics to viruses pathogenic for humans. The most widely investigated bacteriophages are the F^+-specific RNA coliphages that infect *E. coli* expressing sex pili on their surface. Coliphages can be measured using plaque assays (plating concentrated water samples on lawns of susceptible host bacteria, see *Figure 8.1*) or with group-specific oligonucleotide probes. Some success has been achieved in distinguishing animal and human sources and in establishing that they may have survival times similar to the human viruses when exposed to UV light under various conditions. More recently, bacteriophages of the anaerobe *Bacteroides fragilis* (one of the main members of the human gut flora) have shown considerable promise. However, there is very little epidemiological evidence linking bacteriophage levels with increased health risk, so they have not yet found application as microbial standards.

11.3.8 Direct testing for pathogens

The above discussion suggests that the indicator concept may have basic flaws when applied to marine waters and shellfish. Would it not be better to undertake direct testing for the pathogens themselves? The problem here is that most viruses, the pathogens which cause most concern, are difficult or impossible to culture. The usual method is to inoculate cell cultures of suitable human cells (such as enterocytes) with water samples (after concentration) and monitor for cell lysis or other cytopathic effects. Special safety precautions must be taken in laboratories using human cell culture and the effects, if any, may take several weeks to appear. All attempts to cultivate the most significant shellfish-associated pathogen, Norwalk virus, have failed. Attention has therefore turned to molecular biological techniques. The main approach has been the use of the RT-PCR method, as illustrated in *Figure 11.6*. Suitable primers that target the RNA virus group under study are added to samples of water or shellfish tissue extracts in the presence of the enzyme reverse transcriptase to make a cDNA copy, which is then amplified by *Taq* polymerase. Difficulties with the technique include selection of suitable primers which are representative of virus strains circulating in a particular region and interference with the PCR amplification by inhibitors (especially in shellfish tissue). The methods are still too time-consuming and expensive to be used in routine testing. However, they are proving valuable in special studies such as environmental impact assessments of new sewage disposal schemes and classification of shellfish harvesting areas.

11.4 Heavy metal mobilization

An indirect consequence of the activity of marine microorganisms is the transformation of trace elements, such as mercury, arsenic, cadmium and lead, to organic substances by microbial action in sediments. These can enter the food chain and may cause deleterious health effects. Mobilization of mercury is the best-known example because of its ability to concentrate in animal and human tissues. Mercury is widely used in industrial processes and is released during mining and in the burning of refuse. It is also an important component of electronics equipment and batteries and the safe disposal of out-of-date computers and other electronic products is a growing problem. Mercury is also present in many pesticides. Most mercury enters the sea in the form of the Hg^{2+} ion and adsorbs readily to particles. Several types of bacteria can transform this to methylmercury (CH_3Hg^+), which is very toxic, causing liver and kidney damage in humans. Because methylmercury is soluble, it is concentrated in the food chain, especially in fish. One of the most serious environmental disasters occurred in Japan in the 1970s when thousands of people who had eaten fish from the heavily polluted Minamata Bay were seriously affected by mercury poisoning. Hundreds of deaths and long-term health effects occurred. Because of its sequential concentration at each step of the food chain, high levels of methylmercury occur in top predators such as tuna, shark and marine mammals. Very high levels of mercury have been found in the tissues of Eskimo and Inuit communities who eat

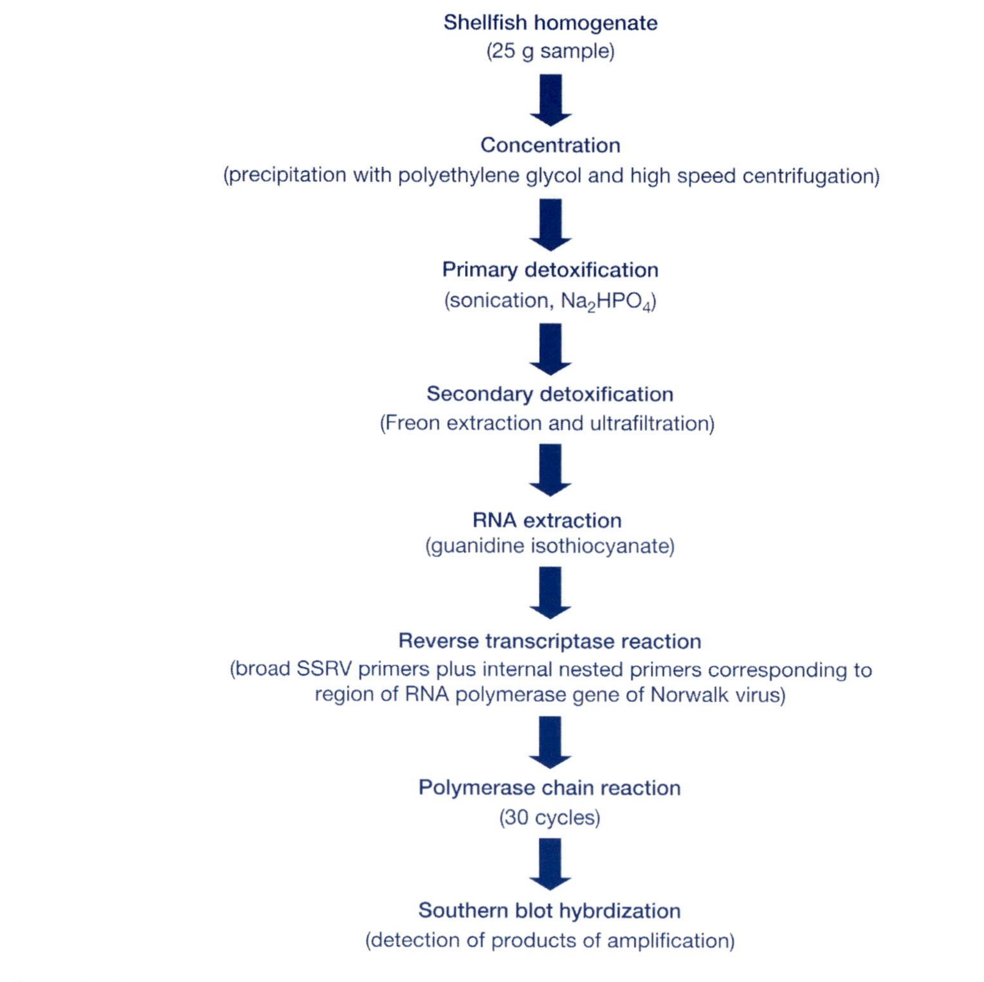

Figure 11.6

Example of scheme for extraction, purification and amplification of enteroviruses from shellfish. A basically similar approach can be used for water testing after concentration (e.g. from 100 liters to 100 microliters) using special filters. Adapted from Henshilwood *et al*. (1998).

large amounts of fish and whale meat. In 2002, health authorities in the UK, USA and Canada issued public health warnings advising weekly limits for the consumption of tuna. There is particular concern about mercury levels in the diet of pregnant women, because of the risks to fetal development.

References and further reading

Boyd, E.F., Davis, B.M., and Hochhut, B. (2001) Bacteriophage–bacteriophage interactions in the evolution of pathogenic bacteria. *Trends Microbiol* **9**: 137–144.

Bradley, G., and Hancock, C. (2003) Increased risk of non-seasonal and body immersion recreational marine bathers contacting indicator microorganisms of sewage pollution. *Mar Poll Bull* **46**: 792–794.

Cabelli, V.J., Dufour, A.P., McCabe, L.J., and Levin, M.A. (1982) Swimming-associated gastroenteritis and water quality. *Am J Epidemiol* **115**: 606–616.

Colwell, R.R. (1996) Global climate and infectious disease: the cholera paradigm. *Science* **274**: 2025–2031.

Colwell, R., and Huq, A. (2001) Marine ecosystems and cholera. *Hydrobiology* **460**: 141–145.

Davis, B.M., and Waldor, M.K. (2003) Filamentous phages linked to virulence of *Vibrio cholerae. Curr Opin Microbiol* **6**: 35–42.

Epstein, P.R. (1993) Algal blooms in the spread and persistence of cholera. *Biosystems* **31**: 209–221.

Faruque, S.M., Albert, J., and Mekalanos, J.J. (1998) Epidemiology, genetics, and ecology of toxigenic *Vibrio cholerae. Microbiol Molec Biol Rev* **62**: 1010–1314.

Fleisher, J.M., Kay, D., Wyer, M.D., and Godfree, A.F. (1998) Estimates of the severity of illnesses associated with bathing in marine recreational waters contaminated with domestic sewage. *Int J Epidemiol* **27**: 722–726.

Godfree, A.F., Kay, D., and Wyer, M.D. (1997) Faecal streptococci as indicators of faecal contamination in water. *J Appl Microbiol* **83**: S110–S119.

Griffin, D.W., Lipp, E.K., McLaughlin, M.R., and Rose, J.B. (2001) Marine recreation and public health microbiology: quest for the ideal indicator. *Bioscience* **51**: 817–825.

Harvell, C.D., Kim, K., Burkholder, J.M., *et al.* (1999) Review: Marine ecology – Emerging marine diseases – climate links and anthropogenic factors. *Science* **285** 1505–1510.

Henshilwood, K., Green, J., Gallimore, C.I., Brown, D.W.G., and Lees, D.N. (1998) The development of polymerase chain reaction assays for detection of small round structured and other human enteric viruses in molluscan shellfish. *J Shellfish Res* **17**: 1675–1678.

Iida, T., Park, K.-S., Suthienkul, O., Kozawa, J., Yamaichi, Y., Yamamoto, K., and Honda, T. (1998) Close proximity of the *tdh, trh* and *ure* genes on the chromosome of *Vibrio parahaemolyticus. Microbiology* **144**: 2517–2523.

Kay, D., Fleisher, J.M., Salmon, R.L., Wyer, M.D., Godfree, A.F., Zelenauch-Jacquotte, Z., and Shore, R. (1994) Predicting likelihood of gastroenteritis from sea bathing; results from randomized exposure. *Lancet* **344**: 905–909.

Linkous, D.A., and Oliver, J.D. (1999) Pathogenesis of *Vibrio vulnificus. FEMS Microbiol Lett* **174**: 207–214.

National Research Council (1991) *Seafood Safety.* National Academy Press, Washington. http://www.nap.edu/books/0309043875/html/ (accessed March 31 2003).

National Research Council (1999) *From Monsoons to Microbes.* National Academy Press, Washington. http://www.nap.edu/openbook/0309065690/html/ (accessed March 31 2003).

National Research Council (2001) *Under the Weather: Climate, Ecosystems and Infectious Disease.* National Academy Press, Washington. http://www.nap.edu/ openbook/0309072786/html/ (accessed March 31 2003).

Pascual, M., Bouma, M.J., and Dobson, A.P. (2002) Cholera and climate: revisiting the quantitative evidence. *Microbes Infect* **4:** 237–245.

Prüss, A. (1998) A review of epidemiological studies from exposure to recreational water. *Int J Epidemiol* **27**: 1–9.

Ruiz, G.M., Rawlings, T.K., Dobbs, F.C., Drake, L.A., Mullady, T., Huq, A., and Colwell, R.R. (2000) Global spread of microorganisms by ships – ballast water discharged from vessels harbours a cocktail of potential pathogens. *Nature* **408**: 49–50.

Wai, S.N., Mizunoe, Y., and Yoshida, S. (1999) How *Vibrio cholerae* survive during starvation. *FEMS Microbiol Lett* **180**: 123–131.

WHO (2001) *World Health Organization. Bathing water quality and public health: faecal pollution. Report of expert consultation.* http://www.water_sanitation_health/ Recreational_water/wsh01_2.pdf (accessed March 31 2003).

Human disease – toxic dinoflagellates and diatoms

12

12.1 'Red tides' and 'harmful algal blooms'

The growth of marine plankton is affected by a wide range of factors which influence their spatial and temporal distribution. Seasonal periodic increases in plankton density (blooms) obviously have great ecological importance in ocean food webs. However, the discussion in this chapter will focus mainly on exceptional blooms of toxin-producing dinoflagellates and diatoms which affect human health. These are frequently referred to as 'red tides', although this colloquial term is something of a misnomer since not all toxic blooms are red in color or may not even reach sufficient densities to discolor the water. A more generally accepted term is 'harmful algal blooms' (HABs). This term is also not entirely satisfactory, because it includes excessive growth of nontoxic algae and because health effects are not always associated with a distinct 'bloom'. Furthermore, the term HAB includes the toxic *Cyanobacteria*, dinoflagellates and diatoms and nuisance macroalgae. Cyanobacterial HABs are mostly a problem in fresh rather than marine waters and problems due to macroalgae are not considered here. As well as a direct effect on human health, HABs can have profound ecological, economic and social effects through their impact on fisheries and aquaculture.

Sudden changes in the appearance of coastal waters have been described for many centuries because of their association with human illness caused by consuming fish or shellfish. For example, it is often surmised that the reference in the Bible to the 'first plague ... and the river [Nile] turned to blood' (thought to be about 1290 BC) is a description of such a bloom. Ancient cultures such as Pacific Islanders and American coastal Indians had scouts who 'read the sea' for changes in color, in order to warn communities about the dangers of harvesting fish and shellfish. Explorers of the 'New World', from the 15th century onwards, made meticulous observations of fish and shellfish poisoning which affected them and their crews during their journeys. Fossil records show that blooms of toxic microalgae have occurred for millennia. Despite their long history, HABs have assumed growing importance during the last decade. This is because: (a) the frequency of incidence appears to be rising; (b) they are occurring in regions where they have not traditionally occurred; and (c) they are having significant economic impact.

It has been estimated that there are about 4000 known species of phytoplankton in the oceans, of which about 200 species have been identified as causing exceptional blooms. Of these, about 70 species are toxic, most of which are dinoflagellates. The main health hazard for humans comes from eating fish or shellfish which have accumulated the toxin in their tissues through feeding in water containing high levels of toxin-producing plankton. Some toxins may also produce disease symptoms as a result of direct contact with contaminated water or aerosols. Properties of the major toxins are shown in *Table 12.1* and representative structures are shown in *Figure 12.1*. As well as human diseases, some of these toxins have significant effects on marine mammals and fish; these are dealt with in Chapters 13 and 14, respectively.

12.2 Shellfish poisoning

There are three well-characterized disease syndromes associated with contamination of shellfish by dinoflagellate toxins and one caused by a diatom toxin. Although the chemical nature

Table 12.1 The major toxins produced by dinoflagellates and diatoms

Toxin	Produced by	Mode of action
Brevetoxins	*Gymnodinium breve* (and others)	Alters threshold for change in potential across cell membrane; makes nerve hyperexcitable
Ciguatoxins, maitotoxins	*Gambierdiscus toxicus* (and others)	Alters threshold for change in potential across cell membrane; makes nerve hyperexcitable
Domoic acid	*Pseudo-nitzschia multiseries*	Acts on glutamate receptor, preventing entry of cations to the nerve cell plus subsequent intracellular processes; nerve is excited
Okadaic acid	*Dinophysis*, *Procentrum* spp.	Inhibits protein phosphatase; carcinogenic (?)
Saxitoxins	*Alexandrium*, *Pyrodinium*, *Gymnodinium* spp. (and others)	Blocks transport of sodium ions, rendering nerve nonfunctional
Pfiesteria toxin (?)	*Pfiesteria piscicida*	Unknown, proposed neurotoxic

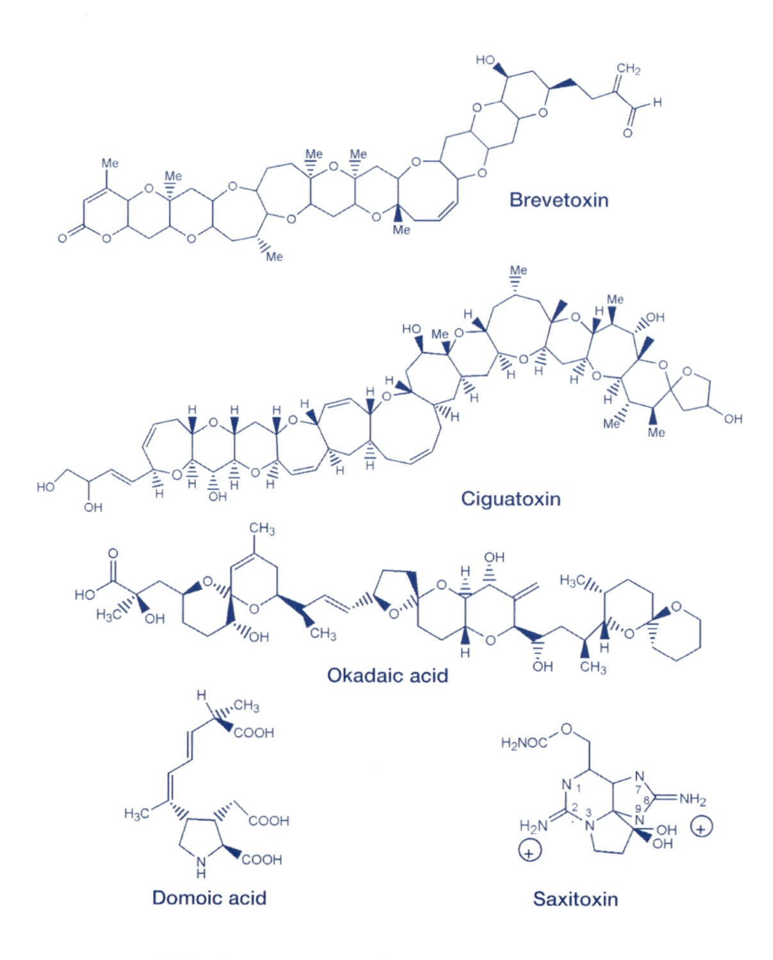

Figure 12.1

Representative structures of major dinoflagellate and diatom toxins. Note that there is considerable variation of structure of brevetoxins, ciguatoxins and saxitoxins.

and precise mode of action of the toxins differs markedly, the diseases mostly share certain characteristics. There is usually a rapid onset of disease symptoms, often within minutes or a few hours of eating contaminated shellfish. Usually, diarrhea and vomiting is quickly followed by neurological symptoms. Most of the toxins inhibit transmission of the nerve impulse by interfering with the passage of sodium ions through the membrane of nerve cells. All of the toxins are nonproteinaceous substances which are still active after cooking. This, together with the usually very rapid onset and neurological symptoms, is an important distinction of HAB intoxications from most bacterial and viral infections associated with consumption of shellfish.

12.2.1 Paralytic shellfish poisoning (PSP)

PSP is the best-known human illness associated with a HAB. It appears to be worldwide in distribution although most occurrences have been described in temperate regions of America, northern Europe, Scandinavia, south-east Asia and Australasia. PSP is usually caused by eating shellfish (especially clams, oysters, mussels and certain species of crabs) in which toxins have built up through filter feeding. Crustacea such as crabs and lobsters can also accumulate the toxin by feeding on contaminated bivalve molluscs. There are at least five species of dinoflagellates associated with PSP, namely *Alexandrium catanella, A. minutum, A. tamarense, Gymnodinium catenatum* and *Pyrodinium bahamense.* All produce one or more toxins known as saxitoxins, of which there are at least 12 different types. PSP has a dramatic onset, often within minutes of consuming contaminated shellfish. Tingling around the lips is quickly followed by numbness of the face and neck, nausea, headache and difficulty in speech. In severe cases, muscular and respiratory paralysis can occur and mortality can be more than 10% unless there is rapid access to medical services. Treatment consists of stomach pumping and administration of charcoal to absorb the toxins. In severe cases, artificial respiration may be required but there are usually no long-lasting effects.

Many countries operate PSP-management programs, which include monitoring the density of toxin-producing dinoflagellates and the assay of shellfish samples for saxitoxins using mouse bioassay or HPLC. Recently, new sensitive methods for assay of PSP toxins using cell culture or receptor-blocking methods have been developed (*Figure 12.2*). Commercial collection of shellfish is prohibited when saxitoxin levels in the tissue exceed a certain threshold (typically, $800\ \mu g\ kg^{-1}$), although many cases of PSP still occur through illegal harvesting or collection of shellfish for personal consumption. Careful analysis has shown that there does not seem to be a straightforward relationship between the presence of toxin-producing dinoflagellates and the level of particular toxins in shellfish tissue. Shellfish may modify the toxins or excrete them differentially according to climatic and physiological conditions, making management of outbreaks more difficult. The closure of fisheries can have devastating economic and social effects, affecting whole communities.

12.2.2 Neurotoxic shellfish poisoning (NSP)

NSP is associated with the toxin brevetoxin, produced by the dinoflagellate *Gymnodinium breve* and a few other species. Blooms of *G. breve* are highly distinctive (a true 'red tide') and are usually seasonal, starting in late summer and lasting for many months. They have been known on the Florida coast for many centuries, but in recent years blooms have occurred in other areas of the eastern US coast and in New Zealand. Like PSP, NSP is caused by consumption of shellfish which have accumulated the toxin, although symptoms are milder. The blooms can be so extensive that disruption of *G. breve* by surf action can lead to aerosols containing the toxin, which can be carried by wind up to 100 km inland, causing shortness of breath and eye irritation. The blooms also cause massive fish kills and the rotting fish can lead to further problems as coastal waters become anoxic. Such red tides have a major impact on fisheries and tourism.

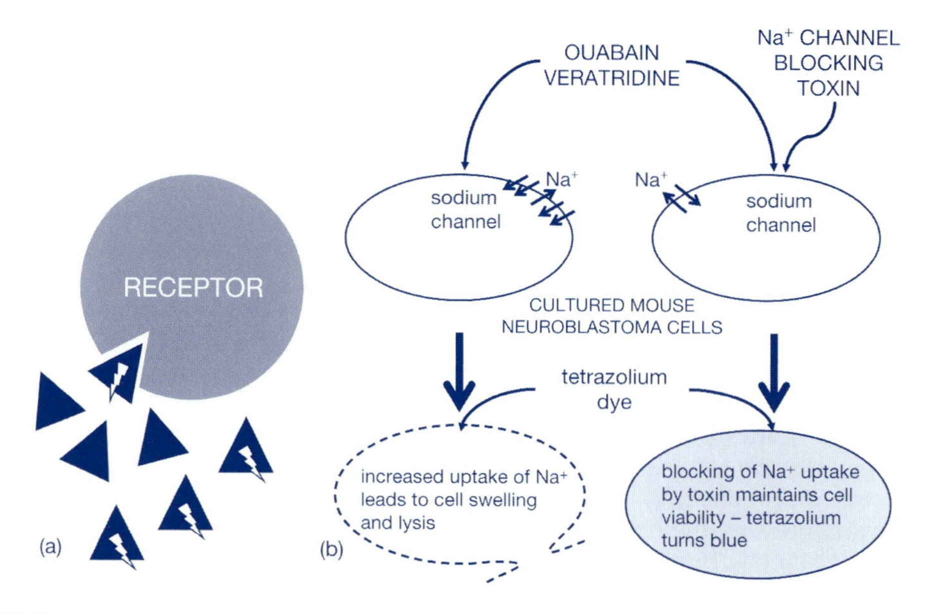

Figure 12.2

Principle of receptor binding and cell culture assays for sodium channel blocking toxins. (a) Native toxin (shown as black triangles) in a standard or the sample to be tested will compete with a radioactive analog of the toxin (shown as triangles with flash) for binding sites on the receptor. By measuring the amount of radioactivity bound, a standard curve is generated which can be used to measure the amount of toxin. Membrane preparations from rat (for PSP) or frog (for ASP) brain are used as a source of receptor; recently, a cloned glutamate receptor for ASP has been expressed in cultured cells. (b) In the MIST® cytotoxicity assay, cultured mouse neuroblastoma cells are treated with the alkaloids ouabain (inhibitor of Na^+/K^+ ATPase) and veratridine (Na^+ activator). In the presence of Na^+ channel blocking toxins, the action of veratridine is inhibited. The intensity of the color in viable cells is inversely proportional to toxin concentration and is read in a plate colorimeter in high-throughput assays employing 96-well microtiter trays.

12.2.3 Diarrhetic shellfish poisoning (DSP) and azaspiracid poisoning

DSP is caused by eating shellfish (usually bivalve molluscs) which contain the toxin okadaic acid (and possibly some other toxins) from dinoflagellates of the genera *Dinophysis* and *Procentrum*. Okadaic acid is an inhibitor of protein phosphatases, and leads to symptoms of vomiting, diarrhea and abdominal pain. Patients usually recover in a day or two. The symptoms resemble those caused by virus infection from shellfish grown in sewage-polluted waters, but a growing body of evidence since the 1980s has shown that DSP is caused by a toxin and can result from consuming shellfish from waters free of sewage contamination. Many outbreaks have been described in Japan and in European countries, particularly on the Atlantic coast of Spain, France and the British Isles. A European Union Directive requires the monitoring of shellfish for the toxin (by HPLC assay) and this has resulted in closure of many traditional shellfish harvesting and aquaculture operations in areas where toxins have been detected (*Figure 12.3*). Unlike the other shellfish poisonings, a DSP risk can occur even in the absence of an exceptional increase in plankton density; *Dinophysis* densities as low as 200 cells ml^{-1} can lead to accumulation of unacceptable levels of toxin in shellfish. Okadaic acid has been shown to be carcinogenic, and long-term exposure has been suggested as a possible cause of cancer in the digestive tract, although there seems to be little epidemiological evidence for this.

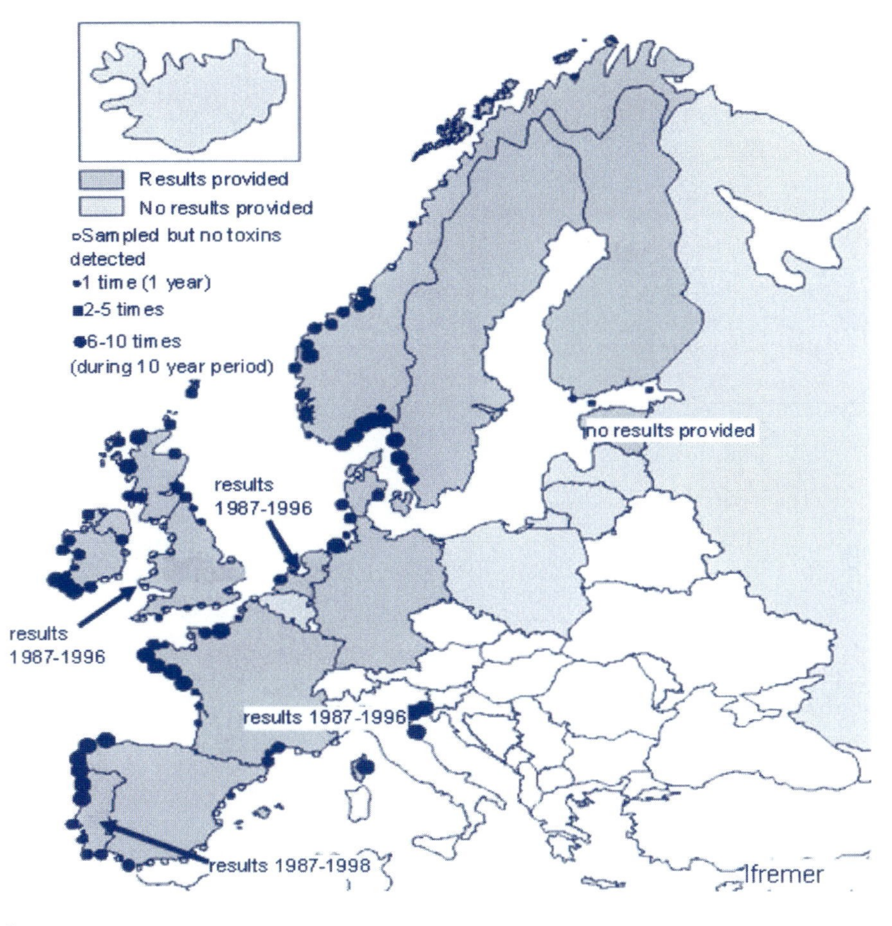

Figure 12.3

Map showing presence of DSP toxins in shellfish samples in Europe, 1990–1999. Image courtesy of Inter-governmental Oceanographic Commission.

Since 1995, a number of intoxications involving groups of people who had consumed mussels harvested from Irish waters have been reported in Europe. A group of closely related toxins called the azaspiracids have been shown to be responsible. Symptoms are severe diarrhea and vomiting, nausea and headache, which may persist for several days. In a mouse bioassay, azaspiracids cause death from serious tissue and organ damage, with neurotoxic effects at high doses. However, their mode of action seems quite different from the other shellfish toxins. Development of improved chromatographic assays means that routine monitoring for azaspiracids is now being introduced with the result that they are much more widespread than previously thought. The toxins seem to be associated particularly with the mussel *Mytilus edulis* and there are no obvious links with bloom events. Rather, it seems as if the toxin accumulates gradually in the tissue of the mussel due to long-term exposure to low levels of toxic microalgae (possibly *Protoperidnium* sp.).

12.2.4 Amnesic shellfish poisoning (ASP)

ASP was discovered in 1987, following an unusual outbreak of seafood poisoning in Prince Edward Island, Canada, in which over 100 people were affected and three died. The symptoms included rapid onset of vomiting, disorientation and dizziness, which did not match those of other known diseases. During the investigation, it emerged that those affected had suffered short-term memory loss, which persisted for many months. This suggested that a new type of neurotoxin was involved, and this was identified as domoic acid produced by the diatom *Pseudo-nitzschia multiseries*. We now know that several other species of *Pseudo-nitzschia* can produce domoic acid and these have been isolated from many parts of the world. Human cases of ASP are rare, but they are almost certainly underreported and in areas where *Pseudo-nitzschia* blooms occur they may be associated with eating planktivorous fish such as anchovies. Domoic acid from *Pseudo-nitzschia* blooms also causes disease in marine birds and mammals (*Box 13.1*).

12.3 Ciguatera fish poisoning (CFP)

In terms of public health, CFP or 'ciguatera' is undoubtedly the most important of the diseases caused by marine toxins, especially for inhabitants of tropical islands which rely on fishing. It is also a well-known hazard among sailors and travelers in the tropics (35°N to 35°S) and was first documented by explorers of the Caribbean and Pacific in the 15th and 16th centuries. In 1774, Captain James Cook gave a detailed account in his ship's log of poisoning which affected him and his entire crew after eating fish in the New Hebrides. The causative agent of CFP was not discovered until 1977 when the toxin-producing dinoflagellate *Gambierdiscus toxicus* was isolated in the Gambier Islands in French Polynesia. The same species occurs worldwide, although isolates vary greatly in the amount and type of toxin produced. The incidence of CFP can be quite localized but unpredictable. Local fishermen will often believe that fish from a particular island or reef will be hazardous whilst others nearby are not. Many travelers have learnt, to their cost, that such claims are not always reliable. The fat-soluble toxins become concentrated as they pass up the food chain and disease is almost always associated with eating large predatory fish. Over 400 species of fish have been implicated in CFP, with moray eel and barracuda being the most common. *G. toxicus* adheres to macroalgae on the surface of dead coral and on the seabed, but little is known about the factors promoting colonization of the macroalgae by the dinoflagellates and its subsequent proliferation and toxicity. Damage to the reef (e.g. by fishing, diving, boat activities, military action, mining or construction work) promotes the initial colonization by the macroalgae. Increased frequency of coral diseases, possibly associated with climate change, may also be leading to higher occurrence of CFP. However, this is only part of the story, as the incidence of toxic fish is very dependent on environmental factors such as rainfall and nutrient run-off from the land. Also, it is likely that there are genetic differences in *G. toxicus* and factors which affect its growth may have a different effect on toxin production.

CFP is characterized by a distinctive sequence of diarrhea, vomiting, abdominal pain and neurological effects within a few hours of eating contaminated fish. Unusual symptoms include numbness and weakness in the extremities, aching teeth and reversal of temperature sensation (cold things feel hot and hot things feel cold). In severe cases, this can progress rapidly to low blood pressure, coma and death. However, the type and severity of symptoms depends on the particular 'cocktail' of toxins consumed, and this varies with species of fish and geographic origin. Doses that cause severe symptoms in one person may be harmless in another. There seem to be important differences in structure between the main CFP toxins implicated in Pacific and Caribbean cases. It is also likely that other ancillary toxins are involved. In some people, neurological problems can persist for many years, and a repeat attack can be triggered by eating any fish (whether or not it contains toxin) or by drinking alcohol. The toxin has also been shown to cross the placenta and affect the fetus and there are even reports of transmission to another person via sexual intercourse. In countries where the disease is endemic, there are a range of local remedies, but these are of unproven value. In severe cases, it is necessary to administer intravenous mannitol which reverses the effect on sodium transport.

The complex ecological, toxicological and physiological factors seen with CFP illustrate dramatically how careful we must be when looking for a simple 'cause and effect' in the etiology of a disease.

12.4 *Pfiesteria piscicida*

The dinoflagellate *Pfiesteria piscicida* was first described in the early 1990s as a cause of mass mortalities of fish in estuarine systems on the eastern seaboard of the USA and was subsequently associated with human illness in people exposed to water containing the organism. The '*Pfiesteria* story' is an exciting example of scientific detective work surrounded by controversy, with important political and social repercussions. *Pfiesteria* is reported to possess a staggeringly complex life cycle and ecology, as shown in *Figure 12.4*. It has both phototrophic and heterotrophic stages. The discovery of closely related *Pfiesteria* strains or species suggests that toxic blooms could involve complex interactions and these dinoflagellates probably exist as a natural component of many estuaries. Fortunately, a particular combination of physical and chemical conditions seems necessary for the development of toxic forms. Analysis by conventional microscopy and taxonomic methods is difficult, but the recent introduction of new molecular techniques, including real-time PCR of sequences from 18S rRNA genes is permitting rapid and accurate assessment of distribution of the *Pfiesteria* complex. There is still much disagreement about the nature of *Pfiesteria* and its effects on human and fish health, as described in *Boxes 12.1 and 14.2*.

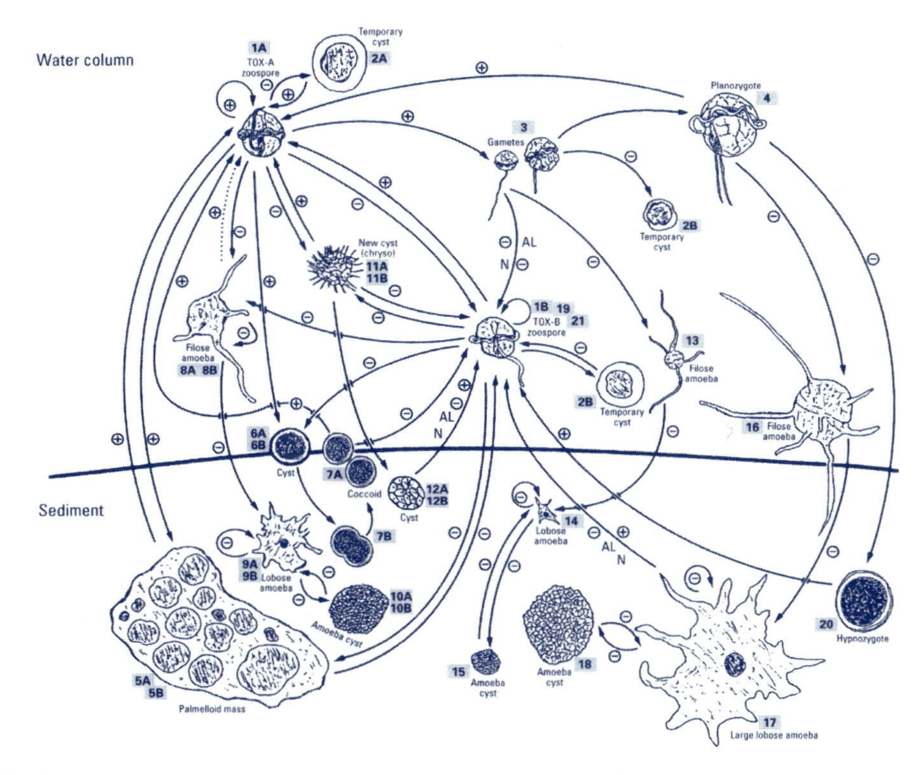

Figure 12.4

Proposed life cycle of *Pfiesteria piscicida*. Image courtesy of Howard Glasgow, Associate Director of North Carolina State University Center for Applied Aquatic Ecology.

Box 12.1 RESEARCH FOCUS

The controversial 'cell from hell'

Human health effects of Pfiesteria piscicida

Clues to the existence of this extraordinary organism initially came from investigation of unusual fish mortalities in brackish water aquaria in the laboratory of JoAnn Burkholder, North Carolina State University in the late 1980s. Death, accompanied by hemorrhagic sores, was associated with the presence of large numbers of a small dinoflagellate in the water. Shortly after death of the fish, the dinoflagellates seemed to disappear until more fish were added to the aquaria. Burkholder and her team were investigating the cause of huge fish kills in the Albemarle-Pamlico estuary of North Carolina. In some outbreaks, up to one million young fish could be killed within the space of a few hours. The discoveries led to intense coverage on television and in newspapers, and the unidentified organism was given the nicknames 'phantom dinoflagellate' and 'the cell from hell'. In field studies, the 'phantom dinoflagellate' was seen in large numbers at the peak of the fish kill, but disappeared a few hours later. The organism was eventually described as a new family, genus and species of dinoflagellate which thrives in brackish estuarine water with an optimum temperature around 18°C and was named *Pfiesteria piscicida* (Burkholder *et al.* 1996). Most dinoflagellates contain chloroplasts and are phototrophic, although they may sometimes supplement their nutrition via phagocytosis or absorption of nutrients. By contrast, *P. piscicida* is heterotrophic and has a highly complex life cycle that includes at least 24 flagellated, amoeboid and encysted stages (*Figure 12.4*). The amoeboid stages feed by engulfing bacteria, other protists or bits of fish tissue, while the flagellated cells attach to their prey by a cell extension called a peduncle and suck out the cellular contents (*Figure 7.3*). Although *P. piscicida* lacks the genetic information for synthesis of chloroplasts, it can retain functional chloroplasts in food vacuoles after engulfing algal prey. This process, known as kleptoplastidy, enables the organism to exist for at least part of its life cycle as a phototroph which may enable the organism to survive conditions when prey is not abundant (Lewitus *et al.* 1999). Both flagellated and amoeboid forms are thought to be toxic to fish, and unidentified chemical signals from fish (possibly excreta or secretions) may stimulate the formation of toxin-producing zoospores. The toxin acts within minutes to narcotize and kill the fish, and other stages consume the fish tissue.

The possibility that the organism could have an impact on human health emerged when laboratory workers who had handled dilute cultures of the organisms began to show a range of symptoms, including lethargy, breathing difficulties, memory loss and dramatic changes in personality and cognitive function (Glasgow *et al.* 1995). It was thought that they had been exposed to a putative *P. piscicida* toxin through aerosol formation or skin contact. The symptoms usually disappeared once people stopped working with the cultures, but in one of the primary investigators the symptoms persisted for many months. Some of the effects can recur years later following strenuous exercise and there are suggestions that the toxin might trigger autoimmune responses. With hindsight, there are obvious parallels with neurotoxins from other HABs such as domoic acid (ASP; induces memory loss and disorientation), brevetoxin (NSP; can cause respiratory effects as an aerosol) and ciguatoxin (CFP; recurrent symptoms triggered by re-exposure) and this emphasizes the need for great caution when working with HABs. Further investigations found that fishermen and others who had regular contact with the estuary reported various skin ulcers, mental effects and other health problems similar to those seen in the laboratory workers, suggesting that chronic exposure to the *P. piscicida* toxin might be responsible (Grattan *et al.* 1998). Experiments with rats in a maze have shown that injection of extracts of the dinoflagellate cultures impairs learning ability and memory (Levin *et al.* 2000). Further progress has been hampered by difficulties in working safely with the organism (category 3 biohazard containment facilities with special air filtration are required), but it seems likely that *Pfiesteria*-like organisms produce a number of toxins. Lipid-soluble toxins may be responsible for the dermonecrotic effects in fish and humans and a water-soluble toxin is possibly responsible for the neurological effects. Initial studies suggest that the toxin may bind to a specific receptor in the brain associated with cognitive function. However, considerable controversy has surrounded the epidemiological investigations and the US Centers for Disease Control has not yet recognized *Pfiesteria* as the definitive causative agent of the proposed 'estuary-associated syndrome'. To add to the uncertainty, research published at the end of 2002 has thrown considerable doubt on the claims by Burkholder's group about the nature of *Pfiesteria's* life cycle and the role of a toxin. This is further discussed in *Box 14.1.*

12.5 Why do dinoflagellates and diatoms produce toxins?

The toxins produced by dinoflagellates and diatoms are usually complex secondary metabolites, and diversion of energy to their synthesis implies some ecological benefit. The extreme sensitivity of humans and other animals to these toxins and the fact that most act on similar mechanisms of nerve action must be coincidental. It is usually presumed that toxins have evolved to deter predation by zooplankton. If this is correct, once physical and chemical conditions are favorable for the initial bloom of a toxic species, the production of toxins will deter predation and prolong the maintenance of high densities of the species. It seems that many zooplankton species, such as copepods, can discriminate dinoflagellates with low toxin content and feed selectively, and some flagellates and ciliates appear to be killed by dinoflagellate toxins. However, toxins do not give universal protection against grazing by all species of predator and it may be that the toxins offer a selective advantage to organisms that produce them under low nutrient conditions, by directing predation pressure onto competitors. It is notable that the production of toxin, and its cellular concentration, is very dependent on the supply of nutrients (especially phosphorus). This has led to the suggestion that the toxins may act as a reserve for storage when nutrient supply is unbalanced. Many dinoflagellates have the capacity for both phototrophic and heterotrophic nutrition, and the production of such secondary metabolites may be associated with major switches in metabolic pathways.

One area that has attracted considerable interest is the involvement of bacteria in toxin synthesis. The saxitoxins, for example, are synthesized by phylogenetically diverse genera of dinoflagellates, and similar compounds appear to be produced by cyanobacteria and eubacteria. This has led to the suggestion that dinoflagellates might harbor symbiotic bacteria or that horizontal gene transfer has taken place, as discussed in *Box 12.2.*

12.6 Why are HABs and toxin-associated diseases increasing?

Fossil studies show that periodic blooms have occurred for millennia and that these have been linked to mass mortalities of marine life in the past. Some phenomena like the Florida *G. breve* red tide and human diseases like PSP and CFP have been described for many years, but there is little doubt that the reported frequency and distribution of HABs have increased markedly in the past few years. For example, until the 1970s PSP-producing blooms were known only in the temperate waters of North America, Japan and Europe, but since 1990 PSP outbreaks have been reported increasingly in the southern hemisphere. As noted above, apparently new problems like ASP, DSP, azaspiracid poisoning and *Pfiesteria*-associated illness have emerged in the last few years.

Box 12.2 RESEARCH FOCUS

Do symbiotic bacteria produce marine toxins?

Their role remains uncertain

Early in the study of the origin of PSP, microbiologists observed bacteria-like structures in electron micrographs of dinoflagellate cells. Could these be endosymbionts, and are they involved in toxin synthesis or breakdown? Besides the intrinsic interest in the possible role of symbiosis in providing their eukaryotic host with a possible defensive mechanism (see *Section 12.5*), answering these questions is of importance in the monitoring and mitigation of fish and shellfish toxicity. However, despite a number of studies in this area, there are no definitive answers to these questions. It is generally accepted that certain *Cyanobacteria*, such as *Anabaena flos-aquae* (Negri *et al.* 1997) and *Lyngbya wollei* (Carmichael *et al.* 1997) can synthesize paralytic shellfish toxins (PSTs) in culture, but synthesis by other bacteria is less certain. Many other studies have shown that several bacteria such as *Moraxella, Vibrio, Pseudomonas* and *Alteromonas* species can produce toxins in pure culture, but different research groups are often fiercely critical of the assay methods used by others. For example, the publication of a paper describing synthesis of tetrodotoxin by a *Vibrio* sp. (Lee *et al.* 2000) in the prestigious journal *Applied & Environmental Microbiology*, resulted in the unusual step of a critical letter to the journal refuting the claims (Matsurama 2001).

In a review of the topic, Gallacher and Smith (1999) concluded that experiments to demonstrate the comparative levels of toxin production in dinoflagellate cultures before and after curing of contaminating bacteria (by antibiotic or enzyme treatment) fail to provide definitive evidence. It is unlikely that the methods used in this type of study conclusively demonstrate the production of axenic (i.e. completely bacteria-free) dinoflagellates; this can only be resolved through the use of specific gene probes. Little progress has been made in studying the genetics of toxin production in dinoflagellates. However, Plumley *et al.* (1999) have begun to identify bacterial genes involved in the production of PSTs and/or the establishment and maintenance of relationships between bacteria and algae. Using transposon mutagenesis and complementation with broad-host-range plasmids of various bacterial strains thought to be involved in the production of PSTs, they isolated a strain of *Pseudomonas stutzeri* which could be transformed when grown under appropriate conditions. This bacterium accumulated PSTs and increased toxin production when added to axenic cultures of *Alexandrium lusitanicum*.

Improved methods are beginning to resolve these issues. Smith *et al.* (2002) used the sensitive receptor-binding MIST® assay (*Figure 12.2*) to measure sodium channel-blocking agents, and found a number of specific bacteria associated with toxigenic strains of *A. tamarense* which were not present in a nontoxigenic strain. Some isolates were capable of transforming PSTs via oxidase activity or reductive elimination. Hold *et al.* (2001) used sequencing of 16S rRNA genes to demonstrate that the dinoflagellates *Alexandrium* and *Scrippsiella* spp. have distinctive communities of members of the α- and γ-*Proteobacteria* and CFB groups. Evidence seems to point towards the role of surface-associated bacteria rather than endosymbionts in PST production and transformation.

Blooms occur when a particular combination of physical (temperature, sunlight, water stratification and circulation) and chemical (nutrient and O_2 levels) conditions occur. Species dispersal due to large-scale water movements is important, and unusual currents and storms may account for appearance of atypical blooms. For example, in some recent years, the *G. breve* red tide off Florida has moved much further north due to formation of unusual circulation patterns in the Gulf Stream. The occurrence of some HABs has been closely correlated with the El Niño Southern Oscillation phenomenon. The increased incidence of HABs is often cited as evidence of disturbance to our oceans and atmosphere due to the 'greenhouse effect'.

Nutrient enrichment of coastal waters is often believed to be responsible for the increased incidence of HABs. Natural upwellings of nutrients into cold oligotrophic waters are responsible for many blooms, such as those that occur regularly off the California coast. Such natural phenomena may now be overlaid with anthropogenic sources of nutrients and some studies have provided clear evidence for this. It may be that eutrophication from sewage and run-off from agricultural land is responsible for a general increase in plankton growth and primary productivity. Although few long-term studies exist and it is difficult to compare the results from different areas, it seems likely that nutrient input into coastal waters with limited exchange with the open ocean can result in a significant increase in plankton growth. For example, regular blooms of nontoxic microalgae such as *Phaeocystis* and *Emiliana* have become a regular occurrence in the North Sea and English Channel (see *Figure 7.6*). Could eutrophication also lead to an increase in toxic species? One widely held idea is that nutrient inputs from sewage or agricultural land run-off alter the ratio of particular nutrients, as well as the total loading, and that this may change the balance of the different plankton groups. Sewage is rich in nitrogen and phosphorus, but has a low content of silicon. Since diatoms specifically require silicon, it is argued that this would favor the selective growth of dinoflagellates. In the case of the emergence of *Pfiesteria* as a problem in the North Carolina and Maryland estuaries, there appears to be a very strong association with increased nutrient levels due to an expansion in poultry and pig farming on the surrounding land. However, not all *Pfiesteria* outbreaks have occurred in nutrient-enriched waters, and not all areas of the estuaries with increased nutrients show high populations of *Pfiesteria*. The expansion of mariculture, such as high-density salmon or shrimp culture, has been directly implicated in the increase in frequency of HABs due to nutrient enrichment from uneaten food and excreta. For example, the increased incidence in Scotland during the 1990s of high levels of PSP toxins in shellfish is widely believed to be due to poorly sited salmon pens in lochs, bays and inlets with limited water exchange.

One obvious explanation for the apparent increase in HABs and associated diseases is increased awareness of toxic species and more extensive and effective monitoring. This is exacerbated by the increased use of coastal waters for aquaculture of shellfish and finfish. These activities act as sensitive indicators of potential problems and undoubtedly account for at least part of the apparent spread of PSP and DSP. Mass mortalities in fish farms can be due to the effects of HABs, although these do not always involve toxic species (see *Section 14.6*).

Finally, a further explanation for the increased geographic distribution of some HABs is transport of the causative organisms via ship movements. Modern large vessels pick up many thousands of liters of water as ballast, which can be transported to completely different geographic regions. Several studies have shown that the cysts of toxic dinoflagellates and other HAB organisms can be transported in this way. For example, the introduction of PSP into southern Australia in the 1980s is suspected to be due to transport from Japan and Korea. In conclusion, although the factors that affect plankton blooms are numerous and complex, a combination of natural and anthropogenic effects is undoubtedly responsible for a worldwide increase in HABs, a wider range of toxic species and the emergence of 'new' human diseases.

12.7 Monitoring and control of HABs

Regular surveys by microscopic examination of the dynamics of phytoplankton populations within an area may give advance warning of an increase in particular species, which may sometimes precede a toxic bloom. Such surveys are time-consuming and it is often difficult to distinguish toxic species or strains using morphological criteria. Some improvements can be achieved using ELM or FCM, after labeling the target species with specific fluorescent antibodies. Unfortunately, background fluorescence is often a problem. Remote sensing can be used by equipping satellites with spectral scanners that detect chlorophyll and other pigments in surface waters. When coupled with physical measurements such as sea surface temperatures and current flows, satellite images are especially useful in tracking the development and movement of blooms. As genetic sequence data for toxic dinoflagellates and diatoms become available, gene probes are being increasingly used. A common method is to amplify microalgal DNA encoding 18S rRNA using eukaryote-specific primers. The resulting PCR products can be cloned and sequenced, leading to a specific oligonucleotide probe that will hybridize with DNA of the organism in water samples. Increased sensitivity and real-time assays can be achieved using molecular beacons. It would be useful to develop probes for specific genes involved in toxin biosynthesis. In the near future, advances in this method will probably permit automatic sensing devices using gene probe technology to be placed on offshore buoys, and information could be beamed to satellites and integrated with improved remote sensing signal detection systems linked to powerful computer models. This could lead to reliable real-time 'HAB forecasts'.

Such information can be used to limit the economic and health impact of HABs, but whether we can use it to control their development or spread is questionable. HABs often occur over huge areas. The effects of weather, ocean currents and the many physical and biological factors that determine the development and eventual demise of a bloom are unpredictable. It is possible that chemical treatments, such as adding agents which promote agglutination and sinking of microalgal cells, could be applied locally in particularly sensitive areas, but the ecological impact of such intervention would have to be established. Bacteria and viruses associated with HABs have recently received some attention. One theory for the often sudden collapse of blooms is that it is caused by algicidal bacteria or viruses. Viruses capable of initiating lysis of dinoflagellates and diatoms can be isolated from seawater and have been proposed as a possible biological control agent. Some microalgae contain lysogenic viruses which can be induced into the lytic cycle by particular conditions. Some success in initiating the collapse of bloom populations has been achieved in microcosm experiments (see *Box 8.1*), but much more research and evaluation of the ecological and 'scale-up' issues is needed before biological control becomes a practical proposition.

References and further reading

Anderson, D.M. (1995) Toxic red tides and harmful algal blooms: A practical challenge in coastal oceanography. *Rev Geophys* **33**: 1189–1200.

Bowers, H.A., Tengs, T., Glasgow, H.B., Burkholder, J.M., Rublee, P.A., and Oldach, D.W. (2000) Development of real-time PCR assays for rapid detection of *Pfiesteria piscicida* and related dinoflagellates. *Appl Environ Microbiol* **66**: 4641–4648.

Burkholder, J.M. (1999) The lurking perils of *Pfiesteria*. *Sci Amer* **281**: 42–49.

Burkholder, J.M., and Glasgow, H.B. (2001) History of toxic *Pfiesteria* in North Carolina estuaries from 1991 to the present. *Bioscience* **51**: 827–841.

Burkholder, J.M., Glasgow, H.B., and Hobbs, C.W. (1995) Fish kills linked to a toxic ambush-predator dinoflagellate – distribution and environmental conditions. *Mar Ecol Prog Ser* **124**: 43–61.

Carmichael, W.W., Evans, W.R., Yin, Q.Q., Bell, P., and Moczydlowski, E. (1997) Evidence for paralytic shellfish poisons in the freshwater cyanobacterium *Lyngbya wollei (Farlow ex Gomont) comb. nov. Appl Environ Microbiol* **63**: 3104–3110.

Fleming, L.E., Easom, J., Baden, D., Rowan, A., and Levin, B. (1999) Emerging harmful algal blooms and human health: Pfiesteria and related organisms. *Toxicol Pathol* **27**: 573–581.

Gallacher, S., and Smith, E.A. (1999) Bacteria and paralytic shellfish toxins. *Protist* **150**: 245–255.

Glasgow, H.B., Burkholder, J.M., Schmechel, D.E., Tester, P.A., and Rublee, P.A. (1995) Insidious effects of a toxic estuarine dinoflagellate on fish survival and human health. *J Toxicol Environ Health* **46**: 501–522.

Grattan, L.M., Oldach, D., Perl, T.M., *et al.* (1998) Learning and memory difficulties after environmental exposure to waterways containing toxin-producing *Pfiesteria* or *Pfiesteria*-like dinoflagellates. *Lancet* **352**: 532–539.

Grattan, L.M., Oldach, D., and Morris, J.G. (2001) Human health risks of exposure to *Pfiesteria piscicida*. *Bioscience* **51**: 853–857.

Hold, G.L., Smith, E.A., Rappe, M.S., *et al.* (2001) Characterisation of bacterial communities associated with toxic and non-toxic dinoflagellates: *Alexandrium* spp. and *Scrippsiella trochoidea. FEMS Microbiol Ecol* **37**: 161–173.

Inter-governmental Oceanographic Commission of the United Nations. *IOC Harmful Algal Bloom Programme*. http://ioc.unesco/hab/ (accessed March 15 2003).

Lee, M.J., Jeong, D.Y., Kim, W.S., *et al.* (2000) A tetrodotoxin-producing Vibrio strain, LM-1, from the puffer fish *Fugu vermicularis radiatus. Appl Environ Microbiol* **66**: 1698–1701.

Levin, E.D., Rezvani, A.H., Christopher, N.C., *et al.* (2000) Rapid neurobehavioral analysis of *Pfiesteria piscicida* effects in juvenile and adult rats. *Neurotoxicol Teratol* **22**: 533–540.

Lewitus, A.J., Glasgow, H.B., and Burkholder, J. (1999). Kleptoplastidy in the toxic dinoflagellate *Pfiesteria piscicida* (dinophyceae). *J Phycol* **35**: 303–312.

Llewellyn, L.E. (2001) Ecology of microbial neurotoxins. In: Masaro, E.J. (ed.) *Handbook of Neurotoxicology*, vol. 1, pp. 239–255. Humana Press, Totowa.

Matsumura, K. (2001) Letter to the Editor: No ability to produce tetrodotoxin in bacteria. *Appl Environ Microbiol* **67**: 2393–2394.

Morris, J.G. (1999) Harmful algal blooms: An emerging public health problem with possible links to human stress on the environment. *Ann Rev Energy Environ* **24**: 367–390.

Moser, V.C., and Jensen, K. (2000) Rapid neurobehavioral analysis of *Pfiesteria piscicida* effects in juvenile and adult rats. *Neurotoxicol Teratol* **22**: 533–540.

National Office for Marine Biotoxins and Harmful Algal Blooms, Woods Hole Oceanographic Institute. *The Harmful Algae Page*. http://www.redtide/ whoi.edu/hab/ (accessed March 15 2003).

National Research Council (1999) Harmful algal blooms. In: *From Monsoons to Microbes*, pp. 59–70. National Academy Press, Washington. http://www.nap. edu/openbook/0309065690/html/ (accessed March 15 2003).

Negri, A.P., Jones, G.J., Blackburn, S.I., Oshima, Y., and Onodera, H. (1997) Effect of culture and bloom development and of sample storage on paralytic shellfish poisons in the cyanobacterium *Anaebena circinalis. J Phycol* **33**: 26–35.

North Carolina State University. *The Pfiesteria web site*. http://www.pfiesteria.org/ pfiestr/lcycle.html (accessed March 15 2003).

Plumley, F.G., Wei, Z.Y., Toivanen, T.B., Doucette, G.J., and Franca, S. (1999) Tn5 mutagenesis of *Pseudomonas stutzeri* SF/PS, a bacterium associated with *Alexandrium lusitanicum* (Dinophyceae) and paralytic shellfish poisoning. *J Phycol* **35**: 1390–1396.

Richardson, K. (1997) Harmful or exceptional phytoplankton blooms in the marine ecosystem. *Adv Mar Biol* **31**: 302–385.

Smith, E.A., Mackintosh, F.H., Grant, F., and Gallacher, S. (2002) Sodium channel blocking (SCB) activity and transformation of paralytic shellfish toxins (PST) by dinoflagellate-associated bacteria. *Aquat Microb Ecol* **29**: 1–9.

Diseases of marine mammals

13.1 Difficulties of study

In the wild, it is difficult to study diseases of marine mammals except when mass mortalities occur, when animals are stranded, or if they become caught in fishing nets. Obtaining blood or tissue samples from live animals is difficult and post-mortem deterioration happens very quickly, so data are often sparse. Some knowledge comes from captive animals in zoos and marine parks, and the study of diseases of marine mammals has become a specialized branch of veterinary medicine.

13.2 Effects of microalgal toxins

Neuropathological symptoms and mortality in marine mammals are being increasingly associated with HABs (*Figure 13.1*). One explanation for the stranding or beaching of whales and dolphins (*Figure 13.2*) is that they become disoriented because of accumulation of toxins after feeding on contaminated fish or shellfish. In humans, the symptoms of intoxication by microalgal toxins often include nausea, vomiting, diarrhea, temperature reversal effects and paralysis (*Section 12.2*). It is reasonable to conclude that similar effects also occur in marine mammals and this would clearly affect feeding, buoyancy, heat conservation, breathing and swimming. In some studies, high levels of PSP toxin (saxitoxin) have been found in tissues of stranded orcas (killer whales) known to have been feeding on mussels in areas with a toxic *Alexandrium* bloom. Although the levels of toxin recovered from most tissues at post-mortem do not usually seem high enough to account for symptoms it is possible that the toxins become concentrated in the brain. Brevetoxin shows a very high affinity for nerve tissue from manatee, which are often killed off the Florida coast during blooms of *Gymnodinium breve*. Saxitoxin also binds strongly to nerve

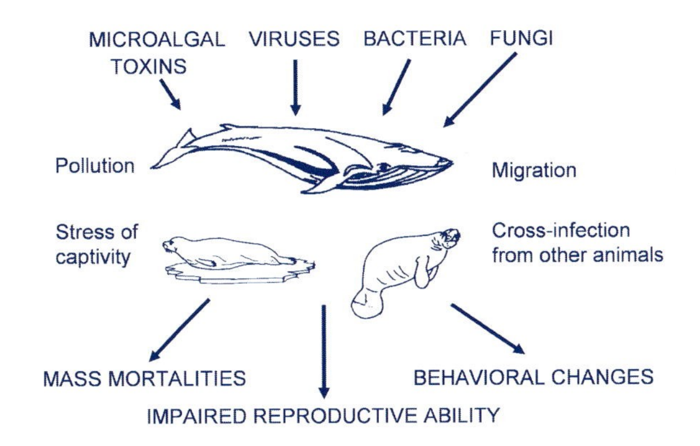

Figure 13.1

Microorganisms as agents of disease in marine mammals. Various factors influence the susceptibility of the host and opportunities for transmission of disease.

Figure 13.2

Mass stranding of pilot whales. Image courtesy of National Fisheries and Marine Service, USA.

tissue from several species of whale. Since the blood flow diverts to the brain during diving, this could deliver toxins absorbed from the gut to the brain, where they could accumulate in high enough concentrations to cause disorientation and other neurological effects, leading to stranding. The toxin domoic acid produced by the diatom *Pseudo-nitzschia* has been linked to illness in marine mammals and birds, as described in *Box 13.1*. The presence of pollutants or infection by viruses or bacteria probably makes marine mammals more susceptible to the effects of toxins.

13.3 Virus infections

In 1988, there were mass mortalities of over 18 000 harbor seals in Northern Europe. At about the same time, an epizootic with similar mortality rates occurred in bottlenose dolphins on the east coast of the USA. In 1992, there were mass mortalities of striped dolphins in the Mediterranean, which reduced the population to 30% of its previous level, and of porpoises in the Irish Sea. These diseases attracted considerable public concern and a number of government-sponsored research projects were set up to investigate the problem. As a result of this work and subsequent studies, we now know that virus diseases are a significant cause of mortality in cetaceans (whales, dolphins and porpoises) and pinnipeds (seals and sea lions). As immunological and molecular biological tests have developed, it is now common practice to test beached cetaceans and pinnipeds for viral diseases, and a worldwide picture of their importance is beginning to emerge. Some knowledge about these diseases has also come from the study of captive animals in zoos. Viruses from nine different families have been linked to diseases of marine mammals, the most important group being the morbilliviruses.

13.3.1 Morbilliviruses

Morbilliviruses are RNA viruses of the paramyxovirus group causing a number of diseases, of which the best-known examples are measles in humans and distemper in dogs. The first clues to the identity of the virus infecting seals came when serum was found to neutralize canine distemper virus (CDV). At first, it was thought that the disease might be linked to an outbreak

Box 13.1 RESEARCH FOCUS

Mad birds, sick sea lions, and choosy otters

The role of HAB toxins in the ecology of marine animals

For many years, occasional mass mortality and erratic behavior of seabirds has been reported in the Monterey Bay area of California. One such incident is thought to have been the inspiration (with some imaginative additions!) for the classic Alfred Hitchcock movie *The Birds,* in which residents of a small town are terrified by the erratic behavior of birds. In 1991, large numbers of brown pelicans and cormorants showed unusual behavior which seemed to indicate some disorder of the central nervous system. Dead birds were tested for heavy metals and pesticides, but these were not found. It was thought that anchovy eaten by the seabirds was the source and toxins from an algal bloom were suspected. Plankton analysis showed that there was a high density of *Pseudo-nitzschia australis.* A few years earlier, a diatom of the same genus had been shown to be responsible for amnesic shellfish poisoning (ASP) in humans (*Section 12.2.4*) due to the formation of the neurotoxin domoic acid. However, this was not measured in the tissues of the birds. In the summer of 1998, seabirds were affected again and over 400 California sea lions died in Monterey Bay, with many others showing neurological symptoms (such as seizures and abnormal behavior). It was notable that over 50% of the deaths were in pregnant females. An extensive research investigation involving scientists from many disciplines was mounted. Scholin *et al.* (2000) reported that the tissue of affected animals contained high levels of domoic acid and pathological examination showed lesions in the brain. High levels were also found in anchovies and sardines on which the sea lions were feeding. These fish were feeding on the *P. australis* plankton bloom, but were apparently not affected by the toxin. In contrast, when filter-feeding mussels were analyzed, they contained very low levels of the toxin. Therefore, it seems that a short pulse of domoic acid entered the food-chain through the planktivorous fish which thenpoisoned the birds and mammals that joined in the 'feeding frenzy'. Bargu *et al.* (2002) have recently shown that krill, a principal component of the diet of squid, baleen whales and seabirds, feed on toxic *Pseudo-nitzschia* and accumulate domoic acid in their tissue. Krill is therefore a likely source for transfer of domoic acid to higher trophic levels in marine food chains.

Another interesting observation concerns the effect of PSP toxins on the ecology of sea otters and butter clams off the coast of Alaska (Kvitek *et al.* 1991). In this area the clams, which are a principal component of the otters' diet, are often contaminated with saxitoxin from regular blooms of *Alexandrium.* In an experimental study of otters kept in captivity, the animals were apparently able to detect and avoid clams with high levels of saxitoxin. It is thought that high levels of saxitoxin might protect some populations of clams from predation and restrict the distribution of otters to exposed areas of the coast where clams are less contaminated.

of distemper in sled dogs in Greenland. However, the seal virus (now called phocine distemper virus, PDV) was shown to be a new species using sequencing of the viral capsid proteins, although it shares some antigenic cross-reactions with CDV. One possible explanation for the sudden epizootic is transfer from another species of marine mammal which migrated from a different geographic region in which the disease is enzootic (constantly present at low levels in the population). Serological studies have shown that PDV-like viruses are present in several species of marine mammals. By contrast, the virus responsible for outbreaks in seals in the Caspian Sea and Lake Baikal in Siberia appears to be identical to CDV, and it almost certainly came from dogs. In seals, the main symptoms are respiratory, gastrointestinal and neurological, often with secondary bacterial infections.

The morbilliviruses isolated from diseased porpoises, dolphins and several species of whale appear to be closely related antigenically and genetically and are now recognized as different strains of cetacean morbillivirus (CMV). Phylogenetic studies show that CMV is close to the ancestor of the morbillivirus group. These viruses have probably infected cetaceans for the several millions of years that they have populated the oceans. Infected animals show pneumonia-like symptoms and disturbance of swimming and navigation ability and this is obviously another possible explanation for stranding events. In the enzootic state, the virus probably has long-term effects on population dynamics, causing mortalities mainly in young animals in which no immunity exists. Morbilliviruses typically lead to either rapid death or recovery with life-long immunity, with no persistent carrier state. They therefore require large populations to sustain themselves through the input of new susceptible hosts. As with seal distemper, epizootics in cetaceans probably occur as a result of cross-infection from different species or animals from other geographic regions once a sufficiently large population of susceptible individuals has built up to allow spread. After the 1988 outbreak in northern Europe, the harbor seal population slowly recovered. In 2002, a new epizootic of phocine distemper in harbor seals emerged, due to the build up of a threshold number of nonimmune animals.

13.3.2 Other viruses

Poxviruses have been implicated as the cause of skin lesions in several species of small cetaceans although they do not usually cause significant mortalities. Papillomaviruses and herpesviruses cause genital warts in porpoises, dolphins and whales. As with their human equivalent, they are sexually transmitted and may affect reproduction and social behavior (sexual 'play' is an important part of cetacean social group interactions). Various types of caliciviruses occur in a wide range of marine mammals, having been first described as San Miguel sea lion virus. From time to time, influenza A virus has been associated with mortalities in cetaceans and pinnipeds. As occurs in humans, animals infected with influenza become very weak and often succumb to secondary bacterial infections. The isolates are closely related to avian influenza in birds and are highly virulent. Seabirds are often associated with marine mammals during feeding at the surface and this favors transmission.

13.4 Bacterial and fungal infections

Several species of bacteria are primary pathogens of marine mammals. The most important is probably *Brucella*, a highly contagious intracellular pathogen that also occurs in cattle, sheep and other farm animals. It infects the reproductive tract, especially the placenta and amniotic sac, leading to abortion. Frequent abortions have been observed to occur in closely monitored dolphins, and the bacterium could therefore have a significant effect on fertility and population dynamics. Several strains have been isolated from many different species, and serological studies show that up to 30% of marine mammals surveyed have evidence of exposure. Leptospirosis, caused by several species of *Leptospira,* occurs in many populations of seals and sea lions. As in the human equivalent (Weil's disease, transmitted by rats), the main symptom is renal failure. Tuberculosis is a chronic multiorgan disease caused by *Mycobacterium tuberculosis* and *M. bovis*. Other mycobacteria such as *M. marinum* and *M. fortuitum* can cause lesions in the skin and lungs. In seals, sea lions and cetaceans, there have been several cases in zoos and marine parks. More recently, pathological and serological evidence for the disease in wild populations has been increasing. Infection by *Burkholderia pseudomallei* can cause septicemia and has been described as the cause of death of 25 dolphins, whales and sea lions over an 11-year period in an oceanarium and sporadically elsewhere. *Erysipelothrix* can cause sudden mortality in captive cetaceans. Other bacterial pathogens include *Streptococcus* sp., *Salmonella* sp., *Bordetella bronchiseptica* and *Pasteurella* spp. In addition to primary pathogens, marine mammals can succumb to a host of opportunistic skin and respiratory infections by bacterial and fungal pathogens (e.g. pneumonia caused by *Aspergillus fumigatus* and *A. terrus*). This is a particular problem in stressed captive animals.

13.5 Effects of environmental pollution on infectious diseases

The mass mortalities of marine mammals observed in the 1980s and 1990s led to widespread speculation that build-up of pollutants in the sea was to blame. Because these carnivorous animals are at the top of the food chain, their tissues have been shown to concentrate environmental pollutants such as mercury, polychlorinated biphenyls (PCBs), dioxins and organochlorine pesticides such as DDT and dieldrin. These compounds accumulate in lipid-rich tissues, especially blubber and milk and can have a variety of deleterious effects. Since they have been shown to impair the function of T- and B-cells in the immune system, organochlorines may increase susceptibility of marine mammals to bacterial or viral diseases. An experimental study in which different groups of seals were fed on fish caught in either highly polluted or less polluted areas showed that accumulation of high levels of pollutants impaired the immune system, as shown by reduction in the natural killer cells that are critical in defense against virus infection. In porpoises, there is a strong correlation between mortality due to infection and levels of PCBs in blubber. Tributyl tin (TBT), which is widely used as an antifouling compound on boats and aquaculture cages, has also been linked with immunosuppression and disease in dolphins. Direct acquisition of microbes of human origin (especially opportunist bacterial infections of wounds) may also occur when mammals swim in sewage-contaminated waters. Because of the problems of conducting surveys and, in particular, controlled experiments, it will always be difficult to conclusively prove that pollution is directly responsible for increased mortality of marine mammals due to infectious disease in marine mammals. However, the circumstantial evidence is strong.

13.6 Zoonoses

Some diseases of marine mammals can be transmitted to humans. Contact between humans and animals such as whales and seals has existed for a very long time in hunting communities and in these cases there have been some well-described zoonoses of microbial origin (infection by metazoan parasites acquired from consumption of whale and seal meat is also important). Bacterial zoonoses include 'seal finger', an extremely painful infection of the hands caused by *Staphylococcus* or *Erisypelothrix* which occurs in seal hunters (and research workers), especially in the Arctic regions of Norway and Canada. Seal bites can also cause serious infections caused by a variety of poorly characterized bacteria, and are notoriously difficult to treat unless antibiotics are used promptly. Large outbreaks of *Salmonella* infection, causing diarrhea and vomiting, have occurred in Inuit communities that have eaten infected whale or seal meat. Research workers and aquarium staff may be at special risk of contracting disease. There have been a few cases of infection by *Mycobacterium* sp. (causing respiratory and skin lesions) and *Brucella* infection (causing severe lethargy, fever and headaches) in such people. Of the viruses, influenza poses the most important risk and several instances of direct transmission from persons in close contact with infected seals have occurred. Evolution of new strains of influenza virus occurs regularly through antigenic changes, due to recombination events in the fragmented genome. It is therefore possible that marine mammals could provide an opportunity for transmission of new variants to humans. Protozoan parasites such as *Giardia* may have a reservoir in marine mammals.

In summary, there is a definite (though probably small) risk of acquiring diseases from marine mammals. Of some concern is the increasing popularity of tourist activities such as 'petting zoos' and swimming with dolphins, which could lead to a rise in transmission of zoonoses. Such activities also increase the likelihood of transmission *from* humans *to* marine mammals, although there is limited evidence of this to date.

References and further reading

Bargu, S., Powell, C.L., Coale, S.L., Busman, M., Doucette, G.J., and Silver, M.W. (2002) Krill: a potential vector for domoic acid in marine food webs. *Mar Ecol Prog Ser* **237**: 209–216.

Barrett, T., Blixenkronemoller, M., Diguardo, G., *et al.* (1995) Morbilliviruses in aquatic mammals – report on round-table discussion. *Vet Microbiol* **44**: 261–265.

Bennett, P.M., Jepson, P.D., Law, R.J., *et al.* (2001) Exposure to heavy metals and infectious disease mortality in harbour porpoises from England and Wales. *Environ Poll* **112**: 33–40.

DeSwart, R.L., Harder, T.C., Ross, P.S., Vos, H.W., and Osterhaus, A.D.M.E. (1995) Morbilliviruses and morbillivirus diseases of marine mammals. *Infect Agents Dis* **4**: 125–130.

Harvell, C.D., Kim, K., Burkholder, J.M., *et al.* (1999) Review: Marine ecology – Emerging marine diseases – Climate links and anthropogenic factors. *Science* **285**: 1505–1510.

Higgins, R. (2000) Bacteria and fungi of marine mammals: A review. *Can Vet J* **41**: 105–116.

Jepson, P.D., Bennett, P.M., Allchin, C.R., *et al.* (1999). Investigating potential associations between chronic exposure to polychlorinated biphenyls and infectious disease mortality in harbour porpoises from England and Wales. *Sci Total Environ* **4**: 339–348.

Kvitek, R.G., Degange, A.R., and Beitler, M.K. (1991) Paralytic shellfish poisoning toxins mediate feeding-behavior of sea otters. *Limnol Oceanog* **36**: 393–404.

Landsberg, J.H. (2002) The effects of harmful algal blooms on aquatic organisms. *Rev Fish Sci* **10**: 113–390.

Ross, P.S. (2002) The role of immunotoxic environmental contaminants in facilitating the emergence of infectious diseases in marine mammals. *Human Ecol Risk Assess* **8**: 277–292.

Scholin, C.A., Gulland, F., Doucette, G.J., *et al.* (2000) Mortality of sea lions along the central California coast linked to a toxic diatom bloom. *Nature* **403**: 80–84.

Trainer, V.L. (2001) Marine mammals as sentinels of environmental biotoxins. In: Masaro, E.J. (ed.) *Handbook of Neurotoxicology*, vol. 1, pp. 349–361. Humana Press, Totowa.

Trainer, V.L., and Baden, D.G. (1999) High affinity binding of red tide neurotoxins to marine mammal brain. *Aquat Toxicol* **46**: 139–148.

Van Bressem, M.F., Van Waerebeek, K., and Raga, J.A. (1999) A review of virus infections of cetaceans and the potential impact of morbilliviruses, poxviruses and papillomaviruses on host population dynamics. *Dis Aquat Org* **38**: 53–65.

Microbial diseases of fish 14

14.1 Importance in wild fish and in aquaculture

The first description of microbial diseases in fish can be traced to the description of 'red pest' in eels in Italy by Canestrini in 1718. This disease, which we know as vibriosis (caused by the bacterium *Vibrio anguillarum*), led to mass mortalities in migrating eels during the 18th and 19th centuries. Such large-scale fish kills in the wild occur occasionally, particularly in estuaries and on coral reefs, and may be caused by a wide range of bacterial, viral or protozoan infections, or by HABs. Apart from these events, it is hard to estimate the normal impact of disease on fish populations in the wild; sick fish do not last long in the natural environment and are quickly removed by predators. One of the few pieces of evidence that infection plays a significant role in controlling natural fish populations comes from observations made in the 1970s that salmon immunized against vibriosis had a 50% greater survival than nonimmunized fish, as shown by the return rate of tagged fish. Fish such as salmon and eels may be particularly susceptible to acute infections due to the pronounced physiological changes that occur during their migration from fresh to salt water, or *vice versa*. Some pathogens are isolated from a high proportion of wild fish; these tend to be those that cause slow-developing chronic diseases.

The development of intensive marine aquaculture (mariculture) in the 1970s led to a rapid growth in the science of fish pathology. Early attempts to farm salmonid fish in intensive off-shore pens and cages were frustrated by large-scale mortality and heavy economic losses. This experience has recurred with many different fish species in all parts of the world and at times has threatened the survival of the industry. The impact of disease should come as no surprise, since it is common in all forms of intensive culture where single species of animals are reared at high population densities. Economic factors demand highly intensive systems and this can lead to stress of the cultured fish, which then succumb to disease transmitted rapidly through the dense populations. In 1997, the economic cost of aquaculture production losses due to disease was estimated by the World Bank to be around US$3 billion per year.

The disease process depends on the interaction of various factors, as shown in *Figure 14.1*. This chapter deals mainly with the microbial component of this diagram, principally the mechanisms of pathogenicity of bacterial and viral infections and the methods for disease control.

14.2 Disease diagnosis

The importance of disease in aquaculture has led to the development of specialized branches of veterinary science and diagnostic microbiology. The need to implement effective control measures following an outbreak puts pressure on investigators to determine the causes of mortality quickly. Careful observation of the stock by the fish farmer, good record keeping and experience play a large part in identifying diseases. Infestation with eukaryotic parasites, such as sea lice and protozoa (e.g. *Icthyobodo, Cryptocaryon* and *Tricodina*), typically manifests as an extended course of mortalities, especially in confined fish. Nutritional deficiency is usually indicated by poor growth rates, skeletal abnormalities and steadily increasing mortalities. Intoxications usually have a very rapid onset, causing mass mortalities within hours. Microbial infection is usually characterized by a rising level of mortality accompanied by characteristic disease signs. These are very varied, but may be broken into three broad categories: (a) bacteremia or viremia, in which

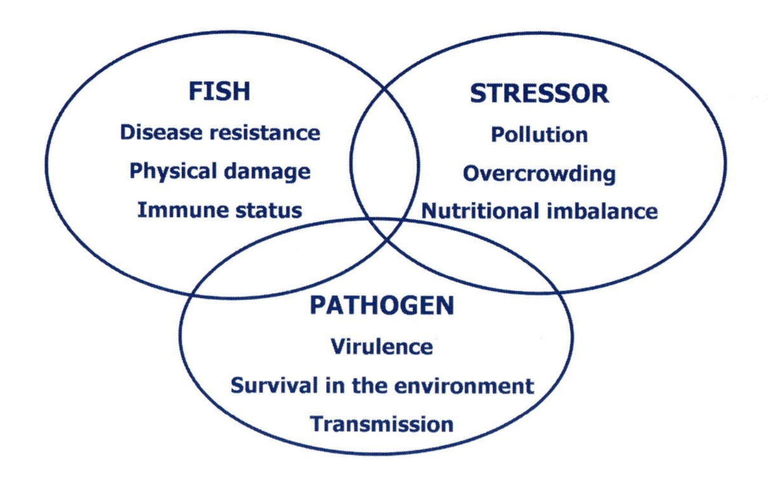

Figure 14.1

Factors affecting the development of disease in fish.

there is rapid growth of the pathogen, often with few external disease signs other than hemorrhaging; (b) skin, muscle and gill lesions; and (c) chronic proliferative lesions.

Post-mortem changes are very rapid due to overgrowth by the normal microbiota of fish, so it is important to examine fish showing signs of infection before they succumb completely. External examination will often reveal the presence of gill and tissue erosion, eye damage, hemorrhages, abcesses, ulcers or a distended abdomen. Internal inspection may reveal organ damage and fluids in the body cavity. To the experienced eye, these signs will often indicate a particular disease agent, but the diagnosis must be confirmed by identification of the pathogen. For bacterial diseases, this is usually achieved by plating tissue samples onto various selective media and performing biochemical tests using diagnostic keys. Not all pathogens are amenable to this approach, as some grow very slowly in culture (notably *Renibacterium* and some mycobacteria). Antibody methods and genomic fingerprinting (*Section 2.6.8*) are often used for precise strain identification, which is especially important in epizootological studies. An alternative approach is the use of methods such as ELISA or FAT (*Section 2.4*) to detect bacterial antigens in the blood or tissues, or to detect a high titer of host antibodies against the pathogen. The diagnosis of viral infections is more difficult and time-consuming because it relies on propagation of the virus in a suitable cell culture. Serological methods are therefore the main method used for rapid identification. There has been some success with 'dipstick'-type kits for rapid diagnosis, based on modifications of the ELISA technique. For many pathogens, there are now accurate molecular diagnostic tests based on PCR amplification and gene probes.

14.3 Bacterial infections

14.3.1 Mechanisms of pathogenicity

The following sections describe some representative major pathogens of marine fish, emphasizing the distinctive features that enable them to colonize their host and produce disease, as summarized in *Table 14.1*. An understanding of these processes is important in the design of control methods, especially vaccines.

14.3.2 *Vibrio* sp.

Species of the genus *Vibrio* are widespread in marine and estuarine waters and in accompanying sediments and are responsible for a range of diseases in humans (*Section 11.2*) and invertebrates

Table 14.1 Stages in bacterial pathogenicity

Stage in pathogenicity	Mechanisms
Initial infection and colonization	Chemotaxis – detection of gradients of exudates or excretion products Attachment to host surfaces
Growth *in vivo* and avoidance of host defense mechanisms	Complement resistance Iron acquisition Avoidance or destruction of phagocytes Intracellular survival and multiplication Antigenic variation
Production of damage	Toxin production Stimulation of cytokines Effects on the cellular immune response

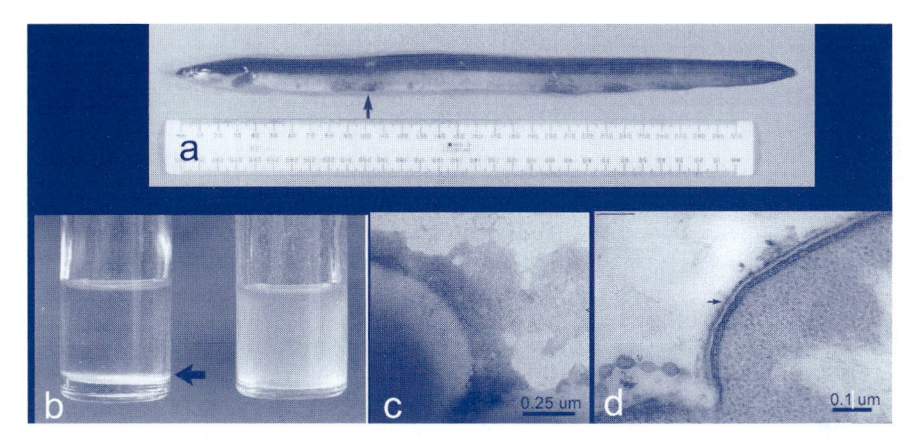

Figure 14.2

(a) Vibriosis in an eel, *Anguilla anguilla,* showing external hemorrhagic lesions. (Image: H Chart & CB Munn, University of Plymouth.) (b–d) The surface A-protein of *Aeromonas salmonicida.* (b) Growth of A^+ (left) and A^- (right) cultures in broth. A^+ cultures autoagglutinate due to the hydrophobic A-protein. (c) Electron micrograph (negatively stained) of an A^+ strain showing regular structure of the surface A-protein. (d) Electron micrograph (section) of an A^+ strain showing the A-protein as an additional layer (arrowed) external to the outer membrane. Note outer membrane vesicles (V). (Images: ND Parker & CB Munn, University of Plymouth.)

(*Sections 15.2.2, 15.3.5, 15.4*), as well as fish. Diagnosis of vibrio infections is based on isolation and identification. Among the characteristic features of vibrios is sensitivity to the chemical 'vibriostat 0/129' (2,4-diamino-6,7-diisopropylpteridine). Immunoblotting and ELISA tests are also used in diagnosis.

As noted above, *V. anguillarum* was the first species to be identified as a fish pathogen causing vibriosis in eels (from which it derives its species name), producing ulcers, external and internal hemorrhages and anemia (*Figure 14.2a*). Vibriosis causes heavy losses in eel culture, especially in Japan. Infections due to *V. anguillarum* also occur in a very wide range of marine species and have caused particular problems in the culture of salmon in North America, sea bass and sea bream in the Mediterranean and yellowtail and ayu in Japan. In all species, the disease is characterized by a very rapid septicemia and, usually, external hemorrhages. Internally, there is extensive hemolysis and organ damage. The major factor in virulence, which explains the extremely rapid growth *in vivo,* is the possession of a 65 Mb plasmid (pJM1) that confers

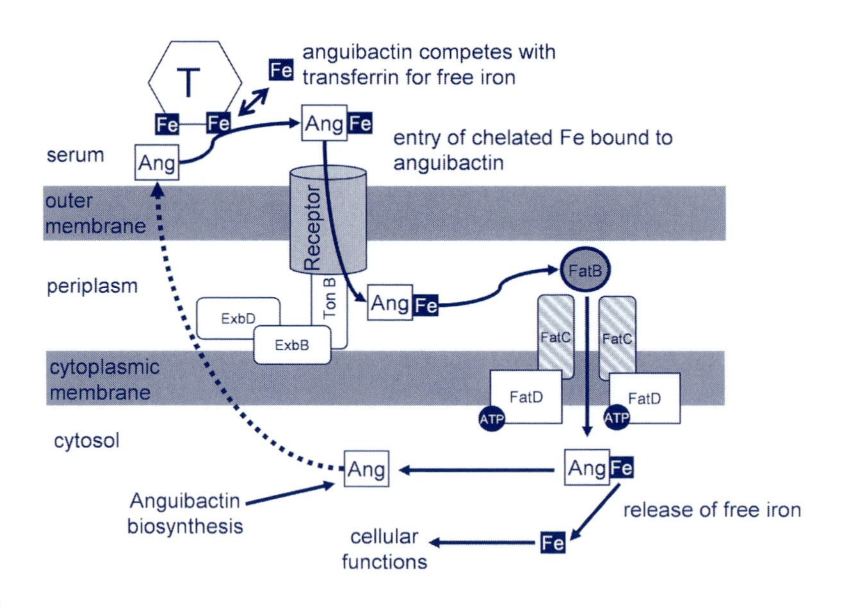

Figure 14.3

Iron uptake in *Vibrio anguillarum.* The biosynthesis of the siderophore anguibactin, the outer membrane receptor and transport proteins are subject to complex regulation at the transcriptional level (not shown). Based on a diagram by J. Crosa, Oregon Health & Science University, http://www.ohsu.edu/som-MicroBio/Faculty/crosa.html.

ability to acquire iron in the tissues of the host. As noted in *Section 4.4.7,* all bacteria require iron for the activity of essential cellular functions. Vertebrate animals possess highly efficient systems for the transport and sequestration of iron. The serum protein transferrin binds iron with extreme avidity, reducing the concentration of free iron in tissues to 10^{-18}M, about 10^{8} times lower than the concentration that bacteria require for growth. During infection, an even more efficient iron-binding protein (lactoferrin) is released and sequestered iron is removed to storage in the liver and other organs. Thus, successful pathogens must compete with the host's iron-binding system in order to obtain sufficient iron for growth and many pathogens achieve this via the production of siderophores and an iron-uptake system. Proof of the key role of the plasmid in virulence of *V. anguillarum* was initially obtained by inducing loss of the plasmid by curing and reintroduction by conjugation (*Section 3.5*) and more recently by directed mutagenesis of key genes. Mutation of any of the genes can result in a reduction of virulence by as much as 10^{5} times. In experimental infections in which iron uptake by transferrin is swamped by injection of excess iron, possession of the plasmid does not confer an advantage. The plasmid pJM1 contains genes for synthesis of the siderophore anguibactin, an unusual hydroxamate–catechol compound, and OM transport proteins and regulators of transcription involved in iron uptake (*Figure 14.3*). Expression of the several components occurs only under low iron conditions, due to negative control at the transcriptional level by the chromosomally encoded Fur protein and at least three plasmid-encoded regulators, the best studied of which is AngR. Outer membrane proteins only expressed under iron-restricted conditions are known as IROMPs. Quorum sensing (QS) via AHL signaling molecules is an important factor in the growth and survival of *V. anguillarum* in its host and in the aquatic environment (see *Section 5.12.3*). There are three separate pathways, which are involved in gene regulation of biofilm formation and protease, pigment and exopolysaccharide production. The hierarchical QS system consists of regulatory elements homologous to those found in both *V. fischeri* (the LuxRI homologs VanRI) and *V. harveyi* (the LuxMN homologs, VanMN, see *Box 16.1*). It is possible that some of the AHLs have effects on eukaryotic host cells, including an immunomodulatory function. Other virulence factors are also involved in *V. anguillarum* pathogenicity, including a metalloprotease of unknown function and a hemolysin, which may contribute to the acquisition of iron by release of hemoglobin from red blood cells. The bacterium is resistant to the bactericidal effects of complement in normal serum; this property depends on structure of the LPS. The flagellum is essential for virulence, as directed mutations in the *flaA* gene result in reduced virulence. LPS on the sheath of the flagellum (genes *virA* and *virB*) are also essential for virulence; these probably contribute to motility across the fish integument and biofilm formation. Vibriosis was one of the first fish diseases for which vaccines were developed and they have generally been very effective.

Vibrio ordalii was originally thought to be a biovar of *V. anguillarum*, but was designated as a separate species, based on a number of biochemical differences and DNA hybridization. It has caused major losses in cage cul-ture of salmon in Pacific coastal waters off Oregon, Washington and British Columbia. Infections with *V. ordalii* tend to be localized in muscle tissue rather than the generalized infections seen with *V. anguillarum*. Virulence has not been so well studied as that of *V. anguillarum*, but complement resistance and a leucocytolytic toxin have been described.

Vibrio salmonicida causes cold-water vibriosis (Hitra disease) in salmon farmed in Norway and Scotland during the winter, when water temperatures drop below 8°C. Disease signs are broadly similar to those of *V. anguillarum*, but the two bacteria are serologically and genetically distinct. *V. salmonicida* is excreted in the feces of infected fish and appears to have good powers of survival in marine sediments, thus causing reinfection even if farm sites are 'fallowed' for a season. Epidemiological studies suggest that there is interchange of the bacterium between populations of cod and salmon. Various strains can be distinguished by different plasmid profiles, but there does not seem to be a close association between plasmid possession and virulence. *V. salmonicida* produces various extracellular enzymes implicated in virulence, and a hydroxamate siderophore and *fur*-regulated iron-uptake system. Interestingly, significant amounts of the siderophore and the IROMPs required for transport are only expressed at temperatures below 10°C, which may explain the increased incidence of disease in the winter. This fact will be very important in vaccine manufacture, since temperature-regulated surface proteins are likely to be important antigens. A project to obtain the genome sequence of *V. salmonicida* has been initiated recently and the results of this will provide further insight into the temperature regulation of virulence.

Vibrio vulnificus causes an infection of eels in Japan. The fish isolate is closely related to clinical isolates associated with human disease (*Section 11.2.2*), but has been designated a separate taxon (biogroup 2) based on phenotypic, cultural and serological properties. Iron acquisition and production of a capsule, hemolysin, protease and other toxins have all been implicated in virulence.

A newly described species, which has also caused salmon infections in northern waters when temperatures fall below 8°C (winter ulcer disease) was originally designated as *Vibrio viscosus*, but has recently been reclassified as *Moritella viscosa*. Ulcerous lesions progress from the skin to the underlying muscle, causing mortalities over 10%.

14.3.3 *Photobacterium damselae* subsp. *piscicida*

The expansion in the 1990s of sea bass and sea bream aquaculture in the Mediterranean, and yellowtail culture in Japan was accompanied by outbreaks of a new disease that has caused heavy losses. The pathogen responsible was originally identified as *Pasteurella piscicida* and the disease is still known as pasteurellosis, although the causative agent has been renamed as above. Pasteurellosis is now a major disease of all areas producing warm water marine fish. When water temperatures exceed 25°C, acute mortalities up to 70% can occur, especially in the larval and juvenile stages. A more chronic condition occurs in older fish. Interest in virulence mechanisms has focused largely on extracellular proteases, adhesive mechanisms and the presence of a polysaccharide capsule. Iron concentration appears to be important in the regulation of expression of superoxide dismutase and catalase, which protect bacteria against reactive oxygen species (*Section 4.6.4*) and may be important in intracellular survival. An iron-uptake system is associated with virulence, but as well as uptake by siderophores the bacterium seems able to acquire iron by direct interaction between hemin molecules and OM proteins. Introduction of rapid diagnostic methods based on PCR technology and a recently introduced vaccine appear to be effective in controlling the disease.

14.3.4 *Aeromonas salmonicida*

A. salmonicida was first described as a pathogen of trout in Europe in 1890, but is now known to have a broad host and geographic range. The taxonomy of *A. salmonicida* has been the subject of much debate over the years and it is now generally recognized that acute 'typical' furunculosis in salmonids is caused by *A. salmonicida* subsp. *salmonicida*. In mariculture, this usually presents as a severe septicemia with acute mortalities. At the peak of the furunculosis outbreaks in Scotland and Norway in the 1980s, total industry losses neared 50% of stock. Externally, the fish show darkening and hemorrhages around the fins and mouth and internally there is extensive hemorrhaging and destruction of the organs. Other subspecies (*achromogenes, masoucida, pectinolytica* and *smithia*) can be distinguished by differences in standard biochemical tests, pigment production and molecular techniques such as gene probes and DNA:DNA hybridization. These subspecies are generally associated with dermal ulcerations in a wide range of other freshwater and marine species (e.g. turbot, halibut and flounder). The name furunculosis derives from a boil-like necrotic lesion seen in a chronic form of the infection in older or more resistant fish. Most strains can be easily isolated on laboratory media although some are fastidious and have a specific requirement for heme.

The virulence mechanisms of *A. salmonicida* are extremely complex, and this pathogen is an example *par excellence* of multifactorial virulence that has attracted the attention of numerous investigators for over 30 years. Different components of the factors responsible for entry, colonization, growth *in vivo* and production of damage interact, as illustrated in *Figure 14.4*. Genes for many of the virulence factors have now been cloned and their properties studied in detail. A key factor in virulence is the 'A-protein', composed of a regular structured array of a tetrahedral

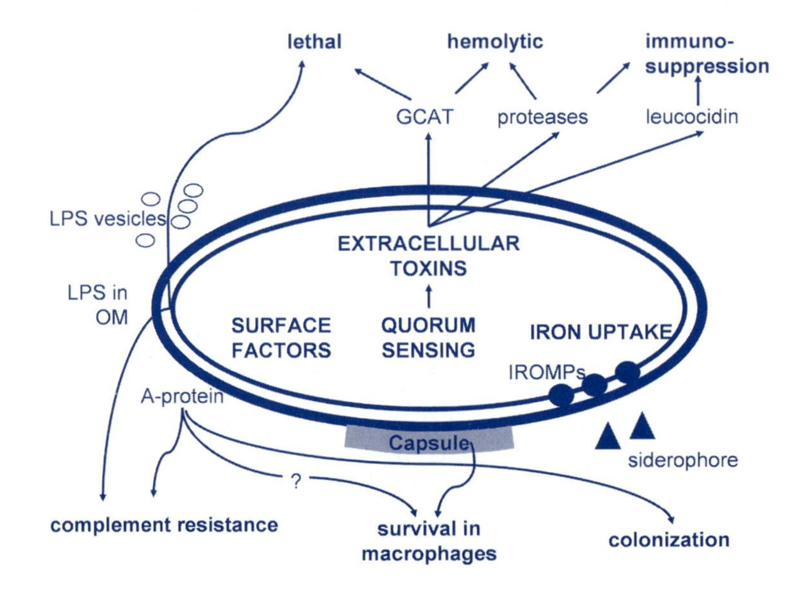

Figure 14.4

Virulence mechanisms in *Aeromonas salmonicida*. The inner line represents the outer membrane (OM) of the cell (cytoplasmic membrane is not shown); the darker outer line represents the S-layer (A-protein). Only part of the *in vivo* capsule and IROMPs (iron-restricted outer membrane proteins) are shown. GCAT = glycerophospholipid:cholesterol acyltransferase.

49 kDa protein (*Figure 14.2c*). It belongs to the family of surface structures known as S-layers that occur in a range of *Bacteria* and *Archaea* (*Section 3.9.3*). Virulent and avirulent strains are usually distinguished by the presence or absence of this layer; the hydrophobic nature of which confers autoagglutinating properties in culture (*Figure 14.2b*). Isolates possessing the A-protein (A^+) can also be distinguished from A^- strains by their growth on agar media containing Coomassie Blue or Congo Red dyes, which the A-protein absorbs. Electron microscopy shows the A-protein to be present as a layer external to the typical Gram-negative OM (*Figure 14.2d*). It is linked to the cell surface via the O polysaccharide side-chain of LPS. The main function of the A-protein is as a protective layer, which contributes to the bacterium's resistance to the bactericidal effects of complement in the serum of the host fish. Because of its hydrophobic nature, it also plays a role in adhesion to host tissue and survival within macrophages. *A. salmonicida* produces a range of toxins and many studies have been carried out on the activity of purified components, although it is now clear that these have synergistic interactions. The enzyme glycerophospholipid: cholesterol acyltransferase (GCAT) and a serine protease are particularly implicated as key virulence factors. GCAT forms a complex with LPS, which is hemolytic, leukocytolytic, cytotoxic and lethal. Serine protease expression is regulated by an AHL-mediated QS mechanism and this enzyme (in synergy with GCAT) is responsible for hemolysis. As in *V. anguillarum*, growth *in vivo* is dependent on the production of a siderophore (2,3 diphenol catechol) and the uptake of sequestered iron via IROMPs. Elucidation of the various components in virulence and their immunological properties was a critical step in the formulation of modern vaccines, which have been largely successful in control of the disease in salmon mariculture since the mid 1990s.

The ecology of *A. salmonicida* has been the subject of much controversy, with some investigators suggesting that it is an obligate fish pathogen and others suggesting that it survives in the environment. It may enter a dormant or VBNC state, during which the cells undergo various morphological changes (see *Box 4.1*). One certainty is that the development of mariculture in enclosed bodies of water such as lochs in Scotland and fjords in Norway has led to a shift in the normal microbiota. Whereas *A. salmonicida* appears to have been primarily a freshwater organism, it has now adapted to life in seawater and recent isolates may have a sodium requirement lacking in earlier strains. Farmed fish can transmit the disease to wild fish around sea cages and these can spread the pathogen to other sites. Wrasse introduced to net pens to remove parasites (sea lice) from salmon can become infected and thus constitute a reservoir of infection.

14.3.5 *Piscirickettsia salmonis*

The rickettsias, which are members of the *γ-Proteobacteria,* have been poorly studied as fish pathogens. Because they are obligate intracellular parasites, they can only be propagated in suitable fish cell lines. *Piscirickettsia salmonis* was identified in 1989 as the etiological agent of a septicemic disease causing severe economic losses in farmed salmon in Chile. Here, it causes predictable annual epizootics, but it has also caused sporadic outbreaks of disease in Norway, Iceland and Canada. The disease progresses rapidly and death often occurs with few or no external symptoms. A range of serological tests has been developed, and genetic heterogeneity among strains can be studied using analysis of the 23S rRNA operon. Most rickettsias have an intermediate host such as ectoparasite, but there is no evidence for this with *P. salmonis* so far. The organism seems to survive well in seawater and this may be sufficient to ensure effective transmission. Antibiotic treatment is ineffective (there are very few agents capable of penetrating host cells without causing damage to the host) and much effort has been applied to the development of a vaccine. A recent recombinant vaccine based on an OM lipoprotein, OspA, shows promise for disease control.

14.3.6 *Renibacterium salmoninarum*

Bacterial kidney disease, caused by the Gram-positive bacterium *R. salmoninarum,* is widely distributed in both wild and cultured salmon in many countries in Europe, North America, Chile,

Iceland and Japan. The expansion of salmon culture through international movement of eggs has assisted the spread of BKD and it causes significant losses in both Pacific and Atlantic salmon. BKD pathology is characterized by chronic, systemic tissue infiltration, causing granular lesions in the internal organs (especially the kidney). External signs include darkening of the skin, distended abdomen, exophthalmia and skin ulcers. Significant changes in blood parameters are consistent with damage to the hematopoietic and lymphopoietic tissues of the kidney, liver and spleen. The pathological signs are the result of the interactions between the host's cellular immune response and the pathogen's virulence mechanisms. Tissue destruction forms a focus of necrosis, due to release of hydrolytic and catabolic enzymes and liberation of lytic agents from the bacteria. Our understanding of the mechanisms of virulence of *R. salmoninarum* is hampered by the fact that the bacterium takes several weeks to grow on culture plates and does not form discrete colonies. Reproducible infection is also difficult to achieve in aquarium experiments. One important virulence factor is a 57 kDa surface protein, whose hydrophobic properties facilitate attachment to host cells. The key feature of *R. salmoninarum* is its ability to enter, survive and multiply within host phagocyte cells. Binding of the complement component C3 to the bacterial surface enhances internalization of the bacterium because phagocytic cells possess a receptor for C3. Why does *R. salmoninarum* encourage uptake by cells that normally kill invading pathogens? The answer appears to lie in the pathogen's ability to survive (at least in part) the intracellular killing mechanisms of the macrophage and to replicate (albeit slowly) within the cells. As well as resistance to reactive oxygen species, *R. salmoninarum* lyses the phagosome membrane in order to escape its strongly antibacterial environment (*Figure 14.5*). In the past, BKD has been difficult to diagnose, but disease management and control are helped by serological techniques (ELISA and FAT), together with recently developed gene probes and techniques for accurate differentiation of clinical isolates (based on PCR

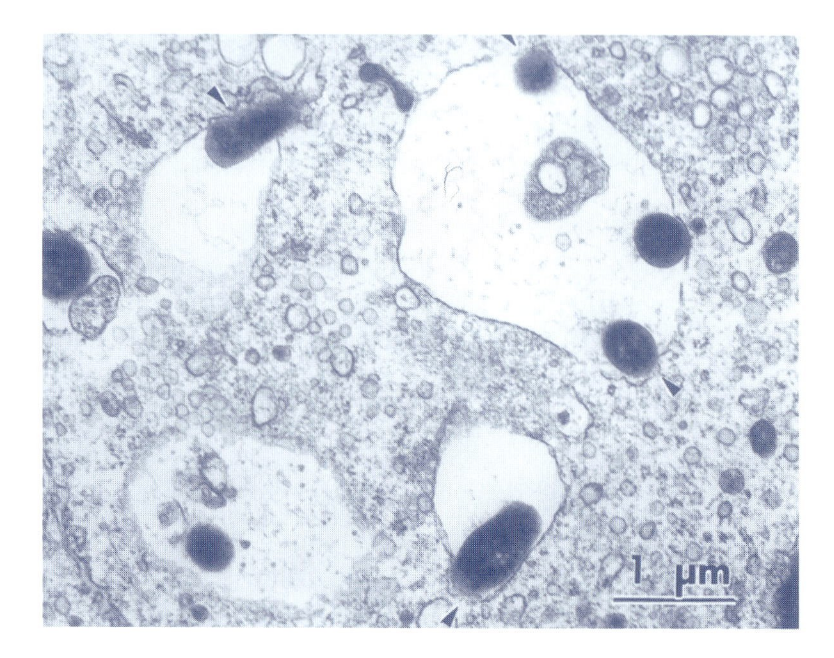

Figure 14.5

Intracellular growth of *Renibacterium salmoninarum*. Electron micrograph shows lysis of the phagosome membrane (arrowheads), prior to entrance of the bacterium into the cytoplasm. Image courtesy of SK Gutenberger and JR Duimstra, Oregon State University: From Evenden *et al.* (1993), reproduced with permission from Elsevier.

amplification of length polymorphisms in the tRNA intergenic spacer regions). These methods are used for certification of broodstock and eggs as disease free and for implementing quarantine procedures to contain disease outbreaks. There are no effective antibiotic treatments and there have therefore been many attempts to develop a vaccine. These are thwarted by the slow growth of the pathogen and recent work has focused on the use of recombinant DNA technology to produce fusion proteins and DNA vaccines (see below). Genes for several virulence factors, including the p57 protein and two hemolysins, have been cloned and expressed in *E. coli* and the nature of the immune response to these is under investigation.

14.3.7 *Tenacibacter maritimus*

Members of the CFB group (*Section 5.19*), many of which are pigmented and show gliding motility, are responsible for a wide range of infections in fish. Mostly, these bacteria are rather weak pathogens that are best described as opportunists. They colonize damaged tissue in fish weakened by stress, especially due to increased water temperature and nutritional deficiency. Most diseases caused by this group occur in fresh water, but *T. maritima* (until recently, classified as *Flexibacter maritimus*) causes a disease called 'marine columnaris', which is characterized by excess mucus production, damage to the gills, tissue necrosis around the mouth and fins, skin lesions and eventual death. The disease responds to antibiotics administered by bath. Some experimental vaccines have been developed, with variable degrees of success.

14.3.8 *Mycobacterium* and *Nocardia*

Like *Renibacterium,* these intracellular Gram-positive bacteria cause chronic, persistent infections. There are a number of related species, of which the most important is *M. marinum.* Mycobacteria are widely distributed in seawater and sediments and infect many species of fish. Disease develops slowly and usually affects mature fish; for this reason, mycobacteriosis is a particular problem in marine aquaria. Most species of fish develop few external disease signs other than emaciation, although histopathological investigation reveals extensive granulomatous lesions with caseous necrotic centers. For this reason, the disease is often called fish tuberculosis, since there is some resemblance to the human condition caused by *M. tuberculosis.* The delayed hypersensitivity reactions and involvement of cell-mediated immunity certainly show some parallels in fish and humans. It should also be noted that *M. marinum* (and possibly other species) can cause an infection in humans known as fish tank or aquarist's granuloma. Again, because of its intracellular nature, there are few antimicrobial agents effective against this pathogen. Valuable aquarium fish are sometimes treated with isoniazid and rifampicin, but this is unwise given the danger of encouraging resistance against these drugs, valuable for the treatment of human tuberculosis. *Nocardia* spp. cause similar chronic granulomatous conditions to mycobacteria, but the organisms can be distinguished in the laboratory.

14.3.9 *Lactococcus* and *Streptococcus*

The Gram-positive bacteria *Lactococcus garvieae* (formerly *Enterococcus seriolicida*) and *Streptococus iniae* have been especially problematic as the cause of disease in both fresh and marine warm-water culture, especially in Japan. Both cause hemorrhagic septicemia with high mortalities and extracellular cytolytic toxins have been implicated in pathogenicity. *S. iniae* has caused epizootics in barramundi culture in Australia and was also implicated in an extensive fish kill in the Caribbean Sea in 1999. *S. iniae* can cause zoonotic infection in humans; infection of wounds in workers handling infected fish leads to severe cellulolytic infection.

14.4 Viral infections

14.4.1 Importance

Although viruses have long been suspected as the causative agents of disease, it was not until the 1960s that development of methods for culture of fish cells led to significant advances. All species of fish are undoubtedly host to one or more viruses, but the importance of this in the natural environment is largely unknown. As with bacterial diseases, the emergence of virus disease problems has accompanied the development of intensive aquaculture. There are no chemotherapeutic agents for virus infections, so development of effective vaccines is a priority.

14.4.2 Infectious pancreatic necrosis virus (IPNV)

IPNV is a member of the birnavirus group of double-stranded RNA viruses, which are widely distributed in fish and invertebrates in fresh, brackish and salt water. IPNV has icosahedral, unenveloped particles with a mean diameter of 60 nm. In mariculture, IPN causes major losses in many species, including salmon, halibut, cod, sea bass and yellowtail. It is mainly a disease of small fry in hatcheries and usually manifests as lack of feeding and corkscrew-like swimming movements. Mortalities can be over 90%. There are often few external or internal signs of disease other than hemorrhaging and exudates in the gut. The virulence of IPNV isolates varies greatly and extensive use is made of serological and molecular methods for disease diagnosis and detection of carriers. The disease is transmitted both horizontally and vertically and has spread to many different countries through the import of salmon and trout eggs. We know little about its entry into the host and subsequent replication. Killed cell vaccines are expensive and of limited efficacy, so attempts are underway to develop recombinant vaccines based on cloned capsid proteins.

14.4.3 Infectious salmon anemia virus (ISAV)

Infectious salmon anemia (ISA) emerged as a new disease of farmed Atlantic salmon in Norway in the 1980s and again in 1996. Outbreaks then occurred in New Brunswick, Canada and it appeared in Scotland and Maine, USA in 1998. European legislation requires that infected stock are killed and the farm site left fallow for at least 6 months, causing great economic hardship to the producers. In Scotland in 1999, 4000 tonnes of fish were slaughtered and losses due to ISA were estimated at £38 million, out of a total industry turnover of £270 million. Similar slaughter policies were adopted in Norway and eastern Canada, but these seem to have been ineffective in preventing the disease from becoming endemic. Wild salmon from these areas now test positive for ISAV. Fish mortalities in sea pens vary from insignificant to moderate and there is significant variation in the severity of ISA depending on virus strain and susceptibility of fish stock. ISA-infected fish are lethargic, with anemia (shown by very pale gills), exophthalmia and hemorrhages in the eye chamber, skin and muscle. ISAV is an enveloped negative-strand RNA virus of the Orthomyxoviridae family, which includes influenza virus. Influenza virus is known for its property of antigenic variation by mutational ('drift') and recombination ('shift') events in the segmented genome, leading to the emergence of new strains. This may be occurring in ISAV as isolates from Norway and North America now seem to be genetically distinct. In the last few years, several institutes and commercial organizations have embarked on an attempt to develop a vaccine using recombinant DNA technology and prototype vaccines show promise in control of the disease in Norway.

14.4.4 Other virus infections

The rhabdoviruses causing infectious hematopoietic necrosis (IHN) and viral hemorrhagic septicemia (VHS) are most commonly known as diseases during the freshwater stage of trout and salmon culture, but they are causing an increasing number of outbreaks in marine net pens. VHSV has now been isolated from a large number of different marine fish species and is responsible for large epizootics in populations of wild herring and sprat and in cultured turbot in northern Europe. Marine isolates of VHSV appear to be genetically different from the freshwater type. There is active research into the development of molecular diagnostics and vaccines for IHN and VHS.

Pancreas disease, caused by a toga-like virus, has caused severe economic losses in salmon culture in Scotland, Ireland and Norway and appears about 6–12 weeks after smolts are transferred to seawater. Although mortalities are low, infected fish become anorexic and emaciated, with internal hemorrhaging and atrophy of the pancreatic tissue.

Lymphocystis is a chronic infection caused by an iridovirus, which results in hypertrophy of the skin cells, causing large 'cauliflower-like' tumors on the body surface. The disease is highly contagious and it has blighted the culture of a wide range of species, especially sea bass and sea bream in Asian and Mediterranean regions. Lymphocystis is widespread throughout the world and signs of the disease can be found in wild fish, where there are occasional reports linking the disease to pollution. Although mortalities are low, the unsightly lesions render the fish unmarketable. A number of other iridoviruses cause anemia and hemorrhagic diseases in fish.

Several herpesviruses can infect fish, and herpesvirus type 2 can cause mortalities up to 30% in coho salmon. The disease is characterized by skin ulcers, fin erosion and papilloma-like tumors around the mouth and gills. A disease known as nervous necrosis, characterized by whirling and erratic behavior due to spongiform lesions in the brain of infected fish, appears to be caused by a virus.

14.5 Control of infectious disease of fish

14.5.1 Husbandry and health management in mariculture

As shown in *Figure14.1*, disease depends on interactions between the fish, the pathogen and the environment and the most important practical measures to prevent or limit diseases are those which reduce stress and maintain good hygiene and the overall health of the stock (*Table 14.2*). Whilst these factors are largely a matter of good husbandry and management practice, the microbiologist has an important role, especially in the development of fast and effective diagnostic procedures.

14.5.2 *Treatment – antimicrobial agents*

Most bacterial pathogens can be killed or inhibited by a wide range of antibiotics and synthetic antimicrobial agents (the intracellular pathogens being an exception). However, the number of effective treatments for fish disease is quite limited for a number of reasons. Firstly, antimicrobials must be proven to be active against the pathogen, but produce minimal side effects in the host. The best chemotherapeutic antimicrobials work by targeting a process present in bacteria that is absent or different in their eukaryotic host. For example, amoxicillin (a semi-synthetic penicillin) targets peptidoglycan synthesis, which is unique to bacteria, whilst oxytetracycline targets the 30S subunit of the prokaryotic ribosome. Secondly, the agent must reach the site of infection in adequate concentrations to kill the pathogen or, more usually, inhibit its growth sufficiently to allow the host's immune system to eliminate the pathogen. The rates of uptake, absorption, transport to the tissues and excretion vary greatly among different fish species. Furthermore, because fish are poikilothermic, these processes are very dependent on temperature. Ideally, therefore, the efficacy of a particular compound should be evaluated for each host–pathogen interaction, under various environmental conditions. A third factor is the need to evaluate the rate of elimination of the drug in order to ensure that there are no unacceptable residues in the flesh of fish intended for human consumption. Again, because fish are poikilothermic, the rate of excretion and degradation depends on temperature, so it is necessary to calculate a 'degree-day' withdrawal period between the last

Table 14.2 Control of diseases in mariculture

Disease control measures	Practices
Design and operation of culture systems	Separation of hatchery and growing-on facilities Good management practices and record keeping
Hygiene	Disinfection of nets Protective clothing and equipment Prompt removal of moribund and dead fish
Nutrition	Careful monitoring of optimal growth rates at all stages of the life cycle Immune stimulants as feed additives Probiotics
Minimizing stress	Avoid netting, grading, overcrowding Maintain good water quality Avoid feeding before handling Use anesthesia during handling Breed 'domesticated' lines of fish
Breaking the pathogen's life cycle	Disinfection of tanks and equipment Separate fish of different ages Fallowing sites for 6 months to 1 year
Eliminating vertical transmission	Test eggs and sperm for pathogen
Preventing geographic spread	Licensing system for egg and larval suppliers Notifiable disease legislation Movement restrictions from infected sites
Eradication	Slaughter policy for notifiable diseases Government compensation
Antimicrobial treatment	Bath, oral or injectable agents Sensitivity testing Limit use to prevent evolution of resistance
Vaccination	Immersion, oral and injectable vaccines Ensure strains used for vaccines are appropriate for local disease experience Well-designed tests for evaluation of efficacy in appropriate species Assess need for re-immunization (boosters)
Genetic improvement of stock	Select for disease resistance traits Transgenics – disease resistance genes

administration of the antimicrobial and the slaughter of the fish for the market. For example, the withdrawal period for oxytetracycline is 400 degree-days (e.g. 40 days at 10°C or 20 days at 20°C). With the growth of aquaculture, government agencies and large retailers in many countries now test farmed fish for antimicrobial residues, in the same way that they test meat or milk. Government regulatory authorities require a considerable amount of testing before licensing a drug for use and the high costs of testing deter pharmaceutical companies from introducing new agents. The range of treatments is therefore very limited. In the UK, only four agents (amoxycillin, trimethoprim/sulfadiazine, oxolinic acid and oxytetracycline), are fully licensed for use in fish intended for human consumption, although enforcement of the regulations is difficult and others are used. Regulatory control is similarly strict in Canada, the USA and Norway, although different agents are approved. By contrast, over 30 agents are licensed in Japan. In some parts of the world, there are no controls at all on antimicrobial usage.

Antimicrobial agents are most commonly administered in medicated feed. Unfortunately, reduction in feeding is often one of the first signs of disease, so infected fish may not receive the appropriate dose of antimicrobial agents from medicated feed. Antimicrobials may sometimes be given

Table 14.3 Examples of the biochemical basis of acquired bacterial resistance to antimicrobials used in aquaculture

Strategy	Example	Mechanism
Modification of the target binding site	Penicillins	Altered penicillin-binding membrane proteins
	Quinolones	Altered DNA gyrase
Enzymic degradation	Penicillins	β-lactamase production
Reduced uptake or accumulation	Tetracyclines	Altered membrane transport proteins active efflux
Metabolic bypass	Sulfonamides	Hyperproduction of substrate (*p*-aminobenzoic acid)

by immersion of infected fish in a bath containing the agent, especially for gill and skin infections. Problems arise with both of these routes of administration because of wastage and contamination of the environment. Injection is rarely used, except for broodstock and aquarium fish.

A major problem with the use of antimicrobial agents is the development of resistance. Bacteria possess three main strategies for resistance, as shown in *Table 14.3*. Individual bacterial isolates often possess more than one resistance mechanism, and individual antimicrobials may be affected by different resistance mechanisms in different bacteria. Bacteria possess *intrinsic* resistance to certain agents because of inherent structural or metabolic features of the bacterial species; this is almost always expressed by chromosomal genes. This type of resistance is relatively easy to deal with, but *acquired* resistance causes major problems in all branches of veterinary and human medicine. The use of almost every antimicrobial leads, sooner or later, to the selection of resistant strains from previously sensitive bacterial populations. This occurs via spontaneous mutations in chromosomal genes (which occur with a frequency of about 10^{-7}) or by the acquisition of plasmids or transposons. Resistance genes carried on conjugative plasmids (R-factors) may spread rapidly within a bacterial population and may transfer to other species. Plasmid-borne resistance to various antimicrobials is frequently encountered in all of the fish-pathogenic bacteria discussed above. Emergence of a resistant strain at a fish site renders particular antimicrobials useless, and the resistance can easily spread until it is the norm for that species. Antimicrobials do not *cause* the genetic and biochemical changes that make a bacterium resistant, but they select for strains carrying the genetic information that confers resistance. The more an antimicrobial is used, the greater the selection pressure for resistance to evolve. If a particular antibiotic is withdrawn from use, the incidence of resistant strains usually declines, because the resistant bacteria now have no advantage and the additional burden of extra genetic information makes them less competitive. (Note, however, that plasmids may confer resistance to several antibiotics.) Resistance causes considerable problems in aquaculture. For example, in Scottish salmon culture, 20–30% of cases of furunculosis have been due at times to *Aeromonas salmonicida* resistant to three or more antimicrobials. Besides the obvious economic losses caused by inefficiency in disease control, many have expressed concerns about the risks of antimicrobial usage in aquaculture to human health and environmental quality. Several studies have shown a build up of resistant strains in sediments underneath sea cages in sites with poor water exchange, due largely to the accumulation of uneaten food. Transfer of resistance genes to marine bacteria is known to occur, and antimicrobial resistant bacteria have been isolated from fish that have escaped from facilities where these agents are used excessively. Experimental studies have shown transfer to human pathogens and commensals such as *Salmonella* and *E. coli* and this raises concerns about risks of transfer of resistance genes into the gut flora of consumers or fish farm workers. The emergence of zoonoses, such as the invasive infections caused by *S. iniae* and mycobacteria, is also a worrying development as infection with resistant strains would be very serious. These concerns about aquaculture are part of a general awareness of the folly of indiscriminate use of antimicrobial agents in medicine, agriculture and everyday products. In many parts of the world, antimicrobial usage is unregulated, and large quantities are used as prophylactic treatments with little regard for testing for sensitivity or proper withdrawal periods. On the other hand, some authorities have concluded that the risks to public health from antimicrobial use in aquaculture are low. Nevertheless, for all of the reasons mentioned above, disease prevention is better than cure and the continued success of mariculture is only possible with the advent of effective vaccines. Indeed, where vaccines have been successfully introduced, antimicrobial usage has dropped sharply. For example, the use of antimicrobials in Norway fell by over 50% in the few years following the introduction of a reliable furunculosis vaccine.

14.5.3 Vaccines, immunostimulants and probiotics

Teleost fish possess an efficient immune response and respond to the administration of microbial antigens by the production of antibodies (B-cell response) and cell-mediated immunity (T-cell response). The most common method of administering vaccines to small fish (up to about 15 g) is via brief immersion in a dilute suspension. Particulate antigens (such as bacteria) probably stimulate immunity after passage across the gills. Intraperitoneal injection is necessary for reliable protection with some vaccines, especially viral vaccines and bacterial vaccines that contain soluble components (e.g. most furunculosis vaccines). Injection vaccines are usually administered with an oil adjuvant, which ensures slow release of the antigen and a heightened immune response. Despite their efficacy, injection vaccines have a number of drawbacks. They cannot be used on fish less than about 15 g and the stress associated with crowding, removal from the water and injection causes mortalities and may even precipitate infection. Despite devices to convey fish from the water onto an injection table and the use of repeater syringes, injection vaccines incur high labor costs. The most desirable form of vaccine is one that can be administered orally. The main difficulty with this approach is that microbial antigens are degraded in the fish's stomach and foregut before reaching the gut-associated lymphoid tissue in the hindgut, where the immune response occurs. This is overcome by microencapsulation of the vaccine in biodegradable polymers such as poly DL-lactide-co-glycolide. Many commercial vaccines using this, or similar, approaches are now becoming available.

In addition, various complex chemical substances induce nonspecific immunostimulation in fish. Indeed, the protection effects observed with many vaccine preparations can be partly due to nonspecific effects rather than specific responses to particular antigens. Complex polysaccharides such as glucans, mannan oligosaccharides and peptidoglycan obtained from yeast and bifidobacteria are particularly effective. Mannan oligosaccharides prevent initiation of infection by interfering with the ability of pathogenic bacteria to attach to host cells. Other compounds attach to specific receptors on the cell surface of phagocytes and lymphocytes, resulting in increased production of enzymes, interferon, interleukins and complement proteins, leading to increased activity of T- and B-lymphocytes. In experimental challenges, salmon fed on diets containing these compounds develop high levels of resistance to vibriosis, furunculosis, BKD and some viral infections.

The use of probiotics in finfish aquaculture has attracted considerable interest in recent years, largely as a result of successes with their use in poultry farming and shellfish culture (*Section 15.3.6*). Human probiotics have also gained in popularity recently. A probiotic is generally defined as a live microbial feed supplement that stimulates health, inducing a beneficial change in the gut flora by competition with harmful bacteria. A wide range of microalgae, yeasts, Gram-positive and Gram-negative bacteria has been investigated for their health-promoting potential. Whether all of these are truly acting as probiotics is a matter of debate. Strictly speaking, probiotic agents are those that colonize the gut transiently or permanently. Many microbial treatments are probably better defined as biocontrol agents, because they are antagonistic to pathogens, or as bioremediation agents if they improve water quality by breakdown of waste products. Some success has been reported in reducing disease with various microbial treatments in experimental infections and in the field. Further research is needed to determine the action of probiotics. They may work by competitive exclusion of pathogens, by production of antibiotics or competition for nutrients, or they may act as nonspecific immune stimulants. Probiotics may also improve general health by producing vitamins, detoxifying compounds or by digesting complex compounds.

Fish immunology and vaccinology have become specialized areas that are beyond the scope of this book. Instead, the remainder of this section will focus on the microbiological aspects of producing vaccines against bacterial and viral infections of fish. The simplest type of bacterial vaccine is a bacterin, which consists of a dense culture of bacterial cells killed by formalin treatment. Although technically simple, careful attention must be given to the quality control of media composition and incubation conditions, in order to ensure that the bacterin contains the appropriate protective antigens. One of the earliest successes in vaccine development was the vibriosis vaccine, which is effective against *V. anguillarum* infection in salmon. The protective antigen in this case appears to be LPS. Commercial vaccines usually incorporate two or more serotypes to allow for antigenic variation and they usually work well in a range of situations. Development of effective bacterins for *V. salmonicida* and *V. ordalii* also proved relatively straightforward, but the early success with the vibrios was not repeated with other diseases. For example, the breakthrough in development of an effective, long-lasting furunculous vaccine was only achieved after recognition of the crucial role of extracellular proteases and IROMPs, and manipulation of the culture and formulation conditions to ensure the correct blend of particulate and soluble antigens. Many bacterial pathogens, notably *R. salmoninarum*, are slow growing and difficult to culture, whilst viruses and rickettsias can only be propagated in cell culture. The cost of production of inactivated vaccines by these methods is very high. Therefore, attention has turned to the use of recombinant DNA technology. After identifying genes important in virulence (e.g. bacterial toxins or surface proteins or virus capsid proteins), they can be cloned and expressed in a recombinant host to produce a subunit vaccine. The most common method of achieving this is via the production of a fusion or hybrid protein in an expression system, as outlined in *Figure 14.6*. This approach has been used to generate vaccines against several bacterial diseases, including furunculosis (based on the serine protease), piscirickettsiosis (ospA membrane protein) and BKD (hemolysin and metalloprotease). Subunit vaccines for the viral diseases ISA and IPN are based on capsid proteins. The development costs of recombinant subunit vaccines are high, but production costs should be little more than conventional bacterins if the subunit can be expressed in *E. coli* or yeast.

Live attenuated vaccines, in which the pathogen is rendered avirulent, are potentially more attractive than killed vaccines, because the bacterium or virus replicates within the host and delivers antigens over a prolonged period. Live vaccines are also better at stimulating mucosal immunity and cell-mediated immunity, and they are more suitable for oral delivery. Many human viral vaccines are based on this principle, although live bacterial vaccines have been less favored because their more complex genomes lead to the possibility of incomplete attenuation or subsequent reversion to virulence by recombination. Recombinant DNA technology allows 'rational attenuation', by which the deletion and replacement of specific genes necessary for virulence and survival *in vivo* ensures a more controlled and targeted approach to attenuation. Such an approach has been used to construct a live vaccine of *A. salmonicida* by deletion of the *aroA* gene, which encodes an essential amino acid biosynthesis pathway, not present in animals. Allelic replacement of this gene and subsequent further attenuation guards against the possibility of reversion. In trials, this vaccine was highly effective and the vaccine strain could be engineered to deliver other antigens. A similar approach has been used recently for *P. damselae* subsp. *piscicida* by removing a siderophore gene. Live vaccines have also been developed for viral hemorrhagic septicemia (VHS), which normally occurs in salmon and trout in the freshwater stage, but is also known in marine turbot. Even with good evidence of protection and no evidence of reversion, live vaccines have not found favor with licensing authorities for use in fish, largely prompted by concerns about deliberate release into the environment.

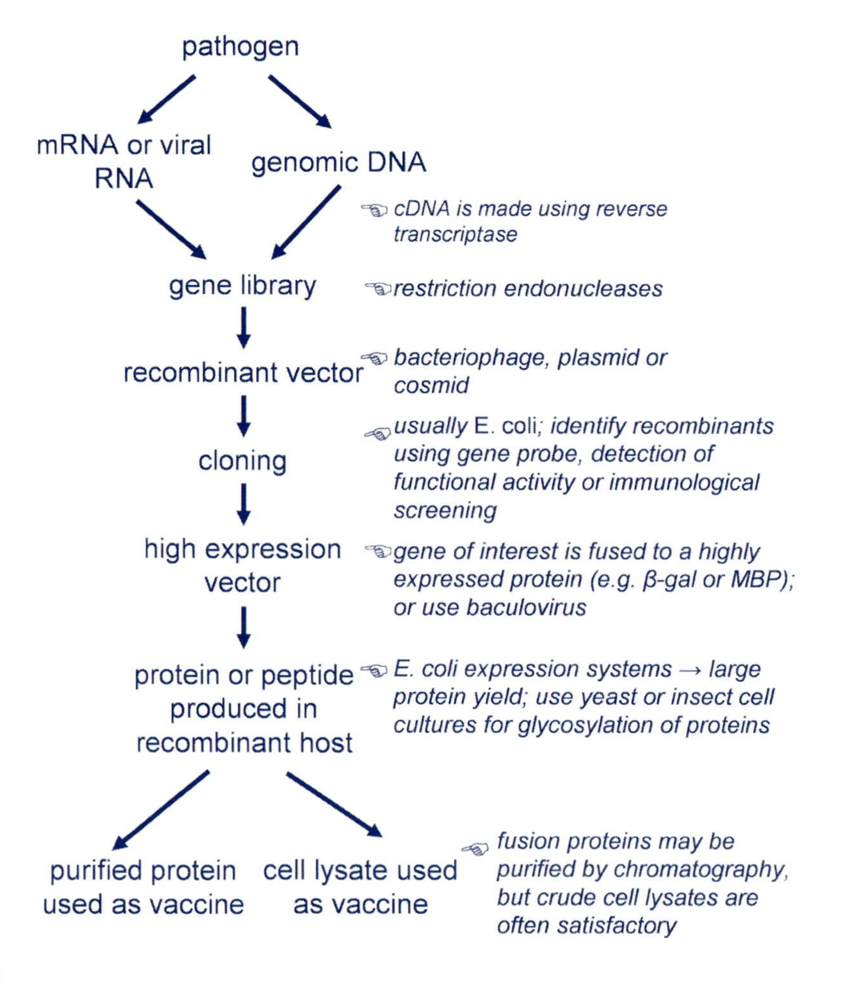

Figure 14.6

Principal steps in the cloning of genes for production of fusion protein vaccines. β-gal = β-galactosidase, MBP = maltose-binding protein. Redrawn from Munn (1994).

Unlike conventional vaccination, which depends on the administration of an antigen, DNA vaccination (also known as genetic immunization) is based on the delivery of naked DNA containing the sequence for the protective epitope of the antigen. Usually, DNA is delivered into the skin or muscle using a 'gene gun', which fires tiny gold particles coated with the DNA into the tissue. The immunogenic DNA has to be incorporated into a plasmid with appropriate promoters; if this is done correctly, the gene will be expressed in the fish host cells. Thus, the fish makes foreign antigens internally and then mounts an immune response to them. An alternative strategy is based on the expression of an antibody fragment inside the cell that can bind to and inactivate the pathogen. Fish cells efficiently express foreign proteins encoded by eukaryotic expression vectors. Although refinement of the vector plasmids and delivery system are required, there is promising evidence of DNA vaccines inducing protective immunity in the diseases VHS and infectious hematopoietic necrosis (IHN). Issues of longevity of protection, stimulation of the correct immune responses and safety need investigation. Unfortunately for the proponents of this technology, public opinion (in Europe at least) is not in favor of such genetic engineering in the food chain. Attempts by the industry to introduce transgenic fish that incorporate genes for faster growth have so far not been acceptable. DNA vaccination is *not*

Box 14.1 RESEARCH FOCUS

A toxic summer of science

Controversy surrounds the lethal effects of Pfiesteria *for fish*

The dinoflagellate *Pfiesteria piscida* was implicated in the 1990s as the causative agent of massive fish kills in estuary systems in the eastern USA (*Section 12.4*). The leading investigators are JoAnn Burkholder and Howard Glasgow, of North Carolina State University's Center for Applied Aquatic Ecology. Considerable controversy has surrounded the pathogen's effects on both fish and humans ever since, as dramatized in a popular book (Barker 1997). The original descriptions of *P. piscicida* referred to it as an ambush predator with a unique 24-stage life cycle (*Figure 12.4*). Burkholder's group propose that a toxin produced by the zoospore stage is responsible for the serious effects in both fish and humans (*Box 12.1*). However, several papers published in the summer of 2002 have thrown considerable doubt on these claims and the issue has provoked heated debate at conferences (Kaiser 2002, Trainer 2002), on the Internet (CAAE website) and in scientific journals. Litaker *et al.* (2002a) used a variety of techniques to investigate the *P. piscicida* lifecycle and were unable to find evidence of many of the reported stages. They concluded that *P. piscicida* has a lifecycle typical of free-living marine dinoflagellates and that the amoeboid stages seen in tanks in which *P. piscicida* is actively feeding on fish are contaminants introduced with the fish. Using *in situ* hybridization techniques (with a recently developed fluorescently labeled peptide nucleic acid probe), Litaker and colleagues showed that the amoeboid stages do not hybridize with a *P. piscicida* probe and that clonal cultures do not contain any unusual stages. Burkholder and Glasgow (2002) criticized this study, reaffirming that they have evidence of transformation to amoebae and cysts, which Litaker *et al.* did not refer to, and claiming that they were working with strains of *P. piscicida* that are noninducible and do not produce toxins. Litaker *et al.* (2002b) presented a vigorous rebuttal of these criticisms. In another study, Berry *et al.* (2002) reported that *P. shumwayae* does not produce toxins. This species is closely related to *P. piscicida* and has also been shown to produce toxins by Burkholder's group. Berry's strain kills fish as rapidly as the allegedly toxigenic strain described by Burkholder, but there was no evidence of toxin production in the work by Berry *et al.* They showed that the lethal effects of cultures could be removed by simple centrifugation and that organic extracts of cultures were nontoxic. Furthermore, attempts to PCR-amplify genes known to be involved in synthesis of polyketides (to which all presently known dinoflagellate fish toxins belong) were unsuccessful. To counter the suggestion that the strain they used was not representative, Berry *et al.* say that they found similar evidence of nontoxigenicity in several isolates of *P. shumwayae, P. piscicida* and other *Pfiesteria*-like organisms. They imply that some of the apparently toxic effects, seen during fish-kills, could be due to a combined effect of other algae, fungal infection and predation by *Pfiesteria.* Vogelbein *et al.* (2002) also found that *P. shumwayae* do not produce toxins and concluded that fish mortality results only when the dinoflagellate is in close contact with the fish, due to micropredatory feeding via a peduncle that sucks fluids from the fish tissue (see *Figure 7.4*). Quesenbery *et al.* (2002) provide a detailed description of an improved assay technique for fish toxicity, coupled with PCR detection, which should be useful in environmental monitoring, but this study does not shed much light on the life cycle issue. Anderson *et al.* (2003) have attempted to clarify the status of the noninducible forms by examining various environmental factors, such as light intensity and prey density. They argue that, because *P. piscicida* is mixotrophic, these conditions will affect phototrophic or phagotrophic nutrition and will have a large effect on the life cycle. As Burkholder and Glasgow (2002) point out, 'absence of evidence does not constitute proof of absence'. It is hoped that future collaboration and exchange of cultures and data will help to resolve the disagreements about the life cycle of *Pfiesteria* and the role of a toxin. This is essential if we are to understand fully the implications of this 'scary' pathogen for both fish and human health.

transgenic, because the DNA is not introduced into the germline, but this is a subtle distinction for those worried about this new technology. It is likely that this situation may change within a few years and DNA vaccines hold considerable potential in aquaculture, especially for control of diseases caused by viruses, eukaryotic parasites and hard-to-culture bacteria like *R. salmoninarum* and *P. salmonis*.

14.6 Protistan infections and HABs

A large number of protozoa can infect wild, farmed and aquarium fish. Often these are free-living or benign parasites and the environmental and nutritional conditions that promote disease are largely unknown. For example, *Paramoeba pemaquidensis* causes sporadic, severe outbreaks in salmon culture in Tasmania. Diplomonad flagellates are commonly found in the gut of fish, where they appear to have little effect, but under some circumstances they can cause systemic infection with high mortalities. Myxosporeans such as *Kudoa* sp. commonly cause muscle infections, causing cysts and softening of the tissue that impairs marketability of the fish. *Loma salmonae* is an obligate intracellular microsporidean parasite infecting the gills of many economically important fish, including both wild and cultured salmon and cod. Protozoan infections are usually controlled by external dip treatments such as formalin, malachite green and chloramines T, but there is concern about the use of these agents on fish intended for human consumption, because of fears of possible toxicity and carcinogenicity.

Excessive growth of phytoplankton in coastal waters can be responsible for mortalities in fish. Some of these blooms are caused by toxin-producing species and the nature of these HABs was introduced in *Chapters 12* and *13* because of their impact on health of humans and mammals, respectively. The most important example of a toxic effect on fish is *Pfiesteria piscida* and related species. *Pfiesteria* was identified as a new organism in the 1990s because of its association with mass fish kills in the Pamlico-Albemarle estuary on the east coast of the USA, and aspects of its life cycle are discussed in *Section 12.4* and *Box 12.1*. The involvement of toxins in fish mortality has provoked considerable controversy and the current status of this debate is presented in *Box 14.1*. Other toxic genera, such as *Alexandrium, Gymnodinium* and *Pseudo-nitzschia* have all been responsible for mass mortality, both in wild and farmed fish. The incidence of fish mortalities from HABs has increased considerably in the last decade and is the cause of major economic losses. There has been a dramatic increase in use of coastal waters for fish and shellfish farming and this has contributed to stimulation of blooms by input of excess nutrients from the aquaculture operations themselves. Caged fish cannot escape the effects.

Nontoxic microalgae can also affect fish. Large blooms reduce light penetration and decrease the growth of seagrass beds, which are often important nursery grounds for the food and young stages of commercially important fish. Clumping, sinking and decay of phytoplankton can generate anoxic conditions. Nontoxic algae can also kill fish directly. For example, the diatom *Chaetoceros convolutus* produces long barbs, which clog fish gill tissue causing excess mucus production, leading to death from reduction in oxygen exchange. Incidents involving the loss of over 250 000 farmed salmon at a time have occurred. Blooms of the flagellate *Heterosigma carterae* can cause mortalities in Pacific coast salmon farms costing the industry US$ several million per year.

References and further reading

Alderman, D.J., and Hastings, T.S. (1998) Antibiotic use in aquaculture: development of antibiotic resistance – potential for consumer health risks. *Int J Food Sci Technol* **33**: 139–155.

Alexander, S.M., Grayson, T.H., Chambers, E.M., Cooper, L.F., Barker, G.A., and Gilpin, M.L. (2001) Variation in the spacer regions separating tRNA genes in *Renibacterium salmoninarum* distinguishes recent clinical isolates from the same location. *J Clin Microbiol* **39**: 119–128.

Austin, B. (1999) The effects of pollution on fish health. *J Appl Microbiol* **85**: 234S–242S.

Austin, B., and Austin, D.A. (1999) *Bacterial Fish Pathogens, Diseases of Farmed and Wild Fish,* 3rd (revised) edn. Godalming, Springer-Praxis.

Barker, R. (1997) *And the Rivers Turned to Blood.* Touchstone Books, New York.

Berry, J.P., Reece, K.S., Rein, K.S., *et al.* (2002) Are Pfiesteria species toxicogenic? Evidence against production of ichthyotoxins by *Pfiesteria shumwayae. Proc Natl Acad Sci USA* **99**: 10970–10975.

Bromage, E.S., Thomas, A., and Owens, L. (1999) *Streptococcus iniae,* a bacterial infection in barramundi *Lates calcarifer. Dis Aquat Org* **36**: 177–181.

Brownie, C., Glasgow, H.B., Burkholder, J.M., Reed, R., and Tang, Y.Q. (2003) Re-evaluation of the relationship between *Pfiesteria* and estuarine fish kills. *Ecosystems* **6**: 1–10.

Burkholder, J.M., and Glasgow, H.B. (2002) The life cycle and toxicity of *Pfiesteria piscicida* revisited. *J Phycol* **38**: 1261–1267.

CAAE Website. North Carolina State University Center for Applied Aquatic Ecology. http://www.pfiesteria.org (accessed April 12 2003).

Cipriano, R.C., and Bullock, G.L. (2001) *Furunculosis and Other Diseases Caused by* Aeromonas salmonicida. National Fish Health Research Laboratory, Leetown, WV. http://www.lsc.usgs.gov/fhb/leaflets/FHB66.pdf (accessed April 12 2003).

Coats, D.W. (2002) Dinoflagellate life-cycle complexities. *J Phycol* **38**: 417–419.

Colquhoun, D.J., and Sorum, H. (2001) Temperature dependent siderophore production in *Vibrio salmonicida. Microb Pathogen* **31**: 213–219.

Colquhoun, D.J., and Sorum, H. (2002) Cloning, characterisation and phylogenetic analysis of the fur gene in *Vibrio salmonicida* and *Vibrio logei. Gene* **296**: 213–220.

Colquhoun, D.J., Alvheim, K., Dommarsnes, K., Syvertsen, C., and Sorum, H. (2002) Relevance of incubation temperature for *Vibrio salmonicida* vaccine production. *J Appl Microbiol* **92**: 1087–1096.

Ellis, A.E. (1999) Immunity to bacteria in fish. *Fish & Shellfish Immunol* **9**: 291–308.

Ellis, A.E. (2002) Mariculture diseases and health. In: **Steele, J.H., Turekian, K.K., Thorpe, S.A.** (ed.) *Enyclopedia of Ocean Sciences,* pp. 1555–1559. Academic Press, New York.

Evenden, A.J., Grayson, T.H., Gilpin, M.L., and Munn, C.B. (1993) *Renibacterium salmoninarum* and bacterial kidney disease – an unfinished jigsaw. *Ann Rev Fish Dis* **3**: 87–104.

Fryer, J.L., and Mauel, M.J. (1997) The Rickettsia: an emerging group of pathogens in fish. *Emerg Inf Dis* 3: 137–144. http://www.cdc.gov/ncidod/EID/vol3no2/ fryer.htm (accessed April 12 2003).

Gatesoup, F.J. (1999) The use of probiotics in aquaculture. *Aquaculture* **180**: 147–165.

Grayson, T.H., Gilpin, M.L., Evenden, A.J., and Munn, C.B. (2001) Evidence for the immune recognition of two haemolysins of *Renibacterium salmoninarum* by fish displaying clinical symptoms of bacterial kidney disease (BKD). *Fish Shellfish Immunol* **11**: 367–370.

Heppell, J., and Davis, H.L. (2000) Application of DNA vaccine technology to aquaculture. *Adv Drug Deliv Rev* **43**: 29–43.

Irianto, A., and Austin, B. (2002) Use of probiotics to control furunculosis in rainbow trout, *Oncorhynchus mykiss* (Walbaum). *J Fish Dis* **25**: 333–342.

Kaiser, J. (2002) Microbiology: The science of *Pfiesteria*: Elusive, subtle, and toxic. *Science* **298**: 346–349.

Kent, M.L., and Poppe, T.T. (2002) Infectious diseases of coldwater fish in marine and brackish water. In: **Woo, P.T.K., Bruno, D.W., and Lim, L.H.S.** (eds) *Diseases and Disorders of Finfish in Cage Culture,* pp. 65–105. CAB International, Oxford. http://www.cabipublishing.org/Bookshop/ReadingRoom/0851994431/ 0851994431Ch3.pdf (accessed May 1 2003).

Kuzyk, M.A., Burian, J., Machander, D., Dolhaine, D., Cameron, S., Thornton, J.C., and Kay, W.W. (2001) An efficacious recombinant subunit vaccine against the salmonid rickettsial pathogen *Piscirickettsia salmonis. Vaccine* **19**: 2337–2344.

Litaker, R.W., Vandersea, M.W., Kibler, S.R., Madden, V.J., Noga, E.J., and Tester, P.A. (2002a) Life cycle of the heterotrophic dinoflagellate *Pfiesteria piscicida* (Dinophyceae). *J Phycol* **38**: 442–463.

Litaker, R.W., Vandersea, M.W., Kibler, S.R., Noga, E.J., and Tester, P.A. (2002b) Reply to comment on the life cycle and toxicity of *Pfiesteria piscicida* revisited. *J Phycol* **38**: 1268–1272.

Masada, C.L., LaPatra, S.E., Morton, A.W., and Strom, M.S. (2002) An *Aeromonas salmonicida* type IV pilin is required for virulence in rainbow trout *Oncorhynchus mykiss. Dis Aquat Org* **51**: 13–25.

Milton, D.L., Chalker, V.J., Kirke, D., Hardman, A., Camara, M., and Williams, P. (2001) The LuxM homologue VanM from *Vibrio anguillarum* directs the synthesis of N-(3-hydroxyhexanoyl) homoserine lactone and N-hexanoylhomoserine lactone. *J Bacteriol* **183**: 3347–3357.

Munn, C.B. (1994) The use of recombinant DNA technology in the development of fish vaccines. *Fish Shellfish Immunol* **4**: 459–473.

Murray, A.G., Smith, R.J., and Stagg, R.M. (2002) Shipping and the spread of infectious salmon anaemia in Scottish aquaculture. *Emerg Inf Dis* **8**: 1–5.
Quesenberry, M.S., Saito, K., Krupatkina, D.N., *et al.* (2002) Bioassay for ichthyocidal activity of *Pfiesteria piscicida*: Characterization of a culture flask assay format. *J Appl Phycol* **14**: 241–254.
Seng, L.T., and Colorni, A. (2002) Infectious diseases of warmwater fish in marine and brackish waters. In: **Woo, P.T.K., Bruno, D.W., and Lim L.H.S.** (eds) *Diseases and Disorders of Finfish in Cage Culture,* pp. 193–230. CAB International, Oxford. http://www.cabipublishing.org/Bookshop/ReadingRoom/ 0851994431/0851994431Ch7.pdf (accessed May 1 2003).
Stork, M., Di Lorenzo, M., Welch, T.J., Crosa, L.M., and Crosa, J.H. (2002) Plasmid-mediated iron uptake and virulence in *Vibrio anguillarum*. *Plasmid* **48**: 222–228.
Trainer, V.L. (2002) Marine biology – Unveiling an ocean phantom. *Nature* **418**: 925–926.
Vogelbein, W.K., Lovko, V.J., Shields, J.D., Reece, K.S., Mason, P.L., Haas, L.W., and Walker, C.C. (2002) *Pfiesteria shumwayae* kills fish by micropredation not exotoxin secretion. *Nature* **418**: 967–970.
Weiner, D.B., and Kennedy, R.C. (1999) Genetic vaccines. *Sci Am* **281**: 34–41.
Winton, J.R. (1998) Molecular approaches to fish vaccines. *J Ichthyol* **14**: 153–158.

Diseases of invertebrates 15

15.1 Introduction

The marine invertebrates are the most diverse group of animal life on Earth, with hundreds of thousands of species. Although microbial diseases have been described in only a tiny fraction of these species, a huge bank of knowledge has been collected by zoologists, especially for the many protozoan parasites that infect invertebrates. Rather less is known about the impact of bacterial and viral diseases in the ecology of marine invertebrates, except in a few situations where our attention has been drawn to animals of particular interest to humans. Thus, over many years we have built up knowledge of microbial diseases of invertebrate animals (mainly bivalve molluscan and crustacean shellfish) used for food, either harvested from the sea or cultured in traditional extensive systems. Here, a number of microbial diseases that affect fishery production have been described. In the last 20 or so years, the intensification of shellfish culture has increased rapidly, accompanied by a dramatic rise in disease incidence. Another area that has expanded greatly is the recent attention given to diseases of corals. Coral reefs are being seriously degraded in all parts of the world. Scientists are rightly concerned about the loss of diversity and the likely ecological impact of their demise, but concerns about adverse effects on tourism are a major economic factor also. This chapter therefore discusses some of the main bacterial and viral diseases of molluscs, crustaceans and corals. The reader should be aware that protozoa have not been included in this treatment, despite their abundance as parasites of invertebrates.

15.2 Bacterial and viral diseases of bivalve molluscs

15.2.1 Viruses

The economically most significant viral disease of bivalve molluscs is gill necrosis in oysters. A major epizootic erupted in France in the 1960s in Portuguese oysters (*Crassostrea angulata*), which demonstrated gill necrosis accompanied by the presence of large inclusion bodies containing icosahedral iridovirus particles in the tissues. This disease spread rapidly along the Atlantic coast of Europe, virtually destroying the European oyster fishery. Pacific oysters (*C. gigas*) were introduced to replace the lost stock and these too are infected by the agent, although mortality is much lower. Another group of viruses, known as birnaviruses. are widely distributed as disease agents in marine invertebrates and fish, including IPNV (*Section 14.4.2*). The host specificity of these viruses is largely unknown. Many other types of viruses have been isolated from marine bivalves and are often associated with disease signs, although proof of etiology and study of pathogenesis is usually lacking, except in cultured species.

15.2.2 Bacteria

The major cause of disease in bivalve hatcheries is infection by various species of Gram-negative bacteria, notably *Vibrio* spp. Full identification of the pathogens implicated in disease outbreaks is often incomplete and it is likely that many organisms have been incorrectly identified, as interpretation of biochemical tests and use of diagnostic keys for the vibrios are not always

Box 15.1 RESEARCH FOCUS

Switching on virulence

Phage conversion may explain the pathogenicity of Vibrio harveyi

No satisfactory genetic or protein profiling method exists for the differentiation of virulent and avirulent strains of *Vibrio harveyi.* Pizzutto and Hirst (1995) suggested that virulence is associated with a genetically mobile element, but plasmids are not responsible. Leigh Owens and colleagues at James Cook University, Queensland, have recently provided a fascinating insight into the virulence of this pathogen. Harris and Owens (1999) isolated two lethal exotoxins from virulent strains of *V. harveyi* and suspected that they might be encoded by a bacteriophage. The phenomenon of lysogenic conversion occurs when a temperate phage is able to exist within the host cell as a stable plasmid, or integrate into the host chromosome, where it resides as a prophage. Bacteria containing a prophage are said to be lysogenic, because the lytic cycle of infection can be induced by various treatments (see *Figure 8.4*). An important property of lysogenic strains is that they are resistant to lysis by free phages of the same type as the prophage. Lysogenic conversion by phages can result in the horizontal transfer of large segments of DNA, which are often recognized in genome sequences as 'pathogenicity islands' characterized by a GC ratio different to other parts of the bacterial chromosome. The production of diphtheria toxin by *Corynebacterium diphtheriae* was the first example of virulence factor encoded by phage genes and other cases have now been discovered in human pathogens (Boyd *et al.* 2001). In *Vibrio cholerae,* Waldor and Mekalanoa (1996) and Karaolis *et al.* (1999) showed, from sequence analysis, that two major virulence factors (pili and cholera toxin) are encoded by phages (see *Figure 11.2*). In *V. harveyi,* Oakey and Owens (2000) used mitomycin C treatment to induce lytic cycles in a number of strains. A phage was isolated from a stable toxin-producing strain and identified using electron microscopy as a myovirus-like icosahedral double-stranded DNA virus, which they called VHML. Oakey *et al.* subsequently determined the complete genome sequence of the phage and found a large number of genes with no known function (Oakey *et al.* 2002). Does this phage convert nonvirulent *V. harveyi* to a virulent form? To answer this question, Munro *et al.* (2002) first checked that the nonvirulent strain did not contain prophage by treatment with mitomycin C. They then infected the nonvirulent strain with VHML and showed a rise in virulence, associated with the production of several new extracellular proteins. At present, it is not clear what virulence properties are conferred by the phage genome. One suggestion is that an ADP-ribosylating toxin is present. Another idea is that the phage has a gene for a DNA adenine methyltransferase, an enzyme which is known to alter the expression of bacterial genes. Thus, the novel possibility is that infection by a phage can change a bacterium from nonvirulent to virulent, either by the transfer of a gene for a new virulence factor, or by modifying an existing bacterial gene.

straightforward. (This may be due in part to the phenomenon of bacteriophage conversion, see *Box 15.1*.) Three species have been particularly implicated, namely *V. alginolyticus, V. tubiashi* and *V. anguillarum.* These organisms are all found as members of the normal microbiota of seawater and in association with marine surfaces. Growth is encouraged by accumulation of organic matter, and careful monitoring of water quality and temperature is essential in hatcheries. Larvae of different bivalve species seem quite variable in their sensitivity to vibrios and there are also marked differences in the virulence of bacterial isolates. Extracellular toxins (hemolysin, protease and a ciliostatic factor) are responsible for the larval necrosis. Vibrio infections do not usually occur in adult bivalves, but an exception is 'Vibrio P1', which causes brown-ring disease in cultured Manila clams (*Tapes philippinarum*). The pathogen attaches to the clam tissue, causing abnormal thickening and a characteristic brown ring along the edge of the shell. Mass mortalities have caused severe economic losses in France since the 1980s. *V. harveyi* has been identified as the cause of high mortalities in cultured pearl oysters (*Pinstada maxima*) in north-western Australia. A wide range of other bacteria can occasionally cause infections in bivalves, especially in the larval and juvenile stages. Gliding bacteria of the CFB group can infect the hinge-ligament of the shell, leading to liquefaction via the production of extracellular enzymes and interference with respiration and feeding. As occurs in fish infections, CFB appear to be weak opportunist pathogens and infection is precipitated by poor nutrition of the animals, rising water temperatures, or other environmental stresses.

Juvenile oyster disease results in seasonal mortalities of hatchery-produced juvenile *Crassostrea virginica* raised in Maine and Massachusetts, which first appeared in the late 1980s. In some years, mortalities as high as 90% of total production have occurred. There are several similarities to brown ring disease, suggesting a bacterial etiology. Growth rate is reduced and the shell becomes fragile and uneven, with proteinaceous deposits on the inner shell surfaces, followed by sudden heavy mortality. Recently, it has been found that infected animals are heavily colonized by a previously undescribed species of the *Roseobacter* group of the *α-Proteobacteria,* recently named as *Roseimarina crassostreae.* A key factor in pathogenicity appears to be the attachment of the bacteria to the oyster tissue via a tuft of type IV pili (*Section 3.10*). The development of molecular diagnostic systems will help in elucidating the ecology of this pathogen and guide the timely use of antibiotics to control the disease.

Intracellular rickettsia-like and chlamydia-like pathogens are widespread and have been reported in at least 25 species of bivalves. These infections frequently produce little evidence of tissue damage, and mortality in adult animals is usually low, except in conditions of environmental stress such as sudden temperature change. The larvae are usually very susceptible, and disease can cause problems in aquaculture hatcheries (e.g. of scallops, *Arcopecten irradians*). The morphology of the bacteria within the cells of the digestive gland and gills is very similar to that of known rickettsias and chlamydia, but detailed identification and taxonomy studies are limited, because these are obligate intracellular pathogens that can only be grown in suitable cell cultures.

15.3 Bacterial and viral diseases of crustaceans

15.3.1 Diseases in aquaculture

A number of crustacean diseases are important in both wild and cultured populations of crustaceans such as lobsters, crabs, prawns and shrimp. Intensive aquaculture of prawn and shrimp now accounts for about half of the total aquaculture production of 2.4 million tonnes. The global demand for prawns seems insatiable and much of the development has taken place in Asia (especially Thailand, Vietnam and Indonesia) and Central America (especially Ecuador). The rapid expansion of intensive culture has led to severe problems of habitat destruction and eutrophication of coastal waters. With a better understanding of these problems, infectious diseases are now considered the most important limiting factor for further development of the industry. For example, in tropical Australia, recent attempts to expand the culture of prawns and other valuable species such as the rock lobster are seriously affected by bacterial and viral disease. For most crustacean species, culture still depends on the use of seed produced in hatcheries, mainly from sexually mature females caught in the wild. Thus, genetic selection of disease-resistant animals is not possible, although it is likely that closure of the life cycle for some species will be achieved soon. Disease transmission in these animals is encouraged by their cannibalistic feeding habits, which ensure rapid spread through ponds and holding tanks. Knowledge of the ecology of pathogens and diagnostic methods has been poorly developed until recently. The development of gene probes and real-time PCR techniques will assist in control methods such as selection of seed and stocks and restrictions on movement unless certified as disease-free. Viruses also remain infectious in frozen seafood products, and these methods are used for screening imports and exports to prevent disease transmission. However, many of these techniques are expensive and require highly trained personnel and it will be necessary to develop cheaper and easier methods for use by rural farmers in developing countries.

15.3.2 Viruses

The number of viruses recognized as important in culture of prawns and shrimp (marine and freshwater) has risen from six in 1988 to nearly 20 today, as intensive culture has expanded in both volume and geographic distribution. One of the most devastating diseases in Asia has been white spot syndrome of penaeid prawns. This was first recognized in eastern Asia and has spread worldwide, with current losses estimated at over US$1 billion. A number of closely related viruses are responsible; these are large, enveloped, rod-shaped double-stranded DNA viruses. The genome of one member of this group has been sequenced and proteomic analysis is being used to elucidate the structural and functional relationships of the various proteins in the virus particle and their role in virus replication and virulence. Animals acquire the infection from the water, and show white spots on the inner surface of the shell. Interestingly, the white spots appear to be caused by a chitinase and this enzyme may have been transferred to the viral genome by bacteriophage conversion. High mortalities (up to 80%) result within 2–3 days for juveniles and 7–10 days for adults. Other important viruses include infectious hypodermal and hematopoietic necrosis virus (IHHNV), hepatopancreatic parvovirus (HPV), baculoviral midgut gland necrosis virus (BMNV), baculovirus penaei (BP) and yellow head virus (YHV). The parvovirus spawner mortality virus (SMV) has emerged in the 1990s in Queensland, Australia and has been responsible for mortalities of 25–50% in black tiger prawns (*Penaeus monodon*). SMV originates in wild broodstock, as 25% of female spawners carry the virus. Integrated PCR and ELISA tests are being developed for high-throughput screening in an attempt to control the disease. Sequence analysis shows that there is considerable homology between SMV and some insect viruses. Monodon baculovirus (MBV) is also a problem in *P. monodon* and *P. plebejus*.

15.3.3 Rickettsias and mycoplasmas

The intracellular rickettsias and mycoplasmas can cause epizootics with high mortalities in crustaceans such as crabs, lobsters and penaeid prawns, due usually to infection of the hepatopancreas. Such epizootics can have a marked effect on marine ecology, for example the outbreaks caused by currently unidentified rickettsias affecting crabs (*Cancer* and *Carcinus* spp.) in Europe and the USA. They are particularly important in prawn culture. Diagnosis of disease is difficult, being based largely on histopathology, but the recent introduction of molecular diagnostic techniques is leading to improved understanding of their ecology.

15.3.4 *Aerococcus viridans* var. *homari*

One of the best-known diseases of marine invertebrates is gaffkaemia in American and European lobsters (*Homarus americanus* and *H. gammarus*), caused by the Gram-positive *Aerococcus viridans* var. *homari* (previously known as *Gaffkya homari*). The disease is usually found in holding ponds and, because of the high value of lobsters, is of considerable concern to lobster fishermen. Schemes to restock depleted fisheries have been badly affected by this disease and there is considerable evidence that disease spread is linked to the commercial movement of infected animals caught from infected wild stocks and transported over large distances. It is highly contagious and is probably acquired from a reservoir in wild animals, which show an infection rate of 5–10%. The bacterium gains entry to the lobster hemolymph via abrasions in the shell and multiplies very rapidly at higher water temperatures (>10°C).

This explains differences in severity of the disease observed during summer and winter impoundment. Antibiotics such as oxytetracycline are sometimes used as a preventive measure in holding ponds, but there is concern about antibiotic residues in treated lobsters; a withdrawal period of at least 30 days should be observed (*Section 14.5.2*).

15.3.5 *Vibrio* spp.

Various *Vibrio* spp. cause devastating losses in hatcheries and growing-on stages of tropical shrimp and prawn culture, the most important of these being *V. harveyi, V. penaecida*. Isolates of *V. parahemolyticus* and *V. vulnificus* have also been described, but difficulties with identification of vibrios mean that these isolates are possibly *V. harveyi*. Because of its bioluminescence (*Section 5.12.2*) a large *V. harveyi* outbreak in prawns can result in a spectacular greenish light in infected ponds, leading to the name luminous vibriosis. In hatcheries, vibrios attach to the feeding appendages and oral cavity of the larvae, which become weakened and swim erratically. The virulence of different strains of *V. harveyi* varies greatly. Some have a minimum lethal dose (100% mortality) of as few as 10^2CFU ml^{-1}, whereas other strains are nonvirulent at 10^6CFU ml^{-1}. *V. harveyi* strains infecting fish produce a siderophore that is essential for virulence, but this does not seem to be the case with prawn isolates, perhaps because invertebrates do not have the efficient iron-sequestering system found in vertebrates (*Section 14.3.2*). Virulent strains produce two lethal protein exotoxins, which probably act in the larval intestinal tract on the gut epithelial cells by facilitating passage across the gut and colonization of other tissues. *Box 15.2* describes recent research that suggests that virulence occurs as a result of bacteriophage conversion. If such phage conversion is widespread, it may have significant implications for aquaculture, as it could lead to extensive mixing of genetic information between bacteria. Bacteriophages could be introduced via feed because they can withstand quite severe heat treatment. In addition, it has been suggested that phage could be used as biological control agents in prawn farming; this will need careful evaluation to ensure that the phages used do not lead to lysogeny in bacterial pathogens. In lobsters and crabs, superficial infection by various vibrios causes 'bacterial shell disease', in which pits in the exoskeleton are produced by bacterial chitinase activity. These can erode to form deep lesions and the bacteria may penetrate into the tissues. The exoskeleton is covered by a thin protective lipoprotein layer, and damage to this layer enables bacteria to attach to the chitin exoskeleton. It is possible that biofilm growth of lipolytic bacteria may facilitate penetration.

15.3.6 Control of disease in crustaceans

Huge quantities of antimicrobial agents are used to control bacterial diseases in prawn and shrimp aquaculture, especially in Asia, where use is largely unregulated. It is impossible to obtain accurate estimates of the amounts used but, in Thailand alone, it is thought that the figure is as high as 500 tonnes per annum. Resistance is a serious problem, prompting concerns about long-term environmental impact and threats to human health (see *Section 14.5.3*). Improved hygiene and management practices are needed to reduce this spiralling dependence on antimicrobials. Many companies now market 'probiotic' treatments for use in hatchery tanks and growing ponds. Most of these agents are probably better defined as biocontrol or bioremediation agents, since they work by modifying the bacterial composition of water and sediment, improving the health of the stock by competitive exclusion of pathogens and rapid degradation of waste matter. The most successful approaches involve careful study of the microbial flora of the water, coupled with careful record-keeping of disease incidence, and isolating bacteria that seem to be associated with local conditions in which disease is low and stock health is good. These agents are then cultured in bulk and added to the water. Strains of *Bacillus* are commonly used as general biocontrol agents. In the control of vibriosis caused by *V. harveyi,* some success has been achieved by isolating other nonvirulent *Vibrio* spp. and using these as competitors of the pathogen. The demonstration of genetic exchange between vibrios, including the transfer of virulence, raises some doubts about the wisdom of this approach, as it may encourage the emergence of new virulent strains. Probiotic usage is giving very encouraging results, but a considerable amount of research is needed to evaluate treatments properly, coupled with thorough evaluation of the dynamic changes in microbial community composition. Bacteriophage therapy is also undergoing experimental trials and has achieved some promising results. The recent demonstration of phage-mediated virulence suggests that this approach also needs careful evaluation.

The immune system of invertebrates is very different to that of vertebrates such as fish. Although shellfish possess both humoral and cellular responses to pathogens, they do not produce true specific antibodies and they have substantially less heterogeneity of lymphocytes. There is little evidence of lasting protective immunity. Nevertheless, some 'vaccines' for shellfish are marketed, notably for gaffkaemia in lobsters (a killed cell bacterin). This does seem to provide protection, but this is almost certainly a nonspecific stimulation of defenses rather than true immunity. Many other non-specific 'immunostimulant' feed additives, such as glucans and peptidoglycans (see *Section 14.5.3*), are now being used in prawn and shellfish culture. Anecdotal evidence and limited peer-reviewed research suggest that they are effective against bacterial infections, warranting further investigation.

15.4 Diseases of corals

As discussed in *Section 10.2.3*, many authorities are concerned about the threat to the world's coral reefs and have expressed fears for their long-term survival. There has been increasing recognition in the past few years of the importance of coral diseases, and many novel pathologies are being described. Most studies have been conducted in the Caribbean, but attention is now being focused on other areas. In some cases, such emerging diseases reach epidemic status and dramatically alter both the abundance of specific corals and the overall diversity of reefs. To date, there are five well-described infectious diseases of corals in which the description of diagnostic pathological changes to the coral tissue has been accompanied by the isolation or characterization of consistent microorganisms (*Table 15.1*). There are also many reports of disorders in corals in reefs or those maintained in aquaria, which are poorly documented and for which pathological and etiological evidence is incomplete. It is often difficult to meet the usual criteria (Koch's postulates) needed to prove the etiology of disease. With 'emerging diseases', it is always difficult to decide whether there is a genuine increase in incidence, or whether it simply reflects more intensive observation. In the case of corals, there is no doubt that there has been recent intense interest in monitoring the health of the world's reefs and there is now a large community of subaqua diving

Table 15.1 Infectious disease of corals for which a clear etiology is established

Disease	Host	Pathogen
Aspergillosis	Gorgonians	*Aspergillus sydowii*
Black band	Wide range of scleractinan corals, and gorgonians	Mixed consortium; *Phormidium corallyticum* and SRB are key components
Bleaching	*Oculina patoginica*	*Vibrio shiloi* (=*V. mediterranei*)
Bleaching and tissue lysis	*Pocillopora damicornis*	*Vibrio corallilyticus*
Coral plague	*Acropora*, *Dichocenia* and other scleractinian corals	*Sphingomonas* sp.
White band (tissue necrosis)	*Acropora* spp.	*Vibrio charcharia* (=*V. harveyi*) and others?
White pox	*Acropora palmata*	*Serratia marcescens*

scientists, which partly explains the increased reporting of disease. However, there seems to be a general consensus that disease may be increasing as a result of recent changes to the marine environment caused by anthropogenic effects such as pollution and global climate change. If, as predicted, the effects of global warming escalate during the 21st century, it is vital that we develop a better understanding of the role of microorganisms in coral disease. At least, we can develop rapid diagnostic tools (such as ELISA, gene probes or microarrays) to monitor the emergence and development of disease. The big question will be whether we can also develop technologies to prevent their spread for the future survival of reef ecosystems.

In 1966, one of the most spectacular mass mortality events yet seen in the marine environment occurred in the Caribbean Sea among gorgonian corals (sea fans). A new fungal species, *Aspergillus sydowii* was identified as the causative agent responsible for massive tissue destruction. It was shown to be transmissible by contact with healthy corals. One explanation for the sudden emergence of this epizootic is that the fungus is of terrestrial origin and is associated with airborne dust transported from North Africa and deposited in the Caribbean under a particular set of climatic conditions. *Aspergillus* is also suspected as the causative agent of the virtual eradication of sea urchins (*Diadema antillarum*) in the 1980s.

Black band disease (BBD) of corals is characterized by a black band (from about 10 mm to several cm wide), which migrates over the coral colony by as much as 1 cm per day. Healthy coral tissue is killed and the band moves on, leaving the exposed skeleton, which usually becomes overgrown by turf algae. BBD usually occurs in the summer when water temperatures rise. It can arise on apparently pristine reefs, but is more often associated with reefs receiving sewage and run-off from the coast. BBD can have severe effects because it often attacks large corals such as *Montastrea* spp., which are important in building the framework of reefs. Infection can be transmitted via contact and damaged colonies are more susceptible. Since its first description in the Caribbean in the 1970s, BBD has been observed in a wide range of coral species throughout the world. Microscopic examination of diseased tissue consistently shows the presence of a large gliding, filamentous cyanobacterium identified as *Phormidium corallyticum,* together with numerous other heterotrophic bacteria including SRB, which are responsible for the characteristic black pigment. Thus, it is generally assumed that BBD is caused by this mixed microbial consortium due to anoxic and sulfide-rich conditions at the base of the band resulting in death of the coral tissue. Most infectious diseases typically result from colonization by one organism, the primary pathogen. Generally, mixed infections are much less common but may result when conditions change in necrotic tissue, such as the development of an anaerobic environment. (A familiar example is the mixed community in the mouth responsible for caries and periodontal disease.) At present, circumstantial evidence points to the cyanobacterium *P. corallyticum* as the most likely agent to initiate the disease, but nothing is known about the mechanisms of pathogenicity such as toxins and extracellular enzymes, which would be expected to be involved in the initial stages of tissue necrosis. Recent use of PCR amplification of 16S rDNA separated by DGGE (*Section 2.6.6*) has shown marked differences between the communities inhabiting healthy tissue, diseased tissue and the overlying seawater. Significantly, the microbial communities from the healthy tissue of different coral species are very different, suggesting that corals may contain species-specific microbial communities. Different investigators have found rather different results with respect to the presence of sequences from *Cyanobacteria* and sewage-associated bacteria (from terrestrial run-off) in healthy and diseased BBD tissue, although sequences corresponding to SRB are consistently found, as predicted from cultural and metabolic studies. Thus, the etiology of BBD remains unclear.

Coral plague was first described as a disease of large encrusting and branching corals in the 1970s and to date has been observed only in the Florida Keys in the Caribbean. It starts at the base of the coral and rapidly progresses upwards, with the destruction of tissue. There appear to be two forms, type I and type II, of the disease with the latter being more virulent. A novel *Sphingomonas* sp. has been isolated from coral plague type II. Unlike BBD, there seems to be a sharply defined line between healthy and diseased tissue, suggesting that toxins or enzymes are excreted by the advancing microbial population.

Box 15.2 RESEARCH FOCUS

Bacteria, coral disease and global warming

How temperature affects virulence of coral pathogens

The group led by Eugene Rosenberg of Tel Aviv University, Israel has shown that *Vibrio shiloi* is the causative agent of bleaching in the Red Sea coral *Oculina patagonica* (Kushmaro *et al.* 1996). The increase in water temperature during the summer facilitates the chain of events that leads eventually to damage or loss of the zooxanthellae. In aquarium experiments, Kushmaro *et al.* (1998) inoculated *O. patagonica* with *V. shiloi* at different temperatures. At 29°C, bleaching (defined as a 10% whitening of the tissue) occurred rapidly, whereas at 20 and 25°C the rate of bleaching was slower and less complete. No bleaching at all was observed at 16°C or in uninoculated controls at any temperature. The key factor affected by temperature seems to be in the initial stages of infection, as shown in work by Toren *et al.* (1998). When *V. shiloi* were grown at 25°C they adhered rapidly to coral, but bacteria grown at 16°C did not adhere, regardless of whether the corals were grown at 16 or 25°C. The adhesion was strongly inhibited by D-galactose (or a synthetic analog of this sugar) suggesting that the coral surface (mucus?) contains a receptor for an adhesin produced by the bacterium. Therefore, it is likely that expression of this adhesin is repressed at lower temperatures. Banin *et al.* (2001) showed that the galactoside-containing receptor is indeed located in the mucus, and that zooxanthellae must be present and actively photosynthesizing in order for the receptor to be produced. Adhesion of bacteria was much reduced if coral fragments were stripped of mucus and photosynthesis was blocked by an inhibitor. Banin *et al.* (2000) and Israely *et al.* (2001) provide further insight into the infection process of *V. shiloi*. In the winter, when temperatures drop below 20°C, the bacterium cannot be detected in either healthy or bleached corals, even when the temperature is raised. Using a specific antibody against *V. shiloi*, Israely *et al.* showed that the bacterium is present in a VBNC state. As discussed in *Box 4.1*, many bacteria (including other vibrios) can enter this state, which is usually interpreted as a form of dormancy initiated by starvation. Surprisingly, in this case, *V. shiloi* seems to become VBNC shortly after penetrating the epithelial cells of the host and appear to multiply intracellularly in this form. In addition, VBNC *V. shiloi* can infect healthy corals. The failure to recover the bacteria on culture media probably reflects the absence of a signal for multiplication provided within the coral tissue. An intracellular lifestyle has not previously been reported for other vibrios, so this finding is somewhat surprising. In other intracellular bacteria, there are specific adaptations permitting survival and multiplication within the host cell, such as escape from the phagocytic vacuole or resistance to reactive oxygen and other antibacterial mechanisms. Little is yet known about how *V. shiloi* multiplies within host cells, although a superoxide dismutase may be important. It will be interesting to see if *V. shiloi* contains pathogenicity islands or bacteriophage-mediated virulence (see *Box 15.1*) which confer ability to grow intracellularly. The final piece of evidence linking *V. shiloi* infection to bleaching is the observation that the bacterium secretes a small peptide (12 amino acids) which rapidly inhibits photosynthesis and lyses zooxanthellae in the presence of ammonia (Ben-Haim *et al.* 1999, Banin *et al.* 2001). Again, this factor is only produced at the higher temperatures implicated in bleaching. Higher molecular weight proteins are probably also involved as toxins. Extrapolating from the *O. patagonica–V. shiloi* model to a general hypotheis that bacteria cause bleaching is difficult because the temperature shifts involved are extreme (16–29°C), whereas mass bleaching events in the Caribbean, Pacific and Indian oceans typically involve temperature elevation of just a few degrees.

The discovery of another temperature-dependent vibrio which causes bleaching and necrosis in *Pocillopora damicornis* is of more general significance because this coral is widely distributed and susceptible to bleaching throughout the world. The organism, originally described as *V. coralyticus* by Ben-Haim and Rosenberg (2002) has been identified as a new species named *V. corallilyticus* by Ben-Haim *et al.* (2002). As with *V. shiloi*, laboratory experiments show that infection and tissue lysis proceeds rapidly at 29°C, but no tissue damage occurs at 25°C or lower. In another recent study, Martin *et al.* (2002) investigated the occurrence of vibrios on gorgonians (sea fans) demonstrating tissue necrosis. Vibrios belonging to the species *V. splendidus, V. pelagius* and *V. campbellii* were able to induce tissue necrosis in a few days, but only at higher temperatures. At present, it is not known whether the effect of temperature is on the expression of bacterial virulence factors, but this is now being investigated.

The role of viruses as disease agents in corals has hardly been investigated, because of the difficulties of study in the absence of coral cell cultures for propagation of suspected agents. Clues to the possible role of viruses come from studies by Willie Wilson and Simon Davy (Marine Biological Association, Plymouth), who used elevated temperature to induce a transferable infectious agent believed to be a virus from zooxanthellae of the sea anemone *Anemonia viridis*. Zooxanthellae could harbor latent viruses that are induced by exposure to elevated temperatures and preliminary evidence for this in hard corals has been found by Wilson's group (pers. comm.).

Because of the diversity of hosts and the likely high degree of specificity of coral–microbe associations, unravelling the processes that occur in different species will require considerable further research.

White band disease (WBD) causes tissue to slough off from the base to the tip, especially in *Acropora* sp. It causes loss of over 90% reef cover by these species in parts of the Caribbean. Histopathological examination suggests involvement of bacteria, but their role is not clear. As with plague, there appear to be different variants of the disease. In Type II WBD observed in the Bahamas, diseased corals contain a bacterium identified as *Vibrio charcharia* (now reclassified as *V. harveyi*), but infection experiments have not been carried out.

White pox is characterized by irregularly shaped patches of bare skeleton, from which the tissue peels off unevenly, and patches up to 10 cm^2 can be killed. The disease progresses most rapidly during the summer, when water temperatures rise. Recently, the bacterium *Serratia marcescens* (normally found in the human gut) was identified as the pathogen responsible for this disease in the Florida Keys, using a combination of cultural and molecular techniques. It is likely that the emergence of white pox is linked to expansion of coastal dwellings and the widespread use of septic tanks, which release sewage run-off to the sea.

A disease called red band occasionally affects hard star and brain corals in the Caribbean and Great Barrier Reef. It is similar to BBD, but probably involves different *Cyanobacteria* and other members of the microbial consortium.

Aquarium corals such as *Acropora* spp., *Pocillopora damicornis, Seratopora* spp. and *Euphyllia* spp. frequently show a characteristic 'brown jelly' band of rapidly spreading necrotic tissue. A ciliated protozoan, *Helicostoma nonatum,* is often found in large numbers actively feeding on coral tissue, but appears to be a secondary invader after initial destruction of the tissue by bacterial toxins and enzymes. This author has shown that the protozoan can be isolated from samples of necrotic tissue of *Acropora grandis* and *A. muricata* taken *in situ* from colonies on the Great Barrier Reef (*Figure 15.1*). Diseased tissue also contains characteristic bacteria, which may be involved in initiation of infection; these have not yet been identified but are possibly *Vibrio* spp.

Rapid tissue necrosis (RTN) is a common scourge of aquarists as corals of many types can die within a few days. Vibrios including *V. vulnificus* have been suspected as being responsible for the disease, but results are inconclusive and another hypothesis is that RTN is initiated either by a virus or by an immune-like cross-reaction between corals in close contact in closed aquarium systems. RTN is not observed in the field.

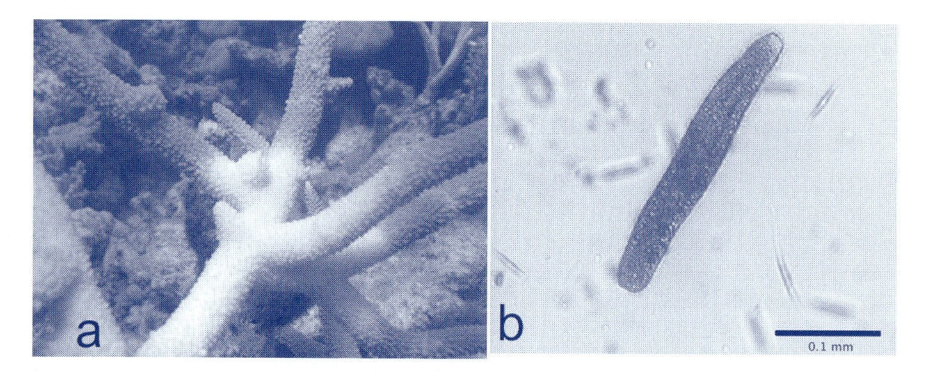

Figure 15.1

Necrotic brown band disease on *Acropora grandis* from the Great Barrier Reef, Australia. (a) Infected branch, showing brown band migrating into healthy tissue, with bare skeleton behind. Image courtesy of Bette Willis, James Cook University, Townsville. (b) Ciliated protozoan, probably *Helicostoma* sp., isolated from the necrotic tissue.

Could bacterial disease be implicated in coral bleaching? Coral bleaching occurs when the symbiotic zooxanthellae lose their photosynthetic ability and/or are eliminated from the host (*Section 10.2.3*). The patchy and spreading nature of bleaching suggests the possible involvement of an infectious agent and *Box 15.2* describes research showing that *Vibrio shiloi* and *V. corallilyticus* infect corals and produce bleaching by processes dependent on elevated temperature. Whether bacteria such as these vibrios can be proved to be responsible for the mass bleaching events observed throughout the world is now a matter of controversy and active investigation. Many coral biologists seem reluctant to accept the involvement of bacterial infection in bleaching, preferring to view the phenomenon as a physiological disorder initiated solely by abnormal temperatures.

References and further reading

Banin, E., Israely, T., Kushmaro, A., Loya, Y., Orr, E., and Rosenberg, E. (2000) Penetration of the coral-bleaching bacterium *Vibrio shiloi* into *Oculina patagonica*. *Appl Environ Microbiol* **66**: 3031–3036.

Banin, E., Israely, T., Fine, M., Loya, Y., and Rosenberg, E. (2001) Role of endosymbiotic zooxanthellae and coral mucus in the adhesion of the coral-bleaching pathogen *Vibrio shiloi* to its host. *FEMS Microbiol Lett* **199**: 33–37.

Ben-Haim, Y., and Rosenberg, E. (2002) A novel *Vibrio* sp. pathogen of the coral *Pocillopora damicronis*. *Marine Biol* **141**: 47–55.

Ben-Haim, Y., Banin, E., Kushmaro, A., Loya, Y., and Rosenberg, E. (1999) Inhibition of photosynthesis and bleaching of zooxanthellae by the coral pathogen *Vibrio shiloi*. *Environ Microbiol* **1**: 223–229.

Ben-Haim, Y., Thompson, F.L., Cnockaert, M.C., Hoste, B., Swings, J., and Rosenberg, E. (2003) *Vibrio corallilyticus* sp. nov., a temperature-dependent pathogen of the coral *Pocillopora damicronis*. *Int J Syst Evol Microbiol* **53**: 309–315.

Boettcher, K.J., Barber, B.J., and Singer, J.T. (2000) Additional evidence that juvenile oyster disease is caused by a member of the *Roseobacter* group and colonization of nonaffected animals by *Stappia stellulata*-like strains. *Appl Environ Microbiol* **66**: 3924–3930.

Boyd, E.F., Davis, B.M., and Hochhut, B. (2001) Bacteriophage–bacteriophage interactions in the evolution of pathogenic bacteria. *Trends Microbiol* **9**: 137–144.

Cooney, R.P., Pantos, O., Le Tissier, M.D.A., Barer, M.R., O'Donnell, A.G., and Bythell, J.C. (2002) Characterization of the bacterial consortium associated with black band disease in coral using molecular microbiological techniques. *Environ Microbiol* **4**: 401–413.

Durand, S.V., and Lightner, D.V. (2002) Quantitative real time PCR for the measurement of white spot syndrome virus in shrimp. *J Fish Dis* **25**: 381–389.

Flegel, T.W. (2002) Emerging shrimp diseases and innovations to prevent their spread. In: Lavilla-Piogo, C.R. and Cruz-Lacierda, E.R. (eds) *Diseases in Asian Aquaculture IV*. Fish Health Section, Asian Fisheries Society, Manila.

Frias-Lopez, J., Zerkle, A.L., Bonheyo, G.T., and Fouke, B.W. (2002) Partitioning of bacterial communities between seawater and healthy, black band diseased, and dead coral surfaces. *Appl Environ Microbiol* **68**: 2214–2228.

Harris, L., and Owens, L. (1999) Production of exotoxins by two luminous *Vibrio harveyi* strains known to be primary pathogens of *Penaeus monodon* larvae. *Dis Aquat Org* **38**: 11–22.

Harvell, C.D., Kim, K., Burkholder, J.M., *et al.* (1998) Review: marine ecology – emerging marine diseases – climate links and anthropogenic factors. *Science* **285**: 1505–1510.

Huang, C.H., Zhang, X.B., Lin, Q.S., Xu, X., Hu, Z.H., and Hew, C.L. (2002) Proteomic analysis of shrimp white spot syndrome viral proteins and characterization of a novel envelope protein VP466. *Molec Cell Proteom* **1**: 223–231.

Israely, T., Banin, E., and Rosenberg, E. (2001) Growth, differentiation and death of *Vibrio shiloi* in coral tissue as a function of seawater temperature. *Aquat Microb Ecol* **24**(1): 1–8.

Karaolis, D.K.R., Somara, S., Maneval, D.R., Johnson, J.A., and Kaper, J.B. (1999) A bacteriophage encoding a pathogenicity island, a type IV pilus and a phage receptor in cholera bacteria. *Nature* **399**: 375–379.

Kushmaro, A., Loya, Y., Fine, M., and Rosenberg, E. (1996) Bacterial infection and coral bleaching. *Nature* **380**: 396.

Kushmaro, A., Rosenberg, E., Fine, M., Ben Haim, Y., and Loya, Y. (1998) Effect of temperature on bleaching of the coral *Oculina patagonica* by Vibrio AK-1. *Mar Ecol Prog Ser* **171**: 131–137.

Lan, Y.S., Lu, W., and Xu, X. (2002) Genomic instability of prawn white spot bacilliform virus (WSBV) and its association to virus virulence. *Virus Res* **90**: 269–274.

Martin, Y., Bonnefort, J.L., and Chancerelle, L. (2002) Gorgonians mass mortality during the 1999 late summer in French Mediterranean coastal waters: the bacterial hypothesis. *Water Res* **36**: 779–782.

McGladdery, S.E. (1999) Shellfish diseases (viral, bacterial and fungal). In: Woo, P.T.K. and Bruno, D.W. (eds) *Fish Diseases and Disorders, Volume 3: Viral, Bacterial and Fungal Infections*. CAB International Publishing, Oxford.

Munday, B.L., and Owens, L. (1998) Viral diseases of fish and shellfish in Australian mariculture. *Fish Pathol* **33**: 193–200.

Munro, J., Oakey, J., Bromage, E., and Owens, L. (2003) Experimental bacteriophage-mediated virulence in strains of *Vibrio harveyi*. *Dis Aquat Org* **54**: 175–186.

Newman, S.G., and Bullis, R.A. (2001) Immune mechanisms of shrimp: form, function and practical application. In: *Proceedings of the Special Session on Sustainable Shrimp Culture, Aquaculture 2001*. World Aquaculture Society, Baton Rouge, LA.

Oakey, H.J., and Owens, L. (2000) A new bacteriophage, VHML, isolated from a toxin-producing strain of *Vibrio harveyi* in tropical Australia. *J Appl Microbiol* **89**: 702–709.

Oakey, H.J., Cullen, W.R., and Owens, L. (2002) The complete nucleotide sequence of the *Vibrio harveyi* bacteriophage VHML. *J Appl Microbiol* **93**: 1089–1098.

Owens, L., Austin, D.A., and Austin, B. (1996) Effect of strain origin on siderophore production in *Vibrio harveyi* isolates. *Dis Aquat Org* **27**: 157–160.

Patterson, K.L., Porter, J.W., Ritchie, K.B., *et al.* (2002) The etiology of white pox, a lethal disease of the Caribbean elkhorn coral, *Acropora palmata*. *Proc Natl Acad Sci* **99**: 8725–8730.

Pizutto, M., and Hirst, R.G. (1995) Classification of isolates of *Vibrio harveyi* virulent to *Penaeus monodon* larvae by protein profile analysis and M13 DNA fingerprinting. *Dis Aquat Org* **35**: 195–201.

Richardson, L.L. (1998) Coral diseases: what is really known? *Trends Ecol Evol* **13**: 438–443.

Ritchie, K.B., Polson, S.W., and Smith, G.W. (2001) Microbial disease causation in marine invertebrates: problems, practices, and future prospects. *Hydrobiology* **460**: 131–139.

Rosenberg, E., and Ben-Haim, Y. (2002) Microbial diseases of corals and global warming. *Environ Microbiol* **4**: 318–326.

Rowher, F., Breitbart, M., Jara, J., Azam, F., and Knowlton, N. (2001) Diversity of bacteria associated with the Caribbean coral *Montastraea franksii*. *Coral Reefs* **20**: 85–91.

Toren, A., Landau, L., Kushmaro, A., Loya, Y., and Rosenberg, E. (1998) Effect of temperature on adhesion of *Vibrio* strain AK-1 to *Oculina patagonica* and on coral bleaching. *Appl Environ Microbiol* **64**: 1379–1384.

Waldor, M.K., and Mekalanos, J.J. (1996) Lysogenic conversion by a filamentous phage encoding cholera toxin. *Science* **212**: 1910–1914.

Williams, E.H., and Bunkley-Williams, L. (2000) Marine major ecological disturbances of the Caribbean. *Infectious Disease Reviews* **2**:110–127.

Wilson, W.H., Francis, I., Ryan, K., and Davy, S.K. (2001) Temperature induction of viruses in symbiotic dinoflagellates. *Aquat Microb Ecol* **25**: 99–102.

Marine microbes and human society

16

16.1 Beneficial and detrimental effects

This chapter reviews some of the activities of marine microbes that have direct consequences for human society. Of course, the ecology of the planet and our very existence depend on marine microbial activities, but these are mostly unperceived by ordinary human experience. What are the more tangible beneficial and detrimental effects on human health, wealth and welfare? Besides their direct impact as pathogens of humans or as the cause of economic losses in aquaculture, marine microbes have some other very important detrimental effects. These are mainly due to the biodeterioration of materials and the spoilage of foods. However, the 'spin-off' from study of these processes is leading to some significant new benefits such as those described in *Box 16.1*. Major economic benefits derive from certain activities of microbes in the environment, especially the bioremediation of pollution, whilst the industrial and biomedical exploitation of microbes isolated from the sea is a major aspect of the growing field of marine biotechnology. Biotechnology is defined broadly as the application of scientific and engineering principles to provide goods and services through mediation of biological agents. In the narrower sense of microbial biotechnology, it encompasses the synthesis of a wide range of useful new products such as enzymes, pharmaceuticals and polymers and new processes such as environmental monitoring, bioremediation and disease diagnostics. The enormous natural diversity of marine microbes and their metabolic activities provides great opportunities for exploitation. As our appreciation of the great variety of marine microbes grows with the study of diverse habitats, so the collection of microbes with unusual properties continues to expand. However, over 99% of microbes in the marine environment are currently unculturable and attempts to bring new species into cultivation in order to study their properties will yield rich rewards (see *Box 2.1*). In addition, advances in environmental genomics, proteomics and informatics now allow us to 'prospect' for interesting genes rather than relying on cultivation. This approach is discussed in *Box 16.2*. The application of new methods for the investigation of community interactions (e.g. in biofilms, plankton and sediments), host–pathogen and host–symbiont interactions holds great promise for exploitable discoveries as well as acquisition of basic scientific knowledge. *Tables 16.1* and *16.2* summarize the principal beneficial and detrimental aspects of marine microbes respectively.

16.2 Biofouling and biodeterioration

16.2.1 Biofilms and biofouling

The surfaces of inanimate objects and living organisms in the sea are colonized by biofilms, which are mixed microbial communities showing complex physical structures and chemical interactions (*Secion 4.5.4*). The process of biofilm formation usually begins with the attachment of bacteria and/or diatoms as primary colonizers, leading to a slimy biofilm up to 500 μm thick (*Figure 16.1*). This is followed by the settlement of other microbes, planktonic algal spores and invertebrate larvae. The complex dense community that develops leads to the biofouling of all types of marine surfaces, including coastal plants, macroalgae, animals, piers, fishing gear, aquaculture cages, engineering materials and boat hulls. The economic effects of biofouling are

Table 16.1 Some beneficial effects of activities or products from marine microbes

Application	Examples
Aquaculture	Disease diagnostics Nutritional supplements Pigments Probiotics Vaccines
Cosmetics	Liposomes Polymers Sunscreens
Environmental protection	Bioremediation of pollution Disease diagnostics Nontoxic antifouling agents Toxicology bioassays Waste processing
Food processing	Enzymes Flavors Preservatives Texture modifiers
Manufacturing industry	Bioelectronics Polymers Structural components
Minerals and fuels	Desulfurization of oil and coal Manganese nodules Oil extraction
Nutraceuticals	Antioxidative compounds Dietary supplements 'Health foods'
Pharmaceuticals	Antibacterial, antifungal and antiviral agents Antitumor and immunosuppressive agents Drug delivery Enzymes Neuroactive agents Self-cleaning implants
Textiles and papers	Enzymes Surfactants

Table 16.2 Detrimental effects of marine microbial activities

Process	Examples
Biodeterioration	Biofouling Corrosion of metals Fish spoilage Sulfide contamination of oil Timber damage
Diseases	Heavy metal mobilization Human infections and intoxications Losses in fisheries and aquaculture

immense. Deleterious effects include deterioration of materials, blockage of pipes, reduced efficiency of heating and cooling plants and interference with the operation of boats. Sailing enthusiasts know the cost (in lost weekends!) of scraping the bottom of their boats, whilst biofouling of large ships is estimated to be responsible for an additional 10% in fuel usage due to

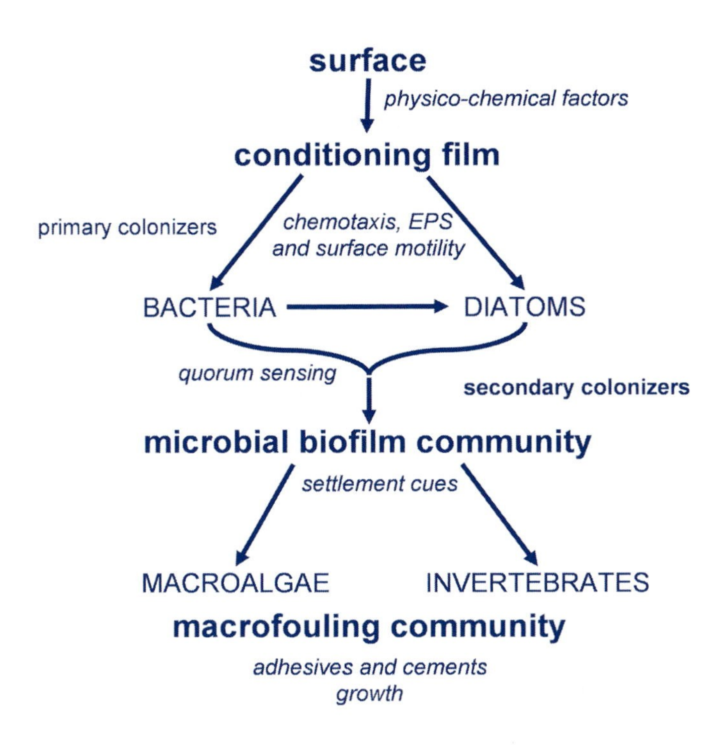

Figure 16.1

Development of biofouling, showing the critical role of microbial activities. Italicized captions indicate processes amenable to disruption, for control of biofouling. Invertebrate macrofouling organisms produce soft fouling (e.g. soft corals, sponges, anemones and tunicates) and hard fouling (e.g. barnacles, mussels and tubeworms). EPS = exopolymeric substances.

viscous drag. The US Navy alone estimates the cost of biofouling at over US$1 billion through losses in fuel efficiency and the cost of antifouling measures.

The prevention of macrofouling of ships' hulls by the use of copper sheets, in use for hundreds of years, led to the development of copper- and tin-based antifouling paints. The use of paints containing tributyl tin (TBT) since the 1970s has led to serious environmental problems, with particular effects on the ecology and reproductive behavior of marine invertebrates and the immunity of marine mammals (*Section 13.5*). TBT will soon be banned in most countries and there is an active search for effective 'environment-friendly' alternatives. Non-toxic 'self-cleaning' silicon coatings are probably the best currently available alternative to antifouling biocides. Adhesion by biofouling organisms to these surfaces is weak and they are easily removed.

Besides these attempts to prevent fouling by macroorganisms, there are a number of points at which the microbial component of biofouling can be disrupted. Indeed, the bacterial/diatom slime alone causes significant viscous drag; experimental removal of just the primary biofilm from ships' hulls can lead to improvements of up to 20% in power usage. Depending on their microbial composition, established biofilms may either promote colonization and larval metamorphosis of algae and invertebrates by the production of chemical cues, or inhibit colonization by producing negative chemotactic repellents. Innovations in the prevention of primary colonization come from the observation that many marine plants and animals have evolved defense mechanisms against colonization by bacterial biofilms. These range from physical barriers (e.g. the almost 'nonstick' surfaces of cetaceans) to the production of chemical inhibitors observed in some coastal plants, seaweeds and animals. Among the most promising developments is the discovery that furanones produced by the red alga *Delisea pulchra*

Box 16.1 RESEARCH FOCUS

Talking to friends, talking to strangers

The importance of quorum sensing in bacterial communication

As described in *Section 5.12.3*, the phenomenon of QS was revealed during investigation of the control of bioluminescence of *Vibrio fischeri* and *V. harveyi.* Instead of thinking of them as simple single cells, it must now be recognized that bacteria can communicate with each other and regulate gene expression so that they can respond in a coordinated fashion to changes in environmental conditions. Many bacteria utilize AHL-dependent QS and circuits resembling the *luxI/luxR* system have been identified in over 30 Gram-negative bacteria (Miller & Bassler 2001). Three unrelated protein families (Withers et al. 2001) can synthesize AHLs. Gram-positive bacteria do not use AHLs, but rely on peptide signals via two-component response regulator proteins resembling those of *V. harveyi.* There is often more than one QS system in the same bacterium. One of the most important examples of the importance of QS is in the interaction of bacteria with plant or animal hosts. The role of QS in the interactions of symbiotic *V. fischeri* with its squid host is described in *Section 10.4.2.* QS also controls the symbiotic association between bacterial symbionts and pathogens of plants, animals and humans. In the case of pathogens, an obvious advantage of QS is that it allows bacteria to delay the induction of virulence factors until they have reached a certain critical density. This means that bacteria can 'hit the host hard' by delaying host defense mechanisms to factors such as enzymes and toxins. One of the most intensively studied roles for QS is in treatment of human infections caused by *Pseudomonas aeruginosa.* Here, two pairs of *luxI/luxR* homologs occur and high cell density promotes expression of several major virulence factors, including elastase, protease and exotoxin. Significantly, QS mutants of *P. aeruginosa* do not develop proper biofilms, which are particularly important in lung infections in cystic fibrosis patients. Can we develop a new class of antimicrobial agents that attack the QS system? The discovery of halogenated furanones in the laboratory of Staffan Kjelleberg and Peter Steinberg at the University of Sydney is an important example of a 'spin-off' from marine microbiology. These compounds were found to be natural antifouling agents in the marine alga *Delisea pulchra* (Kjelleberg *et al.* 1997, Steinberg *et al.* 1998). They prevent colonization by specifically blocking the binding of AHLs to the receptor protein of a transcriptional activator, resulting in inhibition of a broad range of AHL-mediated processes in biofilm formation (de Nys *et al.* 1995). Over 40 derivatives of the furanones have been developed, and they are finding wide applications as chemotherapeutic treatments for biofilm-forming infections such as those caused by *P. aeruginosa* (Hentzer *et al.* 2002, Manefield *et al.* 2002) and as antifouling agents ranging from fishing nets to contact lenses and biomedical implants.

The discovery of the hybrid QS system in *V. harveyi* and the involvement of the LuxS type of autoinducer have stimulated a major research effort by Bonnie Bassler and colleagues at Princeton University. Highly conserved *luxS* homologs occur in many Gram-negative and Gram-positive bacteria. Surette and Bassler (1997) identified 'cross-talk' between different bacteria and Bassler coined the term 'bacterial esperanto' to indicate a common language of gene regulation. Miller and Bassler (2001) suggest that bacteria existing in mixed populations in natural habitats may be able to recognize and respond to multiple autoinducer signals. This would allow the bacteria to monitor the density of its own population and that of other species, regulating genes accordingly. Even more surprising is the recent description by Ian Joint and colleagues at the Plymouth Marine Laboratory of the first evidence of such communication across the bacterial and eukaryotic domains (Joint *et al.* 2002). Biofilms composed of wild-type *V. anguillarum* strongly enhance settlement of zoospores of the alga *Enteromorpha,* but *vanM* mutants (defective in AHL synthesis) do not. *Enteromorpha* detects concentration gradients of the bacterial AHL and swims towards the surface. Interfering with this settlement cue offers an obvious means of controlling this important fouling organism. In summary, QS is an outstanding example of how research in marine microbiology has led to novel discoveries with far-reaching implications.

specifically interferes with chemical signalling within bacterial biofilms by competing with bacterial QS. As well as potential applications in the control of biofouling, this finding is leading to the development of novel pharmaceutical compounds for control of infectious diseases (*Box 16.1*). Antibiotics produced by epibiotic bacteria on the surfaces of algae or corals may have a symbiotic function, working in concert with algal antifouling chemicals. Members of the bacterial genus *Pseudoalteromonas* seem particularly promising in this respect. As well as incorporation into antifouling surface coatings, these agents may also find application as chemotherapeutics. Since these secondary metabolites might only be produced when bacteria grow in mixed biofilm communities, new avenues of research in the search for novel biomolecules are opened by the study of marine microbial ecology.

16.2.2 Biodeterioration of metals and wood

Microbiologically influenced corrosion (MIC) occurs due to the activities of microorganisms within biofilms on metals, alloys and composite materials. SRB have long been known to be the main cause of marine MIC, but the importance of mixed consortia with fermentative acid-producing bacteria (APB) in biofilms is increasingly recognized. Fermentative bacteria produce acid themselves as well as supplying acetate as a substrate for SRB metabolism. MIC has been a particular problem for the offshore petroleum industry, because it causes corrosion of pipelines and oil drilling platforms and SRB also cause 'souring' of crude oil by the production of hydrogen sulfide. As well as interfering with the refining process, the gas can reach toxic levels on drilling platforms that have an SRB contamination problem. Corrosion of steel structures can be limited by the use of cathodic protection, but this can lead to structural weakness unless the electrical potential is carefully applied and monitored; formation of excess hydrogen can cause metal fatigue and embrittlement. Routine monitoring by culturing APB and the judicious application of biocides can help to control the problem. Testing for SRB is more problematic because of the need for anaerobic culture; development of molecular-based test kits could prove useful to the industry. Problems with the use of cathodic protection and environmental concerns about large-scale use of biocides at sea favor a biotechnological solution to the control of SRB/APB biofilms.

Biodeterioration of marine wooden structures (e.g. wharves, jetties, piers and boats) and in the transport and storage of timber at sea cause an immense amount of damage (estimated annual costs in the US alone are US$1 billion). The main cause of damage is penetration by wood-boring invertebrates, of which the most important are the shipworms. These are small bivalves (family *Teredinidae*), which tunnel into wooden structures using their serrated shell. As is common with most animals that feed on wood (termites are a well-studied example), shipworms rely on an obligate symbiotic association with cellulolytic bacteria for the digestion of cellulose. In one shipworm, *Bankia setacea*, the γ-proteobacterium *Teredinibacter turnerae* has been identified as a symbiont of a specialized region of the host gill. In culture, this bacterium synthesizes cellulolytic and proteolytic enzymes for digestion, and fixes atmospheric N_2 via the enzyme nitrogenase. This allows the bivalve host to grow on a diet of wood alone. In the shipworm, the bacterial symbiont is transmitted vertically via the larvae, as shown by specific rRNA probes reveal the bacterium to be present in reproductive tissue and eggs. This ensures a broad distribution of larvae. The use of various preservative wood treatments delays deterioration, but these are expensive, have an adverse environmental impact, and cannot be used on new timber during transport. Identification of the symbionts and their mode of transmission provide a potential weak link in the process of biodeterioration, as it may be possible to identify compounds that inhibit bacterial colonization or metabolism. Some species of catfish eat wood; their gut contains a consortium of symbiotic bacteria that digest wood cellulose. The preservation of archaeological timbers, such as those recovered from shipwrecks, is a highly specialized process. Waterlogged wood is largely protected when buried in anoxic sediments, but is subject to rapid decay by a range of fungi once it is brought to the surface.

16.3 Biodegradation and bioremediation of marine pollutants

16.3.1 Oil pollution

16.3.1.1 Sources of oil in the sea

Petroleum is a complex mixture of thousands of compounds, predominately hydrocarbons. The world economy is totally dependent on petroleum products as a source of energy and raw materials for a vast range of products. As a consequence, over 3 billion tonnes of crude oil are extracted annually, and about 0.1% of this finds its way into the sea during the extraction, transportation and consumption of crude oil and petroleum products. An almost equivalent amount seeps into the ocean naturally from the seafloor. Although oil in the sea has many sources, pollution by oil tankers attracts the greatest public concern. Tankers take on water for ballast and, when this is discharged, considerable local contamination of the sea occurs. Occasionally, tankers collide or run aground, releasing large quantities of oil into the sea. Fortunately, in recent years, greater use of double-hulled tankers has reduced the amount of oil spilt in this way, but major incidents are still occurring. The immediate damaging effect of this on marine wildlife and the economic impact on fisheries and tourism means that environmental agencies are under intense pressure to alleviate the problem quickly and effectively. Apart from immediate effects, there are concerns about toxic residues and long-term disruption of ecosystem communities. Various strategies exist for dealing with spilled oil, including trapping with booms and skimming the surface to remove the oil, using absorbent materials to soak up oil, applying dispersant chemicals or setting fire to oil slicks. The efficacy of these processes is very dependent on the location of the spill and weather conditions and the use of dispersants or burning can have highly damaging effects on marine ecology. In one of the largest recent tanker accidents, in which the grounding of the *Sea Empress* in 1996 led to the release of 72 000 tonnes of crude oil off the coast of Wales, over 500 tonnes of chemical dispersants were sprayed onto the slick by aircraft. This was effective in dispersing the oil and enhancing its degradation, but led to significant changes to the ecology of benthic animals. In contrast, the 85 000 tonnes of oil spilt by the *Braer* off the Shetland Isles in 1993 dispersed within a few hours, due to extreme gales and wave action, and it degraded naturally with few apparent lasting effects.

Fortunately, naturally occurring microbes break down most of the components of petroleum. Large quantities of crude oil have been seeping naturally into the oceans for millions of years, and many organisms have evolved mechanisms to degrade the different hydrocarbon components. When oil is released into the sea as a result of a spillage, it floats on the surface and the low-molecular-weight fractions evaporate quickly. Some components are water-soluble and photochemical oxidation occurs through the action of sunlight. The fate of an oil slick is very dependent on wave action and weather conditions. Sometimes, droplets of oil will become emulsified in the water and disperse quickly; in other cases a water-in-oil emulsion will form and this forms a thick viscous mousse, which takes a very long time to disperse and gives rise to the familiar tar on beaches. Different crude oils vary, but usually over 70% of the hydrocarbons are biodegradable. Only the asphaltenes and resin components are recalcitrant to breakdown.

16.3.1.2 Biodegradation

The biochemical processes involved in the biodegradation of oil have been studied extensively. Aerobic processes are responsible for the efficient biodegradation of oil. The initial step involves incorporation of molecular O_2 into *n*-alkanes by oxygenases, resulting in primary alcohols, which are then further oxidized to aldehydes and fatty acids. The fatty acids are then metabolized to acetyl-coenzyme A which enters primary metabolism via the TCA cycle. In this way, the *n*-alkanes are completely converted to CO_2 and H_2O. There are many different routes for the degradation of the aromatic hydrocarbons, the key step being the cleavage of the aromatic ring; O_2 is again essential for this process. Polyaromatic hydrocarbons (PAHs) such as naphthalene, phenanthrene and pyrene contain four or more aromatic rings and are only degraded by a few groups; this is usually a slow process. Consequently, PAHs that are not degraded in the water column accumulate in sediments and can persist for very long periods. Many harbors and estuaries that are used extensively by shipping have sediments that are chronically polluted with PAHs. In such anoxic sediments, anaerobic degradation occurs with the oxidative processes linked to the reduction of nitrate or sulfate, or the production of methane. Degradation probably begins with carboxylation reactions, followed by ring reduction and cleavage. The activity of burrowing benthic animals can enhance aerobic PAH degradation by microbes, due to the mixing of sediments and introduction of O_2 through the elaborate ventilation systems that they construct; isolation of bacteria from such burrows may prove a rich source of new species.

Many different microbes (at least 160 genera) capable of degrading petroleum hydrocarbons have been isolated from waters all over the world; bacteria appear to be the most important, but yeasts and filamentous fungi have also been described. Most hydrocarbon-degrading isolates are heterotrophs belonging to the *Proteobacteria* (e.g. *Pseudomonas, Acinetobacter, Cycloclasticus* and *Alcanivorax*). Photoautotrophs such as the alga *Ochromonas* and *Cyanobacteria* (e.g. *Agmenellum, Micocoelus* and *Phormidium*) have also been linked with hydrocarbon degradation; some appear to accumulate hydrocarbons within vesicles, but not degrade them. It is possible that these *Cyanobacteria* form consortia with heterotrophs, which leads to breakdown. Some *Archaea* are also known to degrade hydrocarbons as a sole source of carbon and energy. In anoxic environments, SRB and archaeal methanogens are most commonly responsible for this process. The breakdown of oil depends on the activities of a consortium of microorganisms, each responsible for transformation of a particular fraction. Normally, these constitute only a small proportion (less than 1%) of the microflora of seawater, but in a polluted environment they multiply rapidly to as much as 10% of the population. Immobilization of mixed microbial communities on biofilms may be particularly important in efficient biodegradation.

Biodegradation proceeds most quickly when the oil is emulsified into small droplets. Hydrocarbon uptake by microorganisms is stimulated by the production of biosurfactants, surface-active agents containing both hydrophilic and hydrophobic regions which reduce surface tension. Many microbes have hydrophobic surfaces and adhere to small droplets of oil, and many also produce extracellular compounds which disperse the oil. There has been great interest in developing natural biosurfactants as an alternative to chemical dispersants. For example, the bacterium *Acinetobacter calcoaceticus* is particularly effective in biodegradation of oil because it both adheres to hydrocarbons and produces an extracellular glycolipid biosurfactant called emulsan. Emulsan-deficient mutants grow very poorly on hydrocarbons. It may be possible to select naturally occurring strains or use genetic engineering to produce large amounts of biosurfactants, although optimizing the industrial-scale production of these compounds may be difficult. Emulsan is also used to reduce viscosity to aid in the extraction of crude oil.

16.3.1.3 Bioremediation

Bioremediation is usually defined as a biological process to enhance the rate or extent of naturally occurring biodegradation of pollutants, although in a broader sense it can be used for deliberate use of any biological process that reverses environmental damage. (Thus, iron fertilization of the oceans, *Box 9.2*, might be considered the most ambitious bioremediation project of all!) In the case of oil pollution, hydrocarbons provide a carbon source for microorganisms, but oil is deficient in other nutrients (especially nitrogen and phosphorus) and the supply of these is the main factor limiting the rate of degradation. Therefore, the most successful approach to bioremediation is the addition of inorganic or organic nutrients as fertilizers to speed up natural processes. The process of seeding oil-spills with exogenous microorganisms shown to have high degradative activity in the laboratory (bioaugmentation) has been less successful because they are rapidly out-competed by the enrichment of naturally occurring microbes. Furthermore, the idea that genetic modification could create a 'superbug' to digest oil in the marine environment has proved to be misguided. Bioremediation of petroleum products is applied extensively to clean up contaminated soil (e.g. to reclaim land polluted by spillage from oil tanks), and many commercial products have been developed. However, the scientific rigor with which these have been tested for marine bioremediation is questionable. A major problem in testing the effectiveness of bioremediation is monitoring the extent of degradation. Because breakdown proceeds in a progressive fashion, disappearance of compounds such as the alkanes and small aromatic compounds can be measured easily, but monitoring removal of the more recalcitrant compounds is more problematic. One internal standard method is to measure the disappearance of biodegradable components in comparison with the concentration of hopanes, which are highly recalcitrant to breakdown (*Section 3.3*).

Laboratory studies provide little information about how well bioremediation treatments will work in the field. Some mesocosm and controlled release experiments in the field have been performed, but these obviously have to be limited in size and scope, as few authorities are willing to allow the deliberate pollution of coastal waters. Therefore, most of our knowledge about the efficacy of bioremediation comes from studies of opportunity following large-scale spills from tanker accidents in which investigators have little control over the prevailing conditions. Of such incidents, the clean-up after the *Exxon Valdez* spill allowed the first large-scale evaluation of bioremediation and the lessons learnt from that situation have provided a sound basis for future use of the technology. The *Exxon Valdez* ran aground in Prince William Sound, Alaska in March 1989, releasing 11 million gallons of crude oil as a slick that quickly contaminated 500 km of coastline in a pristine wilderness environment. Under the control of the USEPA, a range of approaches to bioremediation was tested. Deliberate experimental contamination of shorelines in Norway and Canada had shown previously that, even under Arctic conditions, oil applied to shorelines would be degraded naturally within a few years, but that addition of agricultural fertilizers like ammonium phosphate, ammonium nitrate or urea increased the initial rate of biodegradation up to 10 times. The best approach to fertilization seems to be to use oleophilic compounds, which stick to the oil and/or release nutrients slowly. One such fertilizer is a proprietary compound called Inipol™ EAP22; this is a microemulsion of urea in brine, encapsulated in an external phase of oleic acid and lauryl phosphate, cosolubilized by butoxyethanol. Although highly effective, the results depend on the nature of the substrate to which the oil is bound. Enhancement of breakdown on pebbles, gravel and large sand particles is much better than on fine sand particles. In the bioremediation of the *Exxon Valdez* spill, Inipol EAP22 was combined with a slow-release fertilizer (Customblen™), which consists of ammonium nitrate, calcium phosphate and ammonium phosphate encapsulated in a coating of polymerized linseed oil. This was better suited to subsurface oil associated with finer sediments. At first, the authorities were very cautious about the amounts of fertilizer that could be applied, for fear of toxicity and unacceptable concentrations of nutrients leaching into the coastal waters, but as the clean-up program continued, they were shown to have no adverse effects. Bioremediation in Prince William Sound was judged very successful; microbial activity was enhanced, and oil biodegradation was stimulated 2–5 fold.

Bioremediation of sediment contamination is more problematic. In terrestrial situations, biodegradation is enhanced by tilling, which introduces O_2, but this is obviously impractical in marine situations. It may be possible to introduce chemical oxidants or alternate electron acceptors, but this requires careful evaluation of ecosystem effects. There have been some successes with the introduction of O_2 by aeration pumps. Bioremediation in sensitive habitats like mangroves and salt marshes is particularly difficult.

To summarize, bioremediation by the application of oleophilic or slow-release fertilizers is now generally accepted to have proved its worth as one component in the response to an oil spill. However, before it is used, careful attention must be given to the nature of the substratum and the degree of penetration of oil into the sediments. Addition of exogenous organisms has not been successful, but further development of surfactant-producing strains and their formulation into products that allow them to compete with the stimulation of indigenous microorganisms may hold promise for the future. Further research on changes in microbial community composition in response to introduction of oil containing different mixtures of hydrocarbons, using tools such as DGGE (*Section 2.6.6*), may also yield valuable information about the best ways to enhance natural processes of degradation.

16.3.2 Persistent organic pollutants and plastics

Persistent organic pollutants (POPs) are very resistant to photochemical, biological and chemical degradation. Most POPs are halogenated and highly soluble in lipids and they accumulate particularly in fatty tissues. Their semivolatile nature allows them to vaporize or to be adsorbed onto atmospheric particles and they are therefore transported over great distances. For example, animals and humans living at the poles have high POP levels in their tissues despite these compounds not being used there. Extensive evidence links POPs to reproductive failure, impairment of the immune system, deformities and other malfunctions in a wide range of marine life (*Section 13.5*). They are highly toxic to humans through ingestion of fish and other routes. POPs include the organochlorine insecticides (e.g. DDT, aldrin and chlordane), industrial chemicals (e.g. polychlorinated biphenyls, PCBs) and by-products (e.g. dioxins and furans). They reach the sea via terrestrial run-off and atmospheric deposition. In numerous countries, the manufacture and use of many of these chemicals has been prohibited, but they are highly persistent in sediments and dump sites. PCBs occur in many electrical products and disposed equipment is a significant source of these chemicals.

Microbial cycling of POPs acquired by plankton in the upper parts of the ocean plays an important role in its distribution through ocean food webs. The bacterioplankton presents a large surface area for the adsorption of POPs. Microbial loop processes release POPs during settlement of plankton debris and organic particles through the water column, but some particles with adsorbed POPs will be buried in sediments and accumulate there. Disturbance of sediments by tides, currents, dredging and the activity of benthic animals can release large quantities of the chemicals into the water. PCBs are highly recalcitrant to degradation and there is an active search for microbes capable of breaking down the chlorine bonds, thus offering a potential use in bioremediation. Many aerobic bacteria can degrade the biphenyl ring in PCBs, but not the heavily chlorinated congenors. Anaerobic degradation of PCBs is known to occur, but isolation and identification of the organisms responsible has been elusive; different bacteria with distinct dehalogenases and congenor specificities occur. DGGE analysis of rRNA from communities, coupled with selective enrichment, has recently been used to isolate new species of PCB degraders. Syntrophic consortia of SRB and dechlorinating bacteria may be responsible.

Every year, society discards thousands of tonnes of plastic debris in the marine environment. Apart from the unsightly aesthetic effects, plastic waste from bottles, rope and netting are harmful to marine mammals, birds and fishes, causing entanglement and suffocation. Also, plastics often contain toxic chemicals such as phthalates. Where does all the plastic go? After breaking up by wave action, smaller debris is ingested by marine invertebrates, but little is known about the consequences of this, nor about the role of microorganisms. Are microbes involved in the biodegradation of plastics in the guts of animals or in sediments? This is a ripe area for investigation.

16.3.3 Other pollutants

Microorganisms (especially SRB, *Bacillus* spp. and *Pseudomonas* spp.) are effective in the removal of heavy metals (including radionuclides) from contaminated sediments. Of course, metals cannot be degraded, but they can be immobilized into a nonbioavailable form or respeciated into less toxic forms (e.g. for the removal of methylmercury, see *Section 11.4*). Bacterially mediated oxidation of soluble Mn(II) to insoluble Mn(III, IV) oxides and oxyhydroxides is particularly important because the products oxidize a variety of organic and inorganic compounds and serve as electron acceptors for anaerobic respiration, leading to immobilization of heavy metals by adsorption. Proteins localized on the surface of dormant spores of *Bacillus* spp. isolated from marine sediments can catalyze the oxidation of metals such as manganese. This bacterial precipitation of metal oxides is also important because it leads, over millennia, to the formation of manganese nodules on the seabed. Besides manganese, these small balls contain a high percentage of other valuable metals, especially nickel and cobalt. They occur in large fields, but harvesting is difficult because these sources are mostly in very deep water (over 5000 m deep), at long distances from major landmasses. If the problems of harvesting can be solved, communities in South Pacific islands could obtain considerable economic returns. Oxidation of Fe(II) by anaerobic bacteria, including anoxygenic phototrophs and nitrate-reducing organisms also occurs in sediments.

Fungi (e.g. *Flavodon*) are effective in the production of lignin-modifying enzymes, which have application in the treatment of paper mill effluents discharged into coastal waters. Yeasts such as *Debaryomyces* sp. and *Trichosporon* sp. have been cultured for use in decomposition of organic polymers in waste water from fish-processing plants.

16.4 Environmental monitoring

The Microtox® system is an established bioassay for the rapid toxicity testing of water, sediment and soil samples. It depends on inhibition of bioluminescence of *Vibrio fischeri* (*Section 5.12.2*), which is supplied as a standardized freeze-dried culture. Light emission is measured using a photometer and is very sensitive to the presence of toxic chemicals at sublethal concentrations. The more recently introduced QuikLite bioassay uses the same principles, but is based on inhibitory effects on bioluminescence of the dinoflagellate *Gonyaulax polyedra* (*Section 7.4.1*). These methods are reported to be many times more sensitive and much easier to carry out than conventional bioassays using fish or amphipods.

The study of natural marine microbial communities is an important aspect of monitoring the effects of pollution, temperature shifts and other environmental disturbance. The DGGE technique for community analysis (*Section 2.6.6*) has found wide application in such studies. Pollution-degrading organisms can be genetically modified to link the expression of degradative genes to reporter systems such as *lux* genes from *Vibrio* (*Section 5.12.2*) or GFP genes from jellyfish (*Section 2.2.4*). These reporter systems have wide applications in cell biology as well as environmental investigations.

16.5 Microbiology of fish and seafood products

Microbial growth and metabolic activities are the major cause of spoilage of fish and shellfish products, resulting in rapid production of discoloration, slime, unpleasant odors and off-flavors. The composition of microbial communities in freshly caught seafood varies considerably and is further affected by the method of processing and storage. Members of the *Vibrionaceae* are particularly prominent as agents of rapid spoilage of unpreserved fish. Psychrophiles such as *Pseudomonas* and *Shewanella* grow even when fish is chilled on ice. The food industry has devised various methods of extending shelf-life in 'value-added' products by inhibiting or delaying spoilage, but many processes used with other foods are unsuitable because the texture and flavor of fish products is easily destroyed by processing, leading to reduced consumer acceptance. Many processes (e.g. salting, pickling and smoking) are based on traditional methods, whilst recent innovations include packaging in a CO_2-modified atmosphere. These methods can shift the ecology of the microbial flora; for example, whilst CO_2 packaging of fresh fish in ice is highly effective, it can suppress the growth of respiratory bacteria so that fermentative *Photobacterium*, lactic acid bacteria (LAB, e.g. *Lactobacillus* and *Carnobacterium*) and enterobacteria become dominant. Spore-forming bacteria (*Bacillus* and *Clostridium*) can survive mild heat treatment (pasteurization) of vacuum-packed products. Specific spoilage organisms are most often associated with the production of ammonia, amines, ketones, organic acids and sulfur compounds that characterize 'off' seafood. Trimethylamine is the most important of these metabolites, produced by the reduction of trimethylamine oxide, naturally present in the tissue of many fish. Most fish spoilage organisms are easily cultured and identified using standard microbiological methods, although there has been some recent use of molecular methods for characterization and early detection of spoilage organisms (e.g. gene probes for *Shewanella putrefaciens*). Bacteria grow rapidly to high levels (over 10^8 CFU g^{-1}) in the nutrient-rich environment of fish tissue, which contains high concentrations of readily utilizable substrates such as free amino acids. Metabolic consortia develop in the mixed microbial community; for example, LAB degrade the amino acid arginine to ornithine, which is further degraded to putrescine by enterobacteria. The production of siderophores is important for the acquisition of iron, since fish tissue contains limiting concentrations (*Section 14.3.2*). LAB characteristically produce bacteriocins, which are highly specific membrane-active peptide antibiotics that inhibit certain other bacteria. As well as spoilage, some fish products are the source of human pathogens. Fish- and shellfish-associated bacterial, viral and toxic diseases were considered in *Chapters 11* and *12,* but other pathogens (e.g. *Salmonella* spp., *Staphylococcus aureus* and *Listeria monocytogenes*) may be introduced during handling and processing. *Listeria* is of particular concern in lightly preserved ready-to-eat products such as cold-smoked fish and shellfish, because of its ability to grow over a very wide temperature range and salinity. Control of these pathogens is an important requirement for commercial processors.

Besides improvements in packaging methods, a number of biotechnological approaches to control of specific organisms has been investigated. Antibiotics such as tetracyclines were once added to ice on board fishing vessels and in markets, to prevent spoilage of fresh fish. This practice is now prohibited because of concerns over antibiotic residues and resistance. However, LAB bacteriocins are generally regarded as safe and are permitted in foods (e.g. nisin is widely used in dairy products). With fish, addition of purified bacteriocins has produced some promising results, as has the addition of nonspoilage LAB as competitors of pathogens or spoilage organisms. It has been shown that AHLs are produced during development of the microbial spoilage community and interference with the QS mechanism (see *Box 16.1*) might offer a potential new method of control.

The 'spoilage' activities of microbes in fish are generally regarded as detrimental, but in some parts of the world, there are a number of ethnic food products that depend on microbial activities for preservation and flavors. Most of these are encountered in Asian countries, especially Indonesia, Thailand, the Philippines and Japan. Many processes are conducted according to traditional recipes, but the microbiology of some has been investigated during commercialization of production for export and pure starter cultures may sometimes now be used. Perhaps the best-known product is nam-pla (fish sauce), which is now widely used in the West because of the popularity of Thai cuisine. Nam-pla is made by fermentation of fish hydrolyzed by a high salt concentration (15–20%), which encourages growth of the extreme halophile *Halobacterium salinarum* (*Section 6.2.4*), leading to characteristic flavors and aromas. This appears to be the only food product that relies on a member of the *Archaea.* Other high-salt products include som-fak, burong-isda and jeikal, a traditional Korean food made from fermented shrimp. Other fermented fish products such as plaa-som employ lower salt concentrations; in these, a microbial flora dominated by LAB and yeasts leads to a characteristic aroma. There must be enough salt present (2–8%) and a final pH of less than 4.5 to inhibit the growth of pathogens. Nevertheless, contamination by *Staphylococcus* can be a problem. Garlic is a major ingredient of some recipes; not only does this serve as a carbohydrate source, but it also has antibacterial activity. Besides direct inhibition, it is interesting that some ingredients of garlic inhibit QS. Ika-shiokara is made from squid and fish guts pickled in 2–30% salt, that depends on growth of the yeast *Rhodotorula;* if this sounds appealing, you will find it as a delicacy in Hokkaido, Japan.

Table 16.3 Some biotechnological applications of extremophilic microbes

Product	Applications
Thermophiles and hyperthermophiles	
Amylases, pullulanases, lipases, proteases	Baking, brewing, food processing
DNA polymerases	PCR amplification of DNA
Lipases, pullulanases, proteases	Detergents
S-layers	Ultrafiltration, electronics, polymers
Xylanases	Paper bleaching
Halophiles	
Bacteriorhodopsin	Bioelectronic devices, optical switches, photocurrent generators
Compatible solutes	Protein, DNA and cell protectants
Lipids	Liposomes (drug delivery, cosmetics)
S-layers	Ultrafiltration, electronics, polymers
γ-linoleic acid, β-carotene, cell extracts	Health foods, dietary supplements, food colors, aquaculture feeds
Psychrophiles	
Ice-nucleating proteins	Artificial snow, frozen food processing
Polyunsaturated fatty acids	Food additives, dietary supplements
Proteases, lipases, cellulases, amylases	Detergents
Alkaliphiles and acidophiles	
Acidophiles	Fine papers, waste treatment
Elastases, keratinases	Hide processing (leather)
Proteases, cellulases, lipases	Detergents
Sulfur-oxidizing acidophiles	Recovery of metals, sulfur removal from coal and oil

16.6 Microbial enzymes

Enzymes are widely used by industry and the current global market is estimated at about US$1.5 billion, growing by 5–10% per year. Many research institutes and commercial organizations have developed culture collections of bacteria (and to a lesser extent, fungi), with i nitial attention being focused on culturable organisms that are easily collected from near-shore habitats. The full potential of the thousands of marine microbes already in culture collections has yet to be investigated. The most commonly exploited enzymes are those that degrade polymers, especially proteins and carbohydrates. Examples of early successes include the production by *Vibrio* spp. of various types of extracellular proteases, some of which are tolerant of moderate salt concentrations and detergents. Another is the extraction of glucanases and other carbohydrate-degrading enzymes from *Bacillus* spp. isolated from muds.

The most successful developments have occurred with the isolation of enzymes from extremophilic microorganisms; some of the products that have resulted are shown in *Table 16.3*. Enzymes from thermophiles and hyperthermophiles have particular attractions in industrial processes, which often require high temperatures. Even at milder temperatures, thermophilic enzymes are beneficial because of their much greater stability. The structural features of thermophilic enzymes that confer stability and function at high temperatures (sometimes >100°C) are discussed in *Section 4.6.2*. The use of proteases and lipases as stain removers in detergents is a particularly important application, since properties that confer high thermostability and activity are often combined with resistance to surfactants, bleaching chemicals and surfactants used in these products. Some washing powders incorporate thermophilic cellulases and hemicellulases, which digest loose fibers and help to prevent 'bobbles' on clothes after washing; cellulases are also important in the manufacture of 'stone-washed' denim. Enzyme production for 'biological' detergents is a very large market accounting for approximately 30% of the total

Table 16.4 Some thermostable DNA polymerases and their sources

DNA polymerase	Organism	Source[1]	Half-life at 95°C	Proof reading
Taq	*Thermus aquaticus*	T (N or R)	40	–
Amplitaq®	*Thermus aquaticus*	T (R)	40	–
Vent™	*Thermococcus litoralis*	M (R)	400	+
Deep Vent™	*Pyrococcus GB-D*	M (R)	1380	+
Tth	*Thermus thermophilus*	T (R)	20	–
Pfu	*Pyrococcus furiosus*	M (N)	120	+
ULTma™	*Thermotoga maritima*	M (R)	50	+

[1]T = terrestrial hot spring; M = marine hydrothermal vent.
(N) = natural; (R) = recombinant.

global production of enzymes produced worldwide. The first source of these enzymes was soil bacteria of the genus *Bacillus*, but enzymes from marine thermophiles have higher temperature optima and superior stability. Modern food processing uses a wide range of enzymes. Almost all processed foods now rely on some form of modified starch product for the improvement of texture, control of moisture and prolonged shelf life. Amylases hydrolyze α-1,4-glycosidic linkages in starch to produce a mixture of glucose, maltooligosaccharides and dextrins. All the remaining α-1,4-glycosidic branches in the products are hydrolyzed by pullulanase. When starch is treated with amylase and pullulanase simultaneously at high temperatures, it yields higher yields of desired end-products. Pullulanases derived from *Thermotoga maritima* have recently been introduced into food processing. A range of other carbohydrate-modifying enzymes have been isolated from marine thermophiles. One novel use proposed for proteases from vent thermophiles is their use in highly acidic, high-temperature treatment of surgical instruments as a method of inactivating the Creutzfeldt-Jacob disease agent on surgical instruments, since the causative prions are resistant to normal sterilization processes.

The discovery of thermostable DNA polymerases used in the PCR (*Section 2.6.3*) has arguably been one of the most spectacular scientific advances since the discovery of DNA itself. The use of PCR in research, disease diagnostics and forensic investigations has led to a huge market; estimated sales of amplification kits in Europe alone were over US$340 million in 2001 and forecast to reach US$790 million by 2008. The original PCR enzyme, *Taq* polymerase, was isolated from *Thermus aquaticus* from a terrestrial hot spring. Although *Taq* is still the least expensive and most widely used enzyme for both research and diagnostic uses of the PCR, a number of alternative enzymes are now available that are superior in certain applications (*Table 16.4*). The choice of enzyme depends on the specific activity and sensitivity required with different amounts and lengths of template, nucleotide specificity and various other factors (including cost!). Enzymes from the hyperthermophilic vent bacteria have greater thermostability and activity at higher temperatures, particularly useful when amplifying GC-rich sequences. They may also have a higher fidelity of replication (due to integral proofreading ability), although careful optimization of the reaction is required; enzymes such as Vent™ or *Pfu* are often used in a mixture with *Taq*. As with many enzymes and other proteins isolated from bacteria, it is sometimes more convenient for manufacturing to clone the genes responsible and express them in *E. coli* or another recombinant host. However, there can be problems with expression and correct folding of thermostable proteins in mesophilic hosts, and if posttranslational modifications are required for enzyme function the recombinant enzyme may not function as required.

Enzymes from psychrophilic microbes have many uses and isolates from the deep sea and polar regions have been exploited for use in food processing applications where low temperatures are required to prevent spoilage, destruction of key ingredients (e.g. vitamins) or loss of texture. Significant savings in energy costs result from the use of cold-water laundry detergents that incorporate proteases and lipases active at temperatures of 10°C or less.

16.7 Microbial polymers

Polymers from marine bacteria are finding increasing application in bioremediation, industrial processes, manufacturing industry and food processing. The best-exploited compounds are extracellular polysaccharides that form the glycocalyx associated with biofilm formation and protection from phagocytosis. Applications in bioremediation were mentioned above. Other potential applications include underwater surface coatings, bioadhesives, drag-reducing coatings for ship hulls, dyes and sunscreens. It has been suggested that the high production (up to 80% cell dry weight under appropriate conditions) of poly-hydroxy-β-hydroxyalkanoates by some marine bacteria as food reserves could be exploited for the production of biodegradable plastics. Oil-derived plastics will continue to dominate the market for some time, but bacterially produced plastics are becoming more competitive because of concerns about sustainability and the polluting effects of nondegradable plastics.

16.8 Biomedical and health products

Natural products provide many compounds used in medicine and health promotion. During much of the 20th century, the pharmaceutical industry was engaged in a continual search for new compounds with biological activities that can be exploited in therapy. Many of our most successful drugs are secondary metabolites obtained from bacteria and fungi isolated in the terrestrial environment (especially soil). However, in the last 20 years, the effort necessary to isolate valuable new compounds became increasingly disproportionate to the returns. Companies moved away from natural-product-based drug discovery and came to rely more on the use of structure–function analysis to chemically modify existing compounds. There is a resurgence of interest in natural products from marine habitats because of the realization that their great biodiversity is likely to yield many novel compounds. Much effort has been applied to the isolation of compounds from marine invertebrates (especially sponges and bryozoans) and a number of antibiotics and antitumor agents have been isolated. Different species of these animals are host to a wide range of symbionts, and it now seems likely that many of the compounds discovered are actually produced by the microbes inhabiting the animals, rather than the host's own metabolism. If this is found to be widespread, it overcomes a major problem, i.e. the need to harvest scarce marine animal life and the consequent disruption of ecosystems. However, although compounds isolated from sponges and bryozoans often share structural similarity with known microbial metabolites, it is not easy to prove that they are indeed of microbial origin in the symbiosis. (*Box 12.2* discussed similar difficulties in the case of the origin of dinoflagellate toxins.) Exploitation of microbial symbionts for pharmaceutical production can follow the methods outlined in *Box 16.2*, i.e. either: (a) isolation or culture of individual bacteria; or (b) identification of genes, cloning and recombinant expression. One notable success is the family of compounds known as bryostatins, isolated from the bryozoan *Bugula neritina*. Bryostatins are cytotoxic macrolides, which are very likely synthesized by an as-yet uncultured γ-proteobacterium. Bryostatin-1 shows great promise for the treatment of certain types of leukemia and esophageal cancer and is currently in phase II clinical trials. Several other antitumor drugs from marine bacteria are at an earlier stage of investigation. For example, *Lyngba majuscule* produces curacins, which inhibit tubulin formation in cell division and have similarities to the successful anticancer compounds taxol and *Vinca* alkaloids. This cyanobacterium also produces microcolin A, which is immunosuppressive for lymphocyte proliferation and *Nostoc* forms cryptophycins, which inhibit solid tumors.

Most of the antimicrobial antibiotics used in treatment of microbial infections are derived from terrestrial fungi (e.g. penicillins) and bacteria, especially actinomycetes (e.g. tetracyclines, aminoglycosides and macrolides). Cephalosporin was obtained from the fungus *Cephalosporium*, isolated near a coastal sewage outfall. Drug discovery companies have begun a renewed search for novel microbes (especially actinomycetes) in the marine environment.

Large-scale rapid-throughput screening programs are underway in several parts of the world, investigating marine sediments, sponges, corals and other habitats. It is likely that the intense competitive pressure found in dense mixed microbial communities, such as those in sponges, will select for microbes producing antimicrobials. Biofilms are likely to be a rich source, and attention should be given to using new types of bioassays to detect metabolites that interfere with cell signalling (see *Box 16.1*) or chemotaxis, as well as those that cause outright growth inhibition in the standard detection methods.

'Nutraceuticals' is a term used to describe the wide range of substances used for health promotion and includes functional foods, probiotics and nutritional supplements. This is an expanding market, currently valued in excess of US$50 billion worldwide. New marine microbial products, especially those from microalgae, are likely to contribute to this area. For example, the PUFA DHA (omega-3) promotes brain growth and is added to infant formula feed. The usual source, fish oils, is increasingly likely to be contaminated with trace amounts of pollutants, so the production of DHA by the marine heterotrophic dinoflagellate *Crypthecodinium cohnii* is finding favor with manufacturers. A spin-off from this work was the development of DHA-enriched diets for larval stages in aquaculture. Newly isolated deep-sea psychrophilic bacteria may also be a good source of PUFAs. It is necessary to prove that microorganisms with no history of use in food products are nontoxic and nonpathogenic and also to develop suitable methods for large-scale cultivation.

A major part of the health food industry is concerned with antioxidative effects. Research with bacteria from coral reefs exposed to very high levels of visible light and UV irradiation suggests that they may have novel mechanisms of reversing the resultant oxidative damage. This could lead to products that overcome some aspects of the aging process. The production of UV-protective pigments by marine bacteria is leading to the development of new sunscreen preparations.

16.9 Biomimetics, nanotechnology and bioelectronics

In materials science and technology, biomimetics is the term given to the process of 'taking good designs from nature'. Nanotechnology involves the construction of materials and functional objects assembled from the basic molecular building blocks, which offers the potential of new products ranging from new computer technology to microscopic machines. Marine microbes are proving to be a rich source for these new technologies. The S-layers of *Bacteria* and *Archaea* (*Section 3.9.3*) are finding applications in nanotechnology because of the ordered alignment of functional groups on the surface and in the pores, which allow chemical modifications and the binding of molecules in a very precise fashion. Isolated S-layer subunits can recrystallize as monolayers on solid supports (e.g. lipid films, metals, polymers and silicon wafers), this has a range of applications in colloid and polymer science and the electronics industry. The uniform size and alignment of pores in S-layers also makes them suitable for use as ultrafiltration membranes. As described in *Section 7.6,* different species of diatoms construct their silica shells in a variety of beautiful forms. Understanding the molecular basis by which diatoms achieve the construction of their frustules may lead to advances in nano-assembly of materials into desired structures. The microscopic rotary motor of bacterial flagella (*Section 3.9*) has attracted the interest of engineers for some years, with suggestions that isolated basal bodies could form the basis of self-propelled micromachines (e.g. for targeted drug delivery systems). The recent discovery of chemotaxis and ultrafast swimming in marine bacteria (*Box 3.1*) could lead to advances in this field, as further research leads to an understanding of the molecu-lar basis of these processes. Magnetotactic bacteria (*Section 5.14.2*) produce magnetic crystals with a uniform structure that is difficult to achieve in industrial processes. These could have important applications in the electronics industry and for biomedical applications (e.g. in the production of magnetic antibodies and in magnetic resonance imaging). Understanding the mechanisms by which bacteria construct the magnetosomes, and introducing changes via genetic modification could be used to display particular proteins.

Box 16.2 RESEARCH FOCUS

Bioprospecting in the biotech gold rush

Genomics in the search for marine microbial products

The development of molecular biology methods is leading to new approaches in the search for novel products from marine microbes (see *Section 2.6.9*). In an analogy with the gold rush, 'bioprospectors' can now use oligonucleotide hybridization probes to 'pan' for genetic sequences of interest in the hope of reaping rich rewards. This enables the screening of communities without the need for culture. For example, comparison of sequence data for genes encoding proteins with a particular function can allow the identification of consensus sequences. These can be used to construct complementary oligonucleotide probes, which are then used to search for genes in DNA amplified from a community of interest. Genes can then be cloned and sequenced. The disadvantage of this method is that it may fail to detect truly novel proteins with sequences unlike those previously undescribed. This limitation is borne out by the results from genome-sequencing projects of marine prokaryotes. In some of the recent sequences published for hydrothermal vent *Archaea*, over 70% of the ORFs in the genomes have no match with existing genes in the database.

An alternative approach is the construction of complex environmental gene libraries by cloning representative gene fragments into suitable vectors. This is the method adopted by the Diversa Corporation (2003), a biotechnology company that specializes in the development of novel genes and gene pathways from diverse environments, including marine sources. Complex microbial communities can contain many thousand different microbial species, with some much more abundant than others and Diversa has developed proprietary technology to facilitate recovery of rare genomes. Powerful sequence analysis techniques are used to screen novel genes and their variants and genes of potential interest are transferred to recombinant hosts for gene expression. High-throughput screening methods employing robotic handling enable thousands of clones to be analyzed in a single day. Diversa have developed FACS (*Section 2.3*) in order to identify DNA sequences or expression of gene products within single cells. Whilst enzymes and other proteins are encoded by single genes (or related genes in the case of subunit proteins), the biosynthesis of secondary metabolites such as antibiotics requires the participation of several or multiple genes encoding enzymes that function within a coordinated biochemical pathway. The genetic information required for such a pathway may be over 25 000 times greater than that required for a simple enzyme. The costs of accommodating, maintaining and manipulating such large amounts of information requires the development of new molecular tools and the development costs mean that these approaches are beyond the scope of most academic research groups and are only possible in commercial laboratories. Diversa have entered into agreements with research institutes in a number of countries for the investigation of marine resources and have engaged in a number of genome sequencing projects, acquiring the exclusive rights to exploitation (see *Box 6.1*).

Initial attention in whole genome sequencing of marine bacteria was focused on extremophiles, but a number of other genome sequences have now been published (*Table 2.3*) and many more are in the pipeline. Improvements in laboratory automation and sequencing technology (capillary array electrophoresis) mean that it is now possible to complete the genome sequencing of a new microbe within a matter of weeks. As the subsequent techniques of bioinformatics and proteomics are increasingly employed and databases grow, we will undoubtedly be able to ascribe properties to many genes of currently unknown function, leading to numerous new biomolecules.

Biomolecular electronics relies on the use of native (or genetically modified) biological molecules such as proteins, chromophores and DNA. One of the best examples to date is the use of bacteriorhodopsin isolated from *Halobacterium salinarum* (*Section 6.2.4*). Bacteriorhodopsin changes its structure every few milliseconds to convert photons into energy. A chromophore embedded in the protein matrix absorbs light and induces a series of changes that change the optical and electrical properties of the protein. Bacteriorhodopsin can store many gigabytes of information in three-dimensional films (holographic memories) and genetic modification produces proteins with various desirable properties. Bacteriorhodopsin has also been used to construct artificial retinas.

References and further reading

Al-Hasan, R.H., Khnafaer, M., Eliyas, M., and Radwan, S.S. (2001) Hydrocarbon accumulation by picocyanobacteria from the Arabian Gulf. *J Appl Microbiol* **91**: 533–540.

Armstrong, E., Yan, L., Boyd, K.G., Wright, P.C., and Burgess, J.G. (2001) The symbiotic role of marine microbes on living surfaces. *Hydrobiology* **461**: 37–40.

Bassler, B.L. (2002) Small talk: Cell-to-cell communication in bacteria. *Cell* **109**: 421–424.

Bertrand, J.C., Bonin, P., Goutx, M., and Mille, G. (1993) Biosurfactant production by marine microorganisms: Potential application to fighting hydrocarbon marine pollution. *J Mar Biotechnol* **1**: 125–129.

Callow, M.E., and Callow, J.A. (2002) Marine biofouling: a sticky problem. *Biologist* **49**: 1–5.

Chung, W.K., and King, G.M. (2001) Isolation, characterization, and polyaromatic hydrocarbon degradation potential of aerobic bacteria from marine macrofaunal burrow sediments and description of *Lutibacterium anuloderans* gen. nov., sp. nov., and *Cycloclasticus spirillensis* sp. nov. *Appl Env Microbiol* **67**: 5585–5592.

Cooksey, K.E., and Wigglesworth-Cooksey, B. (1995) Adhesion of bacteria and diatoms to surfaces in the sea – a review. *Aquat Microb Ecol* **9**: 87–96.

de Nys, R., and Steinberg, P.D. (2002) Linking marine biology and biotechnology. *Curr Opin Biotechnol* **13**: 244–248.

de Nys, R., Steinberg, P.D., Willemsen, P., Dworjanyn, S.A., Gabelish, C.L., and King, R.J. (1995) Broad spectrum effects of secondary metabolites from the red alga *Delisea pulchra* in antifouling assays. *Biofouling* **8**: 259–271.

Diversa Corporation (2003) *Microbial Genomics*. Company website. http://www.diversa.com (accessed April 5 2003).

Gram, L., and Dalgaard, P. (2002) Fish spoilage bacteria – problems and solutions. *Curr Opin Biotechnol* **13**: 262–266.

Harayama, S., Kishira, H., Kasai, Y., and Shutsubo, K. (1999) Petroleum degradation in marine environments. *J Molec Microbiol Biotechnol* **1**: 63–70.

Haygood, M.G., Schmidt, E.W., Davidson, S.K., and Faulkner, D.J. (1999) Microbial symbionts of marine invertebrates: opportunities for microbial biotechnology. *J Molec Microbiol Biotechnol* **1**: 33–43.

Hentzer, M., Riedel, K., and Rasmussen, T.B. (2002) Inhibition of quorum sensing in *Pseudomonas aeruginosa* biofilm bacteria by a halogenated furanone compound. *Microbiol UK* **148**: 87–107.

Holmstrom, C., Egan, S., Franks, A., McCloy, S., and Kjelleberg, S. (2002) Antifouling activities expressed by marine surface associated *Pseudoalteromonas* species. *FEMS Microbiol Ecol* **41**: 47–58.

Joint, I., Tait, K., Callow, M.E., Callow, J.A., Milton, D., Williams, P., and Cámara, M. (2002) Cell-to-cell communication across the prokaryote–eukaryote boundary. *Science* **298**: 1207.

Kjelleberg, S., Steinberg, P., Givskov, M., Gram, L., Manefield, M., and deNys, R. (1997) Do marine natural products interfere with prokaryotic AHL regulatory systems? *Aquat Microb Ecol* **13**: 85–93.

Lebeau, T., and Robert, J.M. (2003) Diatom cultivation and biotechnologically relevant products. Part II: Current and putative products. *Appl Microbiol Biotechnol* **60**: 624–632.

Loferer, H., Jacobi, A., Posch, A., Gauss, C., Meier-Ewert, S., and Seizinger, B. (2000) Integrated bacterial genomics for the discovery of novel antimicrobials. *Drug Discov Today* **5**: 107–114.

Maloney, S. (2003) *Extremophiles: Bioprospecting for Antimicrobials*. Available at: http://www.mediscover.net/Extremophiles.cfm

Manefield, M., Rasmussen, T.B., Hentzer, M., Andersen, J.B., Steinberg, P., Kjelleberg, S., and Givskov, M. (2002) Halogenated furanones inhibit quorum sensing through accelerated LuxR turnover. *Microbiol UK* **148**: 1119–1127.

Miller, M.B., and Bassler, B.L. (2001) Quorum sensing in bacteria. *Ann Rev Microbiol* **55**: 165–199.

National Academies Board on Army Science and Technology (2001) *Opportunities in Biotechnology for Future Army Applications*. National Academies Press, Washington DC. http://books.nap.edu/books/0309075556/html/25.html (accessed April 5 2003).

National Academies Commission on Life Sciences (2000) *Opportunities for Environmental Applications of Marine Biotechnology*. National Academies Press, Washington DC. http://books.nap.edu/books/0309071887/html/index.html (accessed April 5 2003).

National Academies Ocean Studies Board, Marine Board, Transportation Research Board (2003) *Oil in the Sea III: Inputs, Fates, and Effects*. National Academies Press, Washington. http://books.nap.edu/books/0309084385/html/ (accessed April 5 2003).

National Research Council (1999) *Marine-derived Pharmaceuticals and Related Bioactive Agents. From Monsoons to Microbes*, pp.71–82. National Academies Press, Washington. http://www.nap.edu/books/0309065690/html/ (accessed April 5 2003).

Okami, Y. (1993) The search for bioactive metabolites from marine bacteria. *J Mar Biotechnol* **1**: 59–65.

Prince, R.C. (1993) Petroleum spill bioremediation in marine environments. *Crit Rev Microbiol* **19**: 217–242.

Raghukumar, C. (2000) Fungi from marine habitats: an application in bioremediation. *Mycol Res* **104**: 1222–1226.

Ron, E.Z., and Rosenberg, E. (2001) Natural roles of biosurfactants. *Env Microbiol* **3**: 229–236.

Schuler, D., and Frankel, R.B. (1999) Bacterial magnetosomes: microbiology, biomineralization and biotechnological applications. *Appl Microbiol Biotechnol* **52**: 464–473.

Semple, K.T., Cain, R.B., and Schmidt, S. (1999) Biodegradation of aromatic compounds by microalgae. *FEMS Microbiol Lett* **170**: 291–300.

Sipe, A.R., Wilbur, A.E., and Cary, S.C. (2000) Bacterial symbiont transmission in the wood-boring shipworm *Bankia setacea* (Bivalvia: Teredinidae). *Appl Env Microbiol* **66**: 1685–1691.

Steinberg, P.D., and de Nys, R. (2002) Chemical mediation of seaweed surfaces. *J Phycol* **38**: 621–629.

Steinberg, P.D., de Nys, R., and Kjelleberg, S. (1998) Chemical inhibition of epibiota by Australian seaweeds. *Biofouling* **121**: 227–244.

Surette, M.G., and Bassler, B.L. (1998) Quorum sensing in *Escherichia coli* and *Salmonella typhimurium*. *Proc Natl Acad Sci USA* **95**: 7046–7050.

Swannell, R.P.J., Mitchell, D., Lethbridge, G., *et al.* (1999) A field demonstration of the efficacy of bioremediation to treat oiled shorelines following the Sea Empress incident. *Environ Technol* **20**: 863–873.

Valls, M., and de Lorenzo, V. (2002) Exploiting the genetic and biochemical capacities for the remediation of heavy metal pollution. *FEMS Microbiol Rev* **26**: 327–338.

Videla, H.A. (2000) An overview by which sulphate-reducing bacteria influence corrosion of steel in marine environments. *Biofouling* **15**: 37–47.

Weiner, R.M. (1997) Biopolymers from marine prokaryotes. *Trends Biotechnol* **15**: 390–394.

Withers, H., Swift, S., and Williams, P. (2001) Quorum sensing as an integral component of gene regulatory networks in Gram-negative bacteria. *Curr Opin Microbiol* **4**: 186–193.

Wu, Q., Watts, J.E.M., Sowers, K., and May, H.D. (2002) Identification of a bacterium that specifically catalyzes the reductive dechlorination of polychlorinated biphenyls with doubly flanked chlorines. *Appl Env Microbiol* **68**: 807–812.

Concluding remarks 17

This final chapter attempts to provide a brief summary of the major themes that link the various concepts to emerge throughout this book. What have we learned about the importance of marine microbes? What future developments might occur? In looking to the future, I am helped by an article that appeared in the journal *Environmental Microbiology* (Various authors, 2002) in which leading researchers in the field were asked to describe the likely technical and conceptual developments in the next few years. I am hard-pressed to improve on what they have said, so I rely heavily on quotations from this article, which I recommend to students to read in full.

Marine microbiology (and environmental microbiology more generally) has shown dramatic advances in the past few years. This book has described the most significant advances due to the development of many new methods, of which the most important are undoubtedly genomics and bioinformatics. With these methods, it is possible to generate sound biological hypotheses about marine processes and to test them in the field. As Rudolf Amann (Max-Planck Institute of Marine Microbiology) comments, 'environmental microbiology proceeds from *in situ* analysis of microbial diversity into the *in situ* quantification of specific activities of defined populations.' Nevertheless, he emphasizes the fact that we are still far from a good description of those organisms that catalyze essential parts of the biogeochemical cycles. Concerns about global change and the effects of anthropogenic pollutants are the stimulus, and funding source, for much of the recent research in marine microbiology. As Amann says, 'exact numbers and good predictions on the stability of biogeochemical cycles [are not possible] without biologists knowing and understanding the function of the key players, which more often than not will be as yet unknown microbes'. He urges us to 'think big' and initiate major long-term international projects with this goal.

A common theme that has emerged throughout the book is the fact that individual microbes interact with each other and with their immediate environment, so that microscale or nanoscale processes have ecosystem, or even global, effects. Who could have predicted that bacterial motility and extracellular enzyme production affect carbon flux in the oceans, or that viral lysis of algae affects the global climate? Terry J. Beveridge (University of Guelph) argues that interdisciplinary research linking microbiology, molecular biology, chemistry, physics and geology will become increasingly important to understand these processes. Ed DeLong (Monterey Bay Aquarium Research Institute) takes this a step further, highlighting the need for 'integrating analytical capabilities, from remote sensing all the way through to nanoscale interactions'. Students moving into this field as new researchers should heed his comments that 'the challenge to future microbial biologists is that they must become as conversant in Earth science as nanotechnology, as familiar with systems ecology as genomics, and as well versed in global information systems as bioinformatics'. This book can only provide small insights into this continuum.

Despite the successes of genomic approaches in expanding our view of microbial diversity, John A. Breznak (Michigan State University) says that this has 'brought with it the sobering reality that we have little or no understanding of the vast majority of microbes, as ... 0.1% or less have been cultured'. At the time of writing this concluding chapter, a news release has just announced that a team led by Craig Venter (the scientist best known for his role in the human genome sequencing project) has received funding from the US Department of Energy to sequence an entire ecosystem (Genet, 2003). The project aims to sequence the genomes of every organism in the Sargasso Sea. Such ambitious metagenomics initiatives will have dramatic impacts on our understanding of ocean processes, but on their own, they will not tell us everything about microbial habitats and activities. As Breznak states, we must 'mount a steady and sustained effort ... to retrieve the not-yet cultured majority'. Stephen H. Zinder (Cornell University) talks of the 'chilling effect' of the 'frequent use of the term "uncultivable"... on [the] considerable efforts needed to culture organisms from natural habitats'. The successful cultivation in 2002 of SAR11, previously identified as one of the major bacterioplankton groups only from its genetic sequence, stands out as an example of what can be achieved if sufficient effort and technical innovation are applied. Furthermore, as Julian Davies (University of British Columbia) argues, 'to understand microbes and their responses to changing environments ... will require a significant re-invention of [the study of] physiology', a discipline which has declined in importance in microbiological education and research.

This book contains numerous examples of the deterioration in the health of marine ecosystems, with profound implications for biodiversity and for human health and welfare. Striking parallels between the effect of climatic conditions on the epidemiology of diseases in humans and marine animals are emerging. If we are to avoid or mitigate the worst effects of these changes, we must use scientific understanding and human ingenuity to limit further degradation of the environment and to devise methods of monitoring and control of disease and ecosystem disruption. Rita Colwell (University of Maryland and US National Science Foundation) emphasizes the need for understanding how complex systems emerge from the interactions of biological entities at all levels. The fact that most activities of marine microbes are 'unseen' by everyday human experience has been emphasized in this book. It is important that future generations have a sound understanding of marine microbial processes and interactions, which will help human society to make the difficult decisions that lie ahead.

As well as improving our scientific understanding of life processes and Earth systems, I hope that the book has revealed the many economic benefits that have resulted from the applications of marine microbiology. We have been aware of these potential benefits for over 25 years, with one of the most vocal proponents being Rita Colwell, whose work has had a huge influence on the development of applied marine microbiology and marine biotechnology. We are now beginning to see real, significant commercial benefits from the exploitation of marine microbes and their metabolic activities. Unforeseen 'spin-offs' have resulted. Who could have imagined that studying bioluminescent bacteria could lead to a completely new approach to the control of human disease (quorum sensing)? Could we have predicted that studying the adaptation of halophiles to extreme salt concentrations could lead to an artificial eye? Would anyone have thought that studying the microbial communities that develop on whale carcasses on the sea floor could lead to improved laundry detergents? The term 'blue sky research' is often used to describe research undertaken with no obvious foreseeable application. Perhaps 'blue sea research' would be a more appropriate description!

The vastness and complexity of the marine environment is still largely unexplored. At the beginning of the 21st century, we have just discovered previously unknown ocean processes (aerobic anoxygenic phototrophy), new symbiotic associations that are responsible for one of the most important biogeochemical processes in marine sediments (methane oxidation) and an organism that may be the simplest and smallest form of cellular life (*Nanoarchaeum*). What other surprises are out there?

References and further reading

Genet Archive (2003) *News item: Craig Venter to sequence – and patent? – Sargasso Sea organisms.* http://www.gene.ch/genet/2003/Apr/msg00110.html (accessed May 15 2003).

Various authors (2002) Crystal ball. *Environ Microbiol* **4:** 3–17.

Index